D1616075

Recovery Planning for Communications and Critical Infrastructure

For a listing of recent titles in the
Artech House Telecommunications Series,
turn to the back of this book.

Disaster Recovery Planning for Communications and Critical Infrastructure

Leo A. Wrobel
Sharon M. Wrobel

ARTECH
HOUSE

BOSTON | LONDON
artechhouse.com

Library of Congress Cataloging-in-Publication Data
A catalog record for this book is available from the U.S. Library of Congress.

British Library Cataloguing in Publication Data
A catalogue record for this book is available from the British Library.

ISBN-13: 978-1-59693-468-9

Cover design by Yekaterina Ratner

© 2009 ARTECH HOUSE
685 Canton Street
Norwood, MA 02062

10 9 8 7 6 5 4 3 2 1

To my wife of 33 years, Sharon.
As we have grown together in a marriage that has blessed us
with seven children and eleven grandchildren, this book represents
yet another highlight in our journey together. It is Sharon's first as an honest-to-
goodness coauthor. I greatly appreciate her significant contribution to this book,
as well as her impact on my life.

Contents

CHAPTER 3

What Are We Planning For?   55

CHAPTER 4

Case Studies and Examples of Quantifying Risk of Natural Disasters to Critical Infrastructure   77

CHAPTER 5

CHAPTER 8

CHAPTER 9
Other References

CHAPTER 10
Directory of Disaster Recovery Information Sources

Preface

I often wonder if anyone ever specifically sets out to be a disaster recovery or business resumption planner. In my 30 or so years in this business, I have seen many contingency planning professionals, to be sure, but they almost invariably hail from other trades. Some come out of police, fire, and emergency management. A great many planning professionals come out of information technology. Others come out of the military, health care, and a myriad of other professions. Once into the disaster recovery profession, however, they invariably are hooked, just as I was when I entered it. In my case, I came out of the telecommunications industry, which has been my bread and butter since I was in the military 30 years ago. Perhaps that is why I am so enamored with the idea of command and control communications, which is the focus of this book.

You probably bought this book to learn about disaster recovery planning as it relates to your own disaster recovery planning profession or vocation. I am truly happy for you that you did—and not just because the Wrobels and Artech House get to sell a book. I am happy because you will find this a *most* rewarding endeavor. First, you will never learn more about your profession—whatever it is—than you do when you think about all the steps required to pick up the pieces and start all over again after the unthinkable happens. Second, we don't care what anyone says, disasters are interesting! There is a certain morbid curiosity in all human beings about what happened to others and whether it could happen to them or their loved ones. Finally, I believe there are some closet Supermen or Wonder Women among us. It feels good when you are the "super hero" that saves the day, even if it's only on paper or in an exercise. And when you really get down to it, doesn't that sound like a lot more fun than the decades of downsizing, rightsizing, and capsizing in many of our professions and industries? If any of this strikes a chord with you, you are going to have a lot of fun reading this book. The book will not only impart the *technology* required to get your disaster recovery plan in motion, it will also share all the tricks of the trade (those that we know of, anyway) to help you find the resources you need to get it done. Lack of funding is certainly a reality these days, so we will also show you how to plan efficiently and not break your budget in the process. Let's call it our "guerilla warfare" approach to disaster recovery planning, including:

- Inexpensive techniques we have tried that work and how you can use them;
- Where to find resources and talent—internally, externally, and public sources;
- How you can find the money to plan;

• How to quantify vulnerability in terms people understand;

• How to get a plan in place *and* do it *all* on a *budget*.

At the risk of sounding like an infomercial, however: "Wait! There's more!"

For those of you that have not followed our every move these last 20 years (Leo published his first Artech book in 1989), this book asks and perhaps answers a thought-provoking question: Are executives in *public sector companies* responsible for *natural disasters*? The answer will surprise you. In the context of answering that question, we are especially excited about a sizable contribution in this book from the Pacific Disaster Center.

In summary, we must warn you that author Leo grew up in a big Italian-American family. His mother's favorite saying is, "If you leave this house hungry, it's your fault." I hope you don't leave this house hungry. Feel free to corner us for a "second helping" of something if you want it. Both Sharon and I do eventually answer e-mails, and we have included our e-mail addresses here. We appreciate your interest in this fascinating topic and our goal is to make this book as useful, entertaining, and informative as we can.

Leo A. Wrobel: leo@b4Ci.com
Sharon M. Wrobel: sharon@b4Ci.com
Ovilla, Texas
April 2009

CHAPTER 1

Why Do We Need Recovery Plans?

Why Do We Plan At All?

The reasons that justify a disaster recovery plan are as diverse as organizations are from one another. A stockbroker, for example, often deals with recovery time frames measured in seconds following a disaster. Energy companies have different concerns, such as how one well-placed hurricane might mean a doubling of the price of gasoline. For other organizations, such as 911 emergency centers or poison control hotlines, lack of a plan can easily mean loss of life. For this reason, although often recovery plans are not mandated by law, per se, there is at a minimum a moral responsibility to protect public and private assets.

There are many myths and misconceptions about the legal requirements for disaster recovery planning. For a straight story, I asked a lawyer friend of mine exactly what we should be concerned about at least from a legal perspective. He furnished me with some legal insights that were nebulous in some regards but did help place the issues in perspective. I'll also chime in with my technical experience as we explore this issue.

Speaking as a nonlawyer, there are certainly many honest-to-God laws that mandate having a plan. Having said that, it is also my experience as a businessman that people tend to embellish. Sometimes they interpret what they consider "the law" in a way that's favorable to them, rather than as the law truly *is*. As a case in point, my attorney and I found several citations on the Web that said defense contractors must have a disaster recovery plan. That seems to make sense on the surface. (I have published a lot over the last couple of years about how military disaster recovery procedures and processes are now "fashionable" with what is going on today commercially.) Set in that context, the idea that Department of Defense (DoD) contractors *must* have a plan must be true, right? Wrong. When we tried to find a bona fide law, rule, or regulation that actually *stated* this fact, there was nothing to be found.

So what is the deal here? Is the notion of laws that mandate disaster recovery plans the equivalent of an urban legend—a story that is passed along from person to person, without anyone ever tracking back to see if it's true? Or are *all* companies required by law to have a plan? Actually, the answer appears to lie somewhere in between. Even if you try to understand the requirements for your particular industry, legal semantics and terminology come into play. (Hey, anyone up to some antics?) There are myriad terms of law that *might* apply to different cases. Is this confusing enough for you yet? If not, read on and I'll try harder to confuse you.

OK, So Who, Legally, Must Plan?

With the caveats stated previously, let's cover a few of the bona fide laws where there *is* a duty to have a disaster recovery plan. I will try my best, as a nonlawyer, to include the basis for that requirement where there is an implied mandate to do so. The lawyers tell me, not surprisingly, that the best thing to do is to consult your lawyer. At least that part makes sense.

While most of the mandates and laws below are germane to the United States, once again it may be prudent to check with an attorney in your locality to determine what rules and regulations govern the country or principality in which you reside.

Banks and Financial Institutions Must Have a Plan

If you handle money, you need a plan! The Federal Financial Institutions Examination Council (Council) was established on March 10, 1979, pursuant to title X of the Financial Institutions Regulatory and Interest Rate Control Act of 1978 (FIRA), public law 95-630. In 1989, title XI of the Financial Institutions Reform, Recovery, and Enforcement Act of 1989 (FIRREA) established the Examination Council. This Council is empowered to prescribe uniform principles, standards, and reporting for the examination of financial institutions by the Board of Governors of the following:

- The Federal Deposit Insurance Corporation (FDIC);
- The National Credit Union Administration (NCUA);
- The Office of the Comptroller of the Currency (OCC);
- The Office of Thrift Supervision (OTS);
- The Federal Reserve System (FRS).

The Council also makes recommendations to promote uniformity in the supervision of financial institutions. In other words, every bank, savings and loan, credit union, and other financial institution is governed by the principles adopted by the Council.

In March 2003, the Council released its *Business Continuity Planning* handbook designed to provide guidance and examination procedures for examiners in evaluating financial institution and service provider risk management processes. According to our interpretation of all of this, if you are a financial institution, you had better have a plan. We also assume that similar regulations apply to banking systems in other countries that trade cash in one way or another with the United States and with one another.

Securities Brokers Must Have a Plan

The National Association of Securities Dealers (NASD) has adopted rules that require its members to have disaster recovery and business continuity plans. The NASD oversees the activities of more than 5,100 brokerage firms, approximately 130,800 branch offices, and more than 658,770 registered securities representa-

tives. As of June 14, 2004, the rules apply to all NASD member firms. (It is important to note that the rules apply to every company that deals in securities, such as
brokers, dealers, and their representatives. It does *not* apply to the listed companies
themselves. Interestingly enough, we have heard of cases where the inability of callers to get through via phone has actually spawned inquiries by the U.S. Security and
Exchange Commission. Calls cut off in the middle of "sell" orders can sometimes be
construed as suspicious transactions, especially when millions of dollars can be
wired out of the country in less then a second. In any event, the requirements are
specified in Rule 3510:[1]

> 3510. Business Continuity Plans
> (a) Each member must create and maintain a written business continuity plan iden-
> tifying procedures relating to an emergency or significant business disruption. Such
> procedures must be reasonably designed to enable the member to meet its existing
> obligations to customers. In addition, such procedures must address the member's
> existing relationships with other broker-dealers and counter-parties. The business
> continuity plan must be made available promptly upon request to NASD staff.

Electric Utilities Probably Need a Plan

Disaster recovery for the electric utility grid seems pretty self-evident doesn't it?
Indeed, the United States received the equivalent of a shot over the bow five years
ago regarding the potential for massive disruption of the utility grid itself. On
August 14, 2003, a combination of summer heat and a spike in demand, along with
a sagging high voltage power line in Ohio, quickly cascaded into one of the largest
power outages in U.S. history. The U.S.–Canada Power System Outage Task Force
later concluded that electrical power was lost to some 50 million people in a 9,300
square mile area. Full power was not restored for several weeks.

Possibly in response to events like these, and the ever-present specter of deliberate sabotage or terrorism, the industry is making some adjustments. Prior to 2005,
the Federal Energy Regulatory Commission (FERC) really could only *coordinate*
volunteer efforts between utilities. This has changed with the adoption of Title XII
of the Energy Policy Act of 2005 (16 U.S.C. 824o). That law authorizes the FERC to
create an Electric Reliability Organization (ERO). The ERO will have the ability to
adopt and enforce reliability standards for "all users, owners and operators of the
bulk power system" in the United States. The ERO will then begin the process of
establishing reliability standards. It is a pretty safe bet to assume that the ERO will
eventually adopt standards for service restoration and disaster recovery, particularly after such widespread disasters as hurricanes Katrina, Gustav, and Ike. The
new Obama administration also appears to be proactive with regard to infrastructure upgrades, including "smart grid" technology. Therefore if you are an electric
utility, you better get a plan.

---

1. According to Banknet India, an IT-focused banking research company, major banks such as State Bank of
   India, Oriental Bank of Commerce, Canara Bank, and Bank of Maharastra floated tenders to scale up their
   disaster recovery and business continuity systems.

Telecommunications Utilities Should Have Plans—But May Not

Telecommunications utilities are governed on the federal level by the Federal Communications Commission (FCC) and at the state level by the various public utility commissions (PUCs) within each state. The FCC has created the Network Reliability and Interoperability Council (NRIC), whose role is to develop recommendations for the telecommunications industry to "insure optimal reliability, security, interoperability and interconnectivity of, and accessibility to, public communications networks and the Internet." NRIC members include providers and users of telecommunications services together with various commercial telecommunications providers, satellite providers, cable TV providers, wireless providers, computer industry representatives, trade associations, labor and consumer representatives, manufacturers, research organizations, and government organizations. Although there is no *explicit* provision we could find that mandates telecommunications carriers *must* have a disaster recovery plan, I have stated frequently in many books and articles that telecommunications facilities are tempting targets for terrorism. I have not changed my mind in that regard and urge caution. I have further stated that every disaster recovery plan—no matter what type of company or technology or manner of disaster—should begin with *communications*. It is the single most critical component to any recovery. Keep this in mind whether you are a user of telecom services or the purveyor of those services.

As far as users of telecom services, you may also wish to consider what liability a telephone company carries if *it* has a disaster that causes loss to your organization. It's not much. The following is the statement used in most telephone company tariffs with regard to its liability:

> The Telephone Company's liability, if any, for its gross negligence or willful misconduct is not limited by this tariff. With respect to any other claim or suit, by a customer or any others, for damages arising out of mistakes, omissions, interruptions, delays or errors, or defects in transmission occurring in the course of furnishing services hereunder, the Telephone Company's liability, if any, shall not exceed an amount equivalent to the proportionate charge to the customer for the period of service during which such mistake, omission, interruption, delay, error or defect in transmission or service occurs and continues. (Source, General Exchange Tariff for major carrier)

For that reason proceed at your own peril. If the telco screws up, unless it was gross negligence (which is really hard to prove in a court of law), it's kind of like being a professional photographer who sustains a loss when the film processor ruins your film. The value is not determined by the pictures you lost or the loss to your business. The value of your loss is $3 for another roll of film.

All Health Care Providers Will Need a Disaster Recovery Plan

Most persons believe that hospitals must have a disaster recovery plan since they are essential services. That assessment is correct and underscored by recent legislation that mandates health care providers to have plans.

"HIPAA" is an acronym for the Health Insurance Portability and Accountability Act of 1996, or Public Law 104-191. This law, also known as the Kennedy-Kassebaum Act, amended the Internal Revenue Service Code of 1986. It includes a section, Title II, entitled Administrative Simplification, requiring "Improved efficiency in healthcare delivery by standardizing electronic data interchange, and protection of confidentiality and security of health data through setting and enforcing standards." The legislation called upon the Department of Health and Human Services (HHS) to publish new rules that will ensure security standards protecting the confidentiality and integrity of "individually identifiable health information," past, present, or future.

The rules were published by HHS on February 20, 2003, and provide for a uniform level of protection for all health information that is housed or transmitted electronically and that pertains to any individual. The Security Rule requires covered entities to ensure the confidentiality, integrity, and availability of all electronic protected health information (ePHI) they create, receive, maintain, or transmit. The rule also requires entities to protect against any reasonably anticipated threats or hazards to the security or integrity of ePHI, protect against any reasonably anticipated uses or disclosures of such information that are not permitted or required by the Privacy Rule, and to ensure such compliance by their workforce. Required safeguards include application of appropriate policies and procedures, safeguarding physical access to ePHI, and ensuring that technical security measures are in place to protect networks, computers, and other electronic devices. If your organization is a hospital or health care facility, you had better have a disaster recovery plan.

Companies with More Than 10 Employees ... You Guessed It ... Need a Plan!

Stated generically, every CEO of a company with stockholders has the legal and moral right to protect the interests of those stockholders. The chief executive officer and board of directors have an implied responsibility to protect the safety and well being of employees as well. Violation of either of these axioms invites legal action against the corporation, its executives, board of directors, or all of the above.

Above and beyond the fiduciary duties described earlier, the U.S. Department of Labor has also adopted numerous rules and regulations with regard to workplace safety as part of the Occupational Safety and Health Act. For example, 29 USC 654 specifically requires:

(a) Each employer –
(1) shall furnish to each of his employees employment and a place of employment which are free from recognized hazards that are causing or are likely to cause death or serious physical harm to his employees;
(2) shall comply with occupational safety and health standards promulgated under this Act.
(b) Each employee shall comply with occupational safety and health standards and all rules, regulations, and orders issued pursuant to this Act which are applicable to his own actions and conduct.

Other Considerations, Mandates, and Regulations

These laws are not inclusive of all of the laws that mandate a disaster recovery plan—not by a long shot. Some of the other mandates that may affect your organization might include one or more of the following. Consult with your legal professional for more information.

- Contractual performance obligations to customers;
- Document protection and retention laws and requirements;
- Food and Drug Administration (FDA) mandated requirements;
- Foreign Corrupt Practices Act of 1977;
- U.S. Homeland Security Act of 2002;
- Identity theft issues and fraud prevention;
- U.S. Internal Revenue Service (IRS) Law for Protecting Taxpayer Information;
- Pandemic disease prevention (e.g., bird flu);
- Requirements for radio and TV broadcasters;
- ... and *more!*

Suffice it to say, you will need to check with your legal department for specific requirements in your business and industry! Having said all this, however, there are many moral and humanitarian reasons for developing a disaster recovery plan.

Are Private Sector (Commercial) Organizations Responsible for Planning for Natural Disasters?

This is a very interesting question with no simple answer. We will say that the answer appears to be at least in part a *yes*. We base this assessment in part on a recent Homeland Security summit, at which Leo attended and spoke. One of the advantages of these forums is that, as a speaker, it is possible to pose a rhetorical question like this one to the audience as a whole and receive a near instantaneous response. While not a scientific straw poll by any means, the instant response and reaction of such a group, comprised of myriad security and disaster recovery professionals, can sometimes be considered a barometer of things to come.

During his lecture, Leo posed the rhetorical question as to whether contingency planners in commercial and private sector organizations could be held accountable for failing to plan for natural disasters. He offered several possible answers.

1. Answer number one: "Of course not. That's why I pay taxes as a business owner. It's someone else's responsibility."
2. Answer number two: "Of course not. My phone line to God is out. What am I going to do as a business to prevent a tsunami or steer a typhoon away?"
3. Answer number three: "Yes, they are. Sources exist for hurricane, geophysical data, elevational data, and infrastructure vulnerability. These sources should be considered by commercial planners when deciding such matters as where to locate production facilities and when crafting recovery plans."

Much to the speaker's surprise, the audience in this nonscientific straw poll vacillated between blank stares (no opinion or never really considered this question before) and strong opinions that planners *are* responsible at least in part for planning for natural disasters. No one expressed a strong opinion that planners were *not* responsible. This unexpected response resulted in some additional discussion about the history of contingency planning in general.

When one considers the big picture, at least in terms of commercial enterprises, one can see a pattern: (1) a phenomenon whereby technology almost invariably outpaces the ability to back up that technology, and (2) a transition whereby recovery technology matures, becomes adopted by commercial enterprises, and as a result becomes standard practice. It is when the mature technology is available and affordable that it becomes the responsibility of the corporate or private sector contingency planner (see Figure 1.1). Consider, for example, first alert systems, circa 1979, which were in use by the U.S. military. These systems utilized multimillion dollar *autovon* switches, satellite communications, and various types of radio backup systems such as tropospheric scatter systems. All in all, they cost hundreds of millions of dollars.

Since the purpose of these systems was to detect missile attacks and other threats against the United States, they were "mission critical" to the lives and livelihood of hundreds of millions of Americans and their allies. Therefore, they were funded by the government and certainly not required by commercial organizations because the cost was simply out of reach. Today, the capabilities of what used to be a $50 million autovon switch fit into a $2,000 server. Satellite communications is available commercially, affordably, and practically anywhere. While other communication links used by the DoD such as tropospheric scatter radio systems are probably not available in every corner drug store, the pubic switched telephone network (PSTN) is extraordinarily diverse compared with the networks of 30 years ago. Consider the multiple communications paths, which literally are available in every corner drug store in the form of numerous wireless networks, voice over Internet protocol (VoIP) services, and other technologies. Even multilevel preemption enjoyed by the military 30 years ago is available in a fashion under IP version 6. Consider that government emergency telephone service (GETS) for landlines and wireless priority service (WPS) also allow high-priority calls to "bump" or preempt lower-priority calls. While these government-sponsored priority schemes must be arranged in advance, they are certainly much more affordable.

In summary, when the technology for backup and recovery becomes accepted, ubiquitous, and affordable, it is the responsibility of the serious commercial contingency planner to adopt it. Having hopefully made this point, let's return to the central question at hand: Are private sector planners responsible for planning for natural disasters? When set in this context, we believe the answer is *yes*.

1. Commercial organizations *do* have some responsibility for where they locate their facilities. If such a location is on a seismic fault, oceanfront, flood plane, or other risky area, recovery plans must address this fact.
2. Technology for predicting the frequency and intensity of virtually any kind of threat to critical infrastructure is available and affordable if one knows where to look for it and how to interpret it. Critical infrastructure (loosely defined as electrical power, water supplies, fuel distribution, and so on) most

The Evolution of Contingency Planning: Motivations, Causes, and Effects

1948–1968
Military development of the concept of Command, Control, and Communications (3C/3C).

1968–1978
Continued refinement of 3C/3C but Cold War standoff demands decisions faster than humans can make. This calls for the addition of *computers* to the concept of 3C/3C and the concept of 4Ci.

1978–1981
First commercial computer (mainframe) recovery centers as private sector first begins to demand backup, restoration, and recovery.

1981–1986
First Best Practices, also known as Operating and Security Standards, for mainframe computers used in private sector.

1981–1991
First intensive and widespread use of commercial telecommunications to link "mission-critical" mainframe applications. (Private line circuits and later, early packet technologies like Frame Relay and SMDS.)

1990
Publication of *Disaster Recovery Planning for Telecommunications* (Artech House), which was one of the first to document disaster recovery standards for telecom applicable to nonmilitary private sector.

1993–1998
Intensive and widespread use of client server topology to link mission-critical applications. Auditors go wild as mission-critical applications leave the relative security of the mainframe computer and spread directly to end user desktops and servers.

1993
Publication of *Writing Disaster Recovery Plans for Telecommunications Networks and LANS* (Artech House) documents disaster recovery standards for the LAN. (Also in 1993, a bomb explodes in the basement of the World Trade Center in New York, which severely disrupts telecommunications. A chilling harbinger of things to come eight years later.)

1996–2001
U.S. Telecom Reform Act of 1996 allows users the opportunity to buy the piece parts of telecom services such as "dark fiber." Price of telecom services drops dramatically, "native LAN" speed circuits (10 Mbs –> 100 Mbs) replace T1/T3 in the United States, E1/E2 in Europe. Technology such as online televaluating becomes cost-effective, first utilized in the financial service sector.

September 11, 2001
World Trade Centers destroyed. Largest mass activation of computer recovery centers in history. Not everyone makes it into a recovery center. Some businesses never recover. Results in new legislation establishing numerous new agencies such as Department of Homeland Security (DHS) and Transportation Safety Administration (TSA).

2001 to Present
Continued focus on terrorism but also natural disasters. Natural disasters are increasing. No one knows all the reasons why. Awareness of global warming issues. Hurricanes Rita, Katrina, Gustav, Ike. Indian Ocean tsunami causes unprecedented destruction. Earthquakes in Chile, China, and elsewhere. California is overdue.
The private sector realizes that communications is king. It increasingly needs a military level of 4Ci to activate complex recovery plans when the unthinkable happens. Fortunately, the cost of enabling technology (from VoIP to server hardware to all kinds of telecommunications services) drops to accommodate the need. Wireless proliferates and price drops. Local number portability comes into being with the potential capability to relocate entire area codes. Satellite communications become portable and affordable. Commercial companies specializing in 4Ci come into being for both command and control communications as well as backing up complex call centers and other operations.

2009
Disaster Recovery Planning for Communications and Critical Infrastructure (Artech House) is published. First operating and security standards and best practices for 4Ci. This book first suggests that the private sector bears some responsibility for natural disaster prevention.

2009–2012
?

Figure 1.1   High-level evolution.

certainly affects any commercial enterprise. Therefore, the responsibility for having a recovery plan for loss of any of these components rests at least in part with the serious contingency planner.

3. If your U.S.- or U.K.-based organization, for example, has its call centers in India, as many do today, it is no doubt highly dependent on undersea telecommunications cables. These cables are vulnerable to numerous hazards, not the least of which include seismic events or deliberate sabotage.

4. The authors are aware of a significant number of banks and financial services companies in the San Francisco and Los Angeles areas. In deference to the fact that they sit on geologically active real estate, these companies have invested heavily in technologies such as on line televaulting and various other point-in-time recovery systems that back up all data in real time to other remote locations. These will become more the norm than the exception and will become more affordable as the cost of telecom and storage technology continues to fall.

5. Large-scale "point in time" data recovery was totally cost prohibitive from both a telecommunications and storage technology standpoint when Leo wrote his first publications from 1990 to 1995. They were still expensive and utilized only by the financial elite for years after that. Today, they have become ubiquitous and affordable enough that any financial services organization in California that does not use something like them might be considered negligent for their failure to do so.[2]

6. Even elite and leading-edge commercial organizations like those described earlier may not have taken the next logical step of a formal vulnerability analysis of threats to critical infrastructure. We submit to you that the technology to do so is mature and affordable and that the new emphasis on natural disasters and terrorism mandates its consideration. In much the same manner as the military technology described previously, technology for actually computing the probability of and mitigating natural disasters is now available and affordable if one knows where to look for it. What is not yet mature are *operating and security standards* and *best practices* that relate to what constitutes reasonable adoption of such technologies by commercial organizations. Let's put this in terms most commercial organization should remember and understand.

   • *1980.* Most commercial organizations are using large mainframe computers and are just beginning to realize they have outgrown the ability to fall back to paper in the event the mainframe fails. Organizations realize that a fire in the computer room would be very bad, and they will adopt standards for fire-suppression systems and management of combustibles in a computer room. Companies realize that a catastrophic failure of recording media would be very bad; therefore, standards are adopted and written for backup of data and so forth.

---

2. Given the recent meltdown of financial markets worldwide, huge spikes and anomalies in trading patterns can be expected. These in and of themselves can become disastrous if adequate surge capacity is not available or if a disaster occurs at a time of financial stress.

- *1986.* Most commercial organizations have begun to connect the mainframe to branch offices, and service bureaus like electronic data systems (EDSs) spring up. Suddenly dependence on the public telephone network has become acute. In fact, the reaction of the corporate contingency planner of the early 1980s to recovery planning for the phone company was very much like the reaction often received today to the notion of planning for natural disasters. Many planners consider(ed) such an effort to be beyond their scope of responsibility. After all, they paid big phone bills, so why shouldn't the telecom carrier handle its own recovery plan? The truth of the matter was that the phone company could not be everything to everybody. The phone company might be more concerned with restoring pay phones (access by the public to 911 service) than in restoring the T3 line or frame relay network that supported your entire online network of 300 branches. Therefore, much of the 1980s and well into the 1990s was spent writing operating and security standards for telecommunications. A case in point is Leo's 1990 Artech House book, *Disaster Recovery Planning for Telecommunications.*

- *1995.* Just when the typical contingency planner (and also the auditors) felt safe, things changed again. Mission-critical applications moved from the relative safety of the mainframe and computer center and went "out there." "Out *where?*" the auditors asked. The response from the IT department was often neither clear nor comforting. In the eyes of the auditors, applications critical to the well being of the commercial organization had moved from the relative safety and security of the computer room to distributed local area network (LAN) systems that resided who knows where and were being run by who knows who. Again the industry responded with operating and security standards for LANS. Leo again played a role in the transition by publishing another Artech House book in 1995, *Writing Disaster Recovery Plans for Telecommunications Networks and LANS.*

- *September 11, 2001.* No need to really elaborate on this date, but we submit to you that the specter of terrorism, combined with the ever-increasing frequency of natural disasters, would again fundamentally change the mindset of the corporate contingency planner. Planning for natural disasters *is* now arguably in the purveyance of the private-sector planner, much as telecommunications entered the scope of responsibility in the 1980s and 1990s. Just as demand from the private sector, particularly the financial services community, drove innovative new telecommunications solutions and dramatically redefined telecom networks, those same forces will drive and refine responses to natural disasters. Once again, *operating and security standards* and *best practices* will be developed, but this time, for mitigation against natural disasters and terrorism.

What Does This Mean Today?

Disaster recovery planning continues to evolve, even though we can draw parallels with the past. We are literally starting from scratch in many respects with regard to natural disasters. For example, we have sought out a few brain trusts for inclusion in this book that the typical corporate contingency planner may not have heard of yet. One noteworthy contributor is the Pacific Disaster Center (PDC). Formed after Hurricane Iniki devastated the Hawaiian island of Kauai in 1992, the PDC has evolved into "information central" with regard to archival data on disasters, which span the spectrum from earthquakes and seismic events, to hurricanes and typhoons, to tsunamis. If information about disasters was the only requirement, however, the job would be easy. We could all just sit on the Web and mine it for free. But there would be no way to validate which information was truly valid, and then, once it was validated, how to draw conclusions from it that would be believable, and therefore find solutions and improvements that would be *fundable* by executive management. As anyone who has ever done Web-based research knows, when it comes to information it's not how much you have, but how you use it. (Sorry for the gratuitous cliché, but you can see our point.)

What we hope to accomplish in this book is twofold.

First, we believe any recovery plan—and we mean *any* plan—begins with communications. The ability to garner an immediate situational analysis and report to a responsible decision-making executive is paramount to the process. Just ask any military planner, particularly since it is generally acknowledged that "no battle plan survives contact with the enemy."[3] Any plan *will* change, and any change or deviation requires communications. In the same vein, *after* informed command decisions are made, instructions must be *disseminated* in near real time in order to react to the disaster and ultimately recover. That also requires communications.

Every plan begins with communications, and it is the single most indispensable component of any recovery plan. We will show you every technology we can find in this regard, because it is so important to any plan.

The second component of this book is truly new. It is comprised of all the bad things "out there" that could be your organization's undoing, from terrorism to a tsunami.

We must admit that our proposal for this book was met with some slanted grins from a few associates who wondered aloud whether we proposed publishing a handbook for terrorists. We actually thought about that but dismissed the notion for a number of reasons. The biggest reason is that the "bad guys" already know about a lot of what is in this book. Even so, we have downplayed a few items such as specific points of vulnerability for things like water supplies and power grids. If you want more specific information on these points, contact the authors, sign a nondisclosure agreement, and give us a chance to learn who you are. Perhaps some

---

3.   Helmuth Karl Bernhard Graf von Moltke (October 26, 1800–April 24, 1891) was a German Generalfeldmarschall. The chief of staff of the Prussian Army for 30 years, he is widely regarded as one of the great strategists of the latter half of the 1800s and the creator of a new, more modern method of directing armies in the field. He is often referred to as Moltke the Elder to distinguish him from his nephew Helmuth Johann Ludwig von Moltke, who commanded the German Army at the outbreak of World War I. See http://en.wikipedia.org/wiki/Helmuth_von_Moltke_the_Elder.

additional discussion can ensue with us on that basis or we can turn you on to the appropriate people who can help you.

Having said this, nothing here is meant to imply that this book does not have a lot of meat in it, particularly regarding natural disasters, where terrorists and bad guys of all sorts often have little or no control, just like us.[4] We are very excited about this part of the book because we believe it represents some of the first material available to the corporate contingency planner on how to do something proactive about natural disasters.

Summary

Hopefully this first chapter provided you with some of the "whys" for private sector and commercial enterprise recovery planning. We hope we have sold you on the fact that even commercial organizations must consider the issue of natural disasters and terrorism in crafting recovery plans. The rest is up to you. We hope the best practices contained later in this book provide a convenient "check list" format to help you determine not only what is possible to do, but also what is reasonable to do. In this regard, the authors welcome your sincere input, opinions, and academic critique of our books.

---

4.   In some of our earlier books and courses, we made note of the fact that many of the key telecommunications central offices that support U.S. banking and brokerage operations had serious security shortfalls, such as manholes that did not lock. Perhaps due to our efforts, eventually these shortcomings were corrected. Telecommunications vulnerability issues were further strengthened when we published operating and security standards in subsequent books. The authors believe that open publication of such standards, with the obvious exception of things that are clearly classified or directly germane to national security, benefits everyone. Open discussion and academic critique of our conclusion should be encouraged.

CHAPTER 2

The Concept of 4Ci: Command and Control in a Disaster

4Ci Communications—Cornerstone to Any Recovery Plan

CRASH! "*What* in the world was *that*?"

Sharon and Leo have used that phrase a lot over the years, having raised seven children. It usually starts with a large crash or bang originating upstairs with the kids. Leo usually tries to ignore the noise and not look up from the game on TV, but Sharon can almost always be counted on to look up at the ceiling and yell, "What happened?" particularly if sharp cries accompany the aforementioned household anomaly. "What happened?" might in fact just be the most popular phrase in our household, right after "When will the pizza be ready?" Comic relief aside, though, isn't a phrase like this one the first thing anyone asks after an unusual event?

A friend of mine, while speaking at a recent homeland security summit, brought home the point that people like things to be nice and normal. As an example, she stated that if all the lights went out in the meeting room, the first thing people would want is for them to be turned back on. This, she further correctly reasoned, was due to the fact that people don't like things out of the ordinary. After unexpected things happen, they want them to go back to their normal comfort level as promptly as possible. I liked her analogy and example. In fact, I further submit that this return to normalcy that we all crave begins with a statement of surprise, combined with a request for information about what happened in hope that a semblance of normalcy can be quickly restored by virtue of that knowledge. Common examples include:

- What in the world was that?
- What happened?
- What was it?

And right behind that people want to know what *it* affected.

- Are we in danger?
- Can we fix it?
- Will I get in trouble?
- Was I negligent?
- Am I responsible?

Any one of these thoughts or hundreds of others can be going through someone's head at the time they become aware of a sudden change out of the norm. What most responsible people are really seeking at that moment is an *instant situational analysis* of what happened, what it affected, and how it might affect them.

Those in charge of an organization may have the greatest need for such a situation report because they will either have to direct a response or get word to the responsible people so they can direct it. Experience has shown time and again that in the critical first minutes following a potentially catastrophic event, people reach for the phone. If you cannot communicate with your responders after a disaster, you are really up the proverbial creek. (For our foreign readers not familiar with American colloquialisms, that means you are in big trouble!)

As we mentioned in Chapter 1, there is a saying attributed to Helmuth Karl Bernhard Graf von Moltke, a nineteenth-century German general: "No battle plan survives contact with the enemy."

This has been true since the time of cavemen. No matter how well anyone plans, things still go wrong, situations change, and new complexities are introduced. One only needs to imagine terrified cavemen retreating after a well-planned battle while screaming, "They have *clubs*!" Ever since that time, history has been festooned with battle plans that did not go precisely according to plan. Every new development in human history, from bronze to iron to gunpowder to nuclear weapons, has changed the face of battle and the whole paradigm of a battle plan. Those who could quickly communicate and adapt had a better chance to respond and survive. The same is true in disaster recovery planning, where situations and required responses can change just as dramatically.

Surprisingly, few organizations are truly prepared in an area the military refers to as command, control communications computers and intelligence (4Ci). Let's consider each of these terms just for a second.

Command and control is the easiest one to understand. However, in a disaster it is most difficult to be in control much less in command! If you explore the origins of this saying, it originated in the military a long time ago.

Communications is how you stay in control. For example, getting that instant situational analysis when things go pear shaped. Command and control became "command, control and communications" (3C or C3) in the 1950s and 1960s. This was around the time the United States and Russia were keeping a close eye on one another for reasons germane to the Cold War. If one or the other saw a heat bloom in Siberia or North Dakota, they had about 30 minutes to decide whether it was a gas well or something they really had to be concerned about. Set in this context, a "false positive" or a "false negative" was equally catastrophic. If you were told a missile launch was a gas well and did not launch your nation's missile, your retaliatory capability after the first strike would be destroyed. If it was the other way, and you were told a gas well was a missile and launched based on the erroneous information, the end result would be equally bad. Obviously the communications component in any response played a big role when set in a Cold War context. Today it's not quite as dramatic, but reliable, instantaneous communications still play a role, even in commercial or private sector organizations. For example, companies need to know instantly whether to activate their expensive disaster recovery centers (and pay that big activation fee) or hold out for a while to see if their organizations can fix the problems. Sometimes in widespread disasters, recovery centers fill up quickly and people who do not activate them quickly enough may not get in or may have to travel to another recovery center hundreds or thousands of miles away.

Earlier in Chapter 1, we discussed the fact that recovery time for bankers and brokerage companies could in fact be measured in seconds. On September 11, 2001, some 200 companies activated recovery centers according to many sources. Many did not get into their center of first choice. This phenomenon is more pronounced in cases like Hurricanes Katrina, Rita, and Ike. The decision whether or not to evacuate and activate an expensive recovery center (sometimes still called a *hot site*) is based entirely on information received by decision-making executives. Like the military, a false positive (the decision made to activate a hot site unnecessarily) and a false negative (the hot site is activated too late to assure a nearby facility of choice or is not activated at all) are both costly mistakes. But this can be alleviated through quality communications and information.

Even before a decision to evacuate or relocate is made and regardless of whether or not you have a backup center, you have to know whether you even have resources to recover or whether key employees necessary for recovery were casualties of the event. Again, this means an accurate situation analysis early on in the recovery process.

Computers   Like the days of the Cold War, where "launch the missiles" decisions had to be made in minutes, today business decisions must also be made in minutes. That's too fast in many cases for people to do without computers. *Computers* in this context are not the WOPPER 6000 in the 1980s movie *War Games*. Today, computers in the context of 4Ci are all of the items in your pocket, in your car, on your hip, and on your desk that help you think faster. Just having a Blackberry with everyone's phone number in it sets you way ahead of the game when responding to a disaster. In this context, a Blackberry is a computer, a cell phone is a computer, and a two-way pager is a computer. All of them help you think faster and communicate with responders quicker.

Intelligence   Again we are back to needing to know what is going on, what went bang, what it affected, and whether you need to cancel your dinner reservations tonight or wake the CEO. You have to know what's going on in order to react to it. That's what *intelligence* is in the context of 4Ci. The ability to know what happened, with 100% certainty, in minutes, regardless of what happens, in order to be able to coordinate recovery as soon as possible.

Establishing and maintaining 4Ci is the most important first step in *any* recovery plan, for *any* organization, and I dare you to prove me wrong. Try recovering anything with your eyes and ears covered if you don't believe me.

For those of you fond of "table top" disaster recovery exercises, try one sometime and don't allow anyone to speak! This will underscore the importance of 4Ci better than anything else will. Make people use e-mail, cell phones, and instant messaging. It's an interesting twist, and it happens for real when telephone facilities are affected, as described later in this book.

Disaster planners deal with a different kind of enemy than the military does, but one no less likely to morph or change shape and nature as any military opponent does. In order to drive this point home, I have often used stories and examples like those in the following section.

"Houston, We Have a Problem"

Consider that infamous phrase and the importance of communications in responding to an unforeseen catastrophic event. In a disaster, there is no substitute for instant, reliable communications with people who can help. Conversely, there is no more lonely a feeling than when cut off and alone. For those of us who remember that week in April 1970 with the Apollo 13 moon mission (or even if you just saw the movie), consider how important communications were to the safe return of the astronauts. Some of the presumed greatest minds in the world at NASA had to wing it when the unthinkable struck the third lunar mission. Imagine what would have happened if the explosion that crippled the spacecraft also crippled the radio? No CO_2 scrubber (which was fabricated after instructions were relayed from ground controllers) would have been possible, so the astronauts would have suffocated. No reentry angle data (also relayed from the ground) would have been transmitted, so the space travelers would have skipped off into space or burned up on reentry. You can pretty much take an example like this one as far as you want, but it does get people thinking. Imagine Apollo 13 without a radio. Survival would have been impossible, just as survival is impossible today for most organizations that cannot communicate. I watched the entire situation play out as a 14-year-old, and even at that time in my life the importance of communications to those imperiled astronauts was apparent. What was also apparent was that when you give good people the means to coordinate as well as communicate, some extraordinary things can happen.

Now envision a hospital, bank, reservations center, and travel agency, help desk, 911 center, or large retailer. Millions of dollars course through facilities like these every day in cities just like yours. But what if there is an interruption in telephone service? What if there is a fire, an earthquake, a tornado, or terrorism? Even in cases where money is not the main concern, lives may well depend on continuance of services you provide. What is your plan if disaster strikes? Who would you call? How would you contact them? Where would you send your inbound calls? Inbound calls, or, more specifically, inbound communications (inbound calls for the Home Shopping Club, inbound Internet transactions if you are eBay), are the lifeblood of today's companies.

Stated another way, the two most immediate challenges after a disaster involve the following:

1. Alerting the "right" individuals to the disaster;
2. Delivering instructions to orchestrate a safe and effective response.

Corporations, campuses, and government entities must quickly organize and disseminate information to every "first responder" using every conceivable type of technology. Response teams often include executives, building managers, security personnel, fire and emergency medical services personnel, grief counselors, legal personnel, transportation personnel, finance personnel, information technology personnel, telecommunications personnel, and more. The technology used to notify these teams will run the gamut, depending on which communications media survive, including telephones, wireless phones, laptops, PDAs, e-mail, and two-way radio. The technology used depends largely on who is being notified, why, and where they are. But what if an organization plans for one disaster, but a different one occurs? The ability for a plan to change quickly in response to changed circumstances is a major aspect of 4Ci.

Or, as another example, picture yourself spending all of your time planning for a hurricane and getting the most effective plan in place for that particular contingency that your company's money can buy. Shortly after finishing the hurricane response plan, a deranged person enters your building, announces that he is mad at the world, and commits suicide in your lobby. Now you have a different kind of problem altogether. Can your plan change quickly, and does it have a notification system for a whole different set of responders? Rather than calling storm shelters and recovery centers, you will be calling grief counselors and other kinds of responders, as there will certainly be some hysterical people to deal with in the aftermath. That's the way it is with disaster recovery. You plan for a hurricane or some other natural disaster, but the problem turns out to be a disgruntled employee in your computer room—with a permanent magnet. Please don't accuse me of being morbid, but take these examples seriously nonetheless. A lot of people at the U. S. Post Office now are more cautious after the events of a few years ago. (Where do you think the phrase "going postal" came from?) The world is full of various people, so what can happen is equally diverse. The point to take away here is that your 4Ci solution must adapt on the spot to unforeseen circumstances.

"The Right Stuff"

Numerous commercial companies have responded to the need for 4Ci in the private sector. These firms have the advantage of a tailored product geared primarily to business resumption. A good one can actually help you organize your recovery effort and keep you in control by helping you get the word out immediately about any emergency, including severe weather, fire, sabotage and terrorism, system failures, and more. A few of them are highlighted in this section. Some considerations when evaluating the myriad commercial alternatives available today should include several of the following options.

Checklist for "Johnson Space Center" Quality Disaster Recovery

- Can the service redirect inbound telephone numbers instantly to any working telephone number, whether a branch office, home, wireless, VoIP, or satellite phone, *without having to call the phone company*? This obviously saves the time delay of trying to disseminate "new" emergency numbers later.
- Can the service duplicate the same call prompts ("Press 1 for Sales, 2 for Customer Service," and so on) as they exist today in your network? Transparency

to the customer preserves your normal business and projects the mirage that everything is fine to callers, even if they know a disaster has occurred.

- Can the service initiate your emergency response plan through a Web connection, touchtone phone, PDA? Will the service provider do it for you if you are completely cut off? Since it's impossible to say with certainty which telecom technology will serve a disaster, planning for several leaves you "room to live."

- Does the service offer "find me, follow me" features? Whether at home, in the office, in the car, or via pager, PDA, or text message, can you assure you will never lose touch with key first responders? The "find me, follow me" feature tries all these devices until the person is found. There are more phones in service per individual than at any time in history. A service that rings each number for you (home, office, wireless, an so on) saves you the trouble of documenting all those numbers.

- Will "find me, follow me" work by dialing the person's *regular office number* (as opposed to a new one you have to remember and document)? As stated earlier, plan no more than one single number per person, and use technology for the rest. Then keep that number up to date in your secure PDA or wireless phone.

- Will the service broadcast to numerous users? Can emergency notification messages (ENMs) be set up by you and then broadcast to hundreds of responders based on the disaster? This is very much akin to the "launch the missiles" drills used by the military. There is no substitute for getting all your responders online at once in some situations.

- What kind of feature richness can you expect on outbound notification systems? One system we have found actually lets the recipient, upon hearing a recording, to instantly be placed into a voice conference with other members of his or her recovery team, to be connected with a live person, or to hear more detailed instructions. That is a very useful feature in a disaster! Otherwise, you could end up listening to 60 people's voicemail at the same time!

- Can outbound notifications be sent to any network-connected device (phone, PDA, pager, and so on)?

- Can outbound notification systems be recorded with specific information organized by message recipient, recovery team, department, or other criteria, based on their role in the emergency response? For example, some organizations will want to send messages to thousands of employees telling them *not* to come to work, either for safety concerns or because they will just be in the way.

- What if you plan for one disaster, but another disaster happens? Can the caller menus for the commercial solution you selected be changed quickly? Can this be accomplished by a touchtone phone if you can't get to a PC, or vice versa?

- Can voice conference bridges be activated instantly by you, based on department, telephone number, individual persons, or recovery team? Conference calling is enormously helpful in any response.

- Can the system forward faxes? Can it forward faxes to e-mail accounts? Hotel fax machines? Hold them in storage for later retrieval?

- Can the system send faxes including *broadcasting* to hundreds of locations?
- Have you identified hotels in your area with high-speed Internet access? Why send people hundreds of miles to find workspace and lodging? Any nearby hotel with high-speed Internet access can be a recovery center with the right commercial solution.

In one manner of thinking, the 4Ci aspects of disaster recovery have become a lot easier, thanks to VoIP, managed private branch exchanges (PBXs), and some good old-fashioned innovation by commercial providers. These solutions offer a quantum leap in recovery technology on a subscription basis at zero capital expense and a manageable monthly cost. In our opinion, no recovery plan today is complete without an evaluation of what's available in this area.

Truly, in today's organization, "Failure is not an option." You must establish 4Ci, no matter what the circumstances, as the first step in any emergency response. Fortunately, affordable technology exists today to do this and is described in more detail at the end of this chapter.

Every Recovery Plan Starts with 4Ci

Every disaster recovery plan begins with 4Ci, bar none. After an event occurs, the two most immediate challenges involve the following:

1. Alerting the "right" individuals to the disaster;
2. Delivering instructions to orchestrate a safe and effective response.

These are what we can refer to as the "critical first two hours." (After the first hours, 4Ci is still in play but it takes on different dynamics, such as coordination and allocation of resources dedicated to the recovery, which is more logistical in nature.) 4Ci also assumes the creation of disaster recovery teams of one kind or another, in order to assign tasks and designate responsibilities in a disaster. Disaster recovery teams are discussed in greater detail in later chapters. For the moment, suffice it to say that corporations, campuses, or government entities must quickly *organize* and *disseminate* information to every first responder using every conceivable type of technology. This is not only limited to the telephone. As Katrina, Rita, Gustav, and Ike proved to us, oftentimes the phone is not available. In these cases, other technologies may be available, including but not limited to the following:

- Blackberries and other smartphones;
- Wireless phones;
- Laptop computers;
- Instant messaging;
- VoIP phones;
- Satellite phones;
- Satellite and two-way pagers.

Each of these technologies is useful to keep response *teams* in touch with incident commanders responsible for coordinating the recovery and with one another. Use as many of them as you can. Leave yourself room to live.

Multiple "safety nets" work in communications. If telephones are out, wireless might work. If wireless is out, satellite phones and satellite-based text messaging or paging might still work. We have heard of instances where companies planned four different technologies (phone, wireless, satellite, and VoIP) and three out of the four nets broke. In one case, the government took the satellite frequencies after phones and wireless failed. This particular organization used a wireless Internet service provider (WISP) to get IP connectivity back and pull a VoIP dial tone in order to stay in business! The more different and diverse technologies in day-to-day use in your organization, the better the odds you will find one that works in a widespread disruption.

So Who You Gonna Call?

Response *teams* should be designated in advance and included in the recovery plan. Teams often include executives, building managers, security personnel, fire and emergency medical services personnel, grief counselors, legal personnel, transportation personnel, finance personnel, information technology personnel, telecommunications personnel, and more. And as noted earlier, the specific technology used to notify these teams varies with the extent of damage to normal communications infrastructure but includes telephones, wireless phones, laptops, PDAs, e-mail, and two-way radio. The technology used also depends largely on who is being notified, why they are being notified, and where they are when the disaster occurs.

Even small organizations cannot survive a prolonged outage—a telephone cable cut, for example—without a plan. The real question quickly becomes, how can any organization, large or small, remain *in control* of the recovery process and establish 4Ci when phones are down? What if the telephone company is busy elsewhere or has a disaster itself? What if an organization plans for *one* disaster but *a different one* occurs? Can its plan change quickly?

Moreover, while you are in the planning mode, don't make the mistake of assuming that disaster recovery and 4Ci need to be "dead expenses." Actually, quite the contrary is true. If you are a service provider, efforts that help your client establish 4Ci and disaster recovery effectively differentiate your product from your competition and, if properly served up, can make you a lot of money. The following examples and case studies are true stories and underscore this often overlooked fact.

Case Studies and Overview of Commercially Available 4Ci Technologies

Example #1: How We Made $2.5 Million in Airline Business Revenue Because We Provided Disaster Recovery—A True Story!

Over the past 20 years of my career, I have steadfastly maintained that disaster recovery need not be a money-losing proposition. In fact, the right solution can differentiate "commodity" products and earn new business. This is precisely what happened from 1999 through 2005, when Sharon and I owned a 50-state certified telephone company. That company was principally in the business of selling local telecommunications services but added value to those services by addressing disaster recovery as a component.

In 1999, our telecom company was courting a large, Dallas-based airline that had an issue with how customers could communicate with them when the Dallas–Ft. Worth (DFW) airport experienced ice storms. Ice and snow are relatively rare in the DFW area, but they do occur often enough in the winter to present a problem. When it ices outside in North Texas, the call center personnel for this airline could not get to work, since DFW has little, if any, means to deal with winter weather. Anyone who has ever transited DFW airport knows that ice (and thunderstorms) can wreak havoc on airline operations. The airline already had a partial solution for their inbound long-distance (toll-free 800) service that is still useful today. The technology is known as *command routing* (described further in Chapter 7) and is designed to redirect inbound lines from one geographic location to another. In this case, the airline's DFW call center could redirect inbound 800 service to Raleigh, North Carolina, if an ice storm hit DFW. Command routing is nothing new. It's a 25-year-old technology and can move lines in minutes. You should already be using it.

This was all well and good for the 800 numbers. However, the airline also had hundreds of local numbers due principally at the time to the high cost of intra-LATA (nearby) 800 service and to the fact that the call center received overseas calls on these lines (800 service does not work outside North America). In order to move the local lines, it was necessary to call Southwestern Bell (SWBT), the local carrier. Unlike command routing, at the time SWBT had no formal procedure on how to move these lines. SWBT instead resorted to a manual process in the switch—one that resulted in errors and took most of the day to accomplish. In the meantime, ice storms in the United States generally move east. Tomorrow Raleigh would have the ice/snow problem and would wait to move *its* calls to Dallas. It was the devil's own dilemma for the airline.

That was when we had our first chance as a disaster recovery-savvy service provider to try something different. We offered the airline the following proposition:

> *How would you like a disaster recovery service that will move 99 local lines at a time to any number you select in North America from the warmth and comfort of your own living room?*

You have to admit that claim gets a lot more attention than the typical "I can save you money" mantra. The airline was very interested. Here is how we provided this service:

Everyone (in North American anyway) is familiar with the *72 call forwarding feature to forward calls. Not everyone knows that it can be accessed remotely using a feature called remote access to call forwarding (RACF). Using RACF, a caller dials a predetermined ten-digit number and enters a four-digit pin code, along with the number at which they want their calls to ring. Once this info is entered (in seconds), the calls now ring automatically at the new number.

There was still an issue, however. The typical RACF service available under the SWBT tariff only forwarded one line, or *call path*, at a time. This was hardly suitable for an airline with hundreds of inbound lines. Remember, however, as a SWBT competitor at the time, our company had the same kind of certification that SWBT did and had access to the same raw parts and materials. It was kind of like being a gourmet chef with an unlimited cooking pantry. We could cook up items for clients on special diets and did so in this case for the airline. [We are only telling you all this so you know how to handle your own telecom vendors—they can do the same thing if you know how to ask! Call ILEC (incumbent local exchange carrier) competitors like Houston-based CLEC (competitive local exchange carrier) Cypress Telecommunications!]

Getting back to our example, as part of how we addressed this issue for the airline, we did about an hours' worth of research on the SWBT 5ESS and Northern DNS 100 switches that served the airline. Each was capable, according to the manufacturers specifications, of forwarding 99 calls at a time. The issue was not that it couldn't be done, but that SWBT didn't do it. The switch technology could handle the problem. What's more, while hundreds of lines under the SWBT tariff would have been clumsy (imagine trying to restore hundreds of them, one after another, using *72) as well as cost prohibitive (since each was billed essentially as a line), the switch technology allowed for something much cheaper and simpler. That's where our company came in, given that we were a telephone company, too. We did not have to do things the way SWBT did them. As partially explained earlier, we had access to the same "ingredients" SWBT did—something called *unbundled network elements* (UNEs).

In order to understand UNEs from a layperson's perspective, imagine for a moment that the Telecom Act of 1996 (which opened SWBT's market to competition) did not break up SWBT. Imagine instead that it broke up McDonald's. If that were the case, some people in this example would resell completely assembled Big Macs from McDonald's at a discount. In telecom, this is called, not surprisingly, resale. Resale is based on the tariffed rate, and the reseller must follow many of the same rules as the provider.

Other enterprising individuals in the McDonald's analogy, however, would buy the *elements* of a Big Mac. Remember the popular McDonald's jingle of years past?

Two all beef patties
Special sauce
Lettuce, cheese, pickles, onions
On a sesame seed bun.

By buying the elements of a Big Mac, a competitive hamburger provider could cater to people on a special diet. Vegetarians and vegans, for example, might want two soybean patties. My doctor might want me to substitute a cholesterol-free beef patty. Someone else might have an allergy to lettuce or not like the special sauce. Before I make everyone hungry, let's talk about how this related to solving the airline's problem.

In precisely the same manner, a phone company—in this case, a competitive local exchange carrier (CLEC)—could cater to users with special needs and wants. One network element or UNE was called *local switching*. By buying the local switching with the RACF capability needed by the airline, we deployed this service right out of SWBT's own switch. All that was necessary was to "turn on" a feature (which resided in that switch since the day it was purchased from the manufacturer) that told it to turn on 99 paths instead of one. In the process, we gave the airline what it needed, saved them 50% from what they previously paid SWBT, and grossed over $2 million for ourselves. That's because as a facilities-based provider (one that did not have to follow SWBT's rules and pricing like a reseller), we could price the service any way we wanted. Switch features are not inherently expensive. They are, however, often priced hundreds or thousands of times what they cost at the wholesale level by the local carriers. (In its defense, this markup sometimes off-sets other services priced at or below cost.) Since our company did not engage in the same kinds of cross subsidies as SWBT did, and since we were free to price any way we pleased, we did so on behalf of the airline.

And here is the punch line to this true story: Since the feature we ordered to deploy this service for the airline was priced at a "wholesale" rate, not a tariff rate, it was based on something phone companies and regulators call *total element long run incremental cost* (TELRIC). Since the feature *already existed* in the 5ESS and DNS 100 switches, there really was no wholesale TELRIC cost other than to turn it on. This was logically computed to be the amount of time required for a technician to enter the command in the switch to turn it on. That cost was *five cents*. You heard correctly, a nickel. We got over $2 million in business from an airline for a nickel.

In view of that fact, no one has been able to convince Sharon and I that disaster recovery must be a money-losing proposition.

But what if you are dealing with a public sector organization, such as a government office? We offer you another true example of how inexpensive switching components literally saved a city hall here in North Texas that literally burned to the ground.

Example #2: City of Red Oak Fire

Another entity that was a customer while we owned our phone company was the city of Red Oak, Texas.

This particular customer was smaller than the airline, but even so we added RACF to most of their lines because it was still cost effective to do so. Recall from the previous example that the RACF feature cost only five cents to add on to a phone line for all the reasons cited in the previous example. Why not, we reasoned, add it to all lines and package it as a value-added feature? This is precisely what we did. After all, it cost us next to nothing to do so, and it created good will with the client, who perceived they were receiving extra value from us. We installed the feature and gave the client the appropriate personal identification number (PIN) code to activate and deactivate the feature. As often happens, however, personnel changed at the city, and they forgot they had this capability. Now fast forward about two more years to when the city experienced a devastating fire.

Sharon woke up very early one brisk morning and turned on the television set to get the day's news and weather report. In the middle of the broadcast, the anchor-man cut to breaking news: A big fire at Red Oak city hall (see Figure 2.1). Sharon immediately let Leo know (OK, she woke Leo up) about the situation, so that he

Figure 2.1   City of Red Oak fire. How would your organization fare if this was *your* place of business?

and another employee could get down to the city right away to offer assistance. Upon their arrival, the city was in an absolute panic. Everyone knew about the fire. Several television vans had pulled up. Everyone wanted a statement. While the fire was out by that time, the city hall was almost a 100% loss. Most distressingly, the phone lines that thousands of citizens depended upon were terminating in a burned out shell. Calls were not being answered. From a liability and public policy stand-point, the situation was intolerable.

Luckily, Leo and staff were able to send all calls to another location within the hour, due to the RACF feature that was installed three years earlier and subsequently forgotten about at city hall. Calls were redirected to the Ellis County sheriff's department, which had the capacity to take them and which was manned 24 hours. Was the solution perfect? Not exactly. People calling about trash pickup, water bills, or dogs in the yard were surprised to get the sheriff's office. What was maintained, however, was the essential element of 4Ci. People got instructions on what to do in nonemergencies. Emergency calls could be routed. Maybe most important of all, anyone who saw the fire on TV and called city hall (to check per-haps on the status of loved ones who worked there) was answered and remained informed.

Afterwards, our customer thought we hung the moon. All for a feature installed in advance that cost only cents per line to deploy. There is no reason your company should not be doing the same thing. What's even better, however, is that the tech-nology for handling calls in these situations has grown by leaps and bounds. By combining call fowarding technology with other available options, that "Johnson Space Center" level of 4Ci that companies need today is available and affordable. We discuss just these kinds of options as we proceed with this chapter.

Commercially Available Options for Call Forwarding, 4Ci, and Inbound Call Recovery

As we stated earlier in this chapter, a respectable number of firms are beginning to tackle the long-neglected 4Ci issue as well as the issue of inbound call recovery. The vast majority of them are small outfits that have discovered they can slap a few Dialogic boards into an Intel-based server and broadcast lots of emergency messages either in voicemail or e-mail format. I am not being disrespectful to these small firms and believe that they provide some great basic power tools for response. Some of these newcomers are somewhat more sophisticated in providing technology such as a "hosted PBX" solution. In these cases, an organization's PBX (private branch exchange) is located remotely, presumably in a hardened facility such as a carrier collocation. The connection to the user is often provided via a high-speed VoIP connection that is useful in a number of ways. One of the biggest advantages is as follows: if a disaster renders your building inaccessible, wherever you can find an Internet hot spot in addition to some space to recover (such as the Holiday Inn), you can be back in business and still have all of your old phone numbers. This is really cool.

Hosted PBX Systems

A hosted PBX system delivers PBX functionality as a service, rather than having one on site. Instead of buying PBX equipment, users contract for services from the hosted PBX service provider. Many are surprisingly feature-rich and useful in a disaster. Functions such as find me, follow me calling, actually came about in the hosted PBX world before they became available in traditional PBXs. In fact, these days it is possible to get hosted PBX service that includes far more features than you probably have now, if your present PBX is more than a few years old. What makes them appealing in a disaster, however, is they are not in your building like your PBX—they reside off site. With some creative engineering, they become very useful.

A user of a hosted PBX solution does not install any PBX equipment and there is no capital investment. Instead, the hosted PBX equipment is maintained by the service provider, who then shares access to the system among many customers. As with premise-based PBX systems, key functions that can be provided by a basic hosted PBX include, but are not limited to, the following:

- Custom greetings;
- Dial name directories, such as connecting to a specific extension or department or dialing based on a user's name;
- Automatic call distribution (ACD);
- Music on hold while callers wait for an available department or employee;
- Voicemail;
- Call transfer between users or extensions;
- Conference calls.

With just a little forethought, these solutions become very useful in a disaster. Even so, not all of these functions are available from every provider of hosted PBX services. You determine what functions best match your organization's needs. A little effort on your part will pay great dividends in your organization's ability to respond to a disaster and maintain 4Ci. A couple of companies that pop up quickly on a Web search for "Hosted PBX" include http://www.onebox.com and http://www.virtualpbx.com, and there are *many* others. Find one that works for you and improve your recovery plan for very short money and no capital investment!

Outbound Notification Systems

Planning for the essential 4Ci component of your recovery plan consists of getting the word out *immediately* about any emergency, including severe weather, fire, sabotage and terrorism, system failures, and more. (The shooting rampage on April 16, 2007, at Virginia Tech is a noteworthy example.) These technologies are referred to as *mass notification systems*.

The purpose of a mass notification system is to get word out to lots of people quickly, whether the receivers of the emergency messages are residents of a city, employees of a company, first responders like police or fire personnel, students on a campus, or whomever. In the past, someone had to do the work and sit on a phone calling number after number. Technology has introduced equipment such as auto dialers that call a list of numbers and play a recorded message (probably just leaving hundreds of voicemails given the way we do business today). These days, systems exist with very advanced features such as find me, follow me that ring a responder's office, home, mobile, and pager numbers sequentially until the message is delivered. Some ask the recipient to press a key to acknowledge that they received the message, assuring that a critical bulletin is not languishing somewhere on a voicemail system. Mass notification is only one aspect however of a full-blown 4Ci system. An equally important feature in such a system is the ability to restore inbound phone calls. This is much more difficult. Many recovery plans, in the past, responded to changed work locations by publishing "emergency" phone numbers that were different from the original numbers. This in practice just does not work. In the heat of a disaster people do not remember to look up new numbers. Your customers certainly will not know them either. It is therefore essential to redirect the original numbers to the emergency work location to maintain 4Ci in a disaster and assure your customers that everything is OK.

We have highlighted several commercial providers of services that are suitable for 4Ci communications in this chapter. The addresses for several more are available in Chapter 10. This specific section is not intended to endorse any particular product or service; rather, it is designed to familiarize the reader with the specific characteristics of various providers so that informed choices and assessments of technology can be made.

One-Way Mass Notification

One method of communication that most people often overlook is a low-power AM radio station. Useful for airports and traffic control to broadcast information to the public, it is also effective in campus environments and universities. When I was mayor of a small municipality, we bought one of these to augment a siren system purchased at the same time. The idea was that if someone heard the sirens go off, they would tune in to 880 kHz AM to see what was going on, rather than go look out the window and possibly get killed by a tornado or other event. In practice, it worked very well. In the "normal" mode, the system broadcasts community events and announcements. In a disaster, the fire or police department could quickly change the announcements and broadcast emergency instructions. Running at about $15,000, it was relatively cheap, given the peace of mind it provided. (The system has been in for more than 10 years now and is still running, so the actual monthly cost over 120 months is well less than $20—a bargain by any measure for the service it provides!) Obviously, this communication media is only one way to get the word out, but a low-power radio station fits the description of a mass notification system and merits consideration, particularly in campus and other locations.

Soft Switches

Other mass notification systems that fall under this general heading may be based on VoIP or traditional telephone technology. Any system of this type is better than nothing at all, but many are actually quite good. A number of systems are enterprise-focused and function only when the user is in the building or has e-mail access. A few are also not suitable for a residential offering or any type of mass market solution (e.g., "I've fallen and can't get up!"). Many systems have no real-time knowledge of how much network capacity they have at a given moment or call load, resulting in jammed calls and notifications or otherwise thwarting timely delivery for important message notifications. Some have little leeway for user-defined delivery options or other kinds of customization, which is important because every organization is different.

Ideally, a notification system is one that enables an organization to respond appropriately to attention-grabbing emergency events (tornado warnings, hurricane/tsunami warnings, terrorism events, amber alerts, campus alerts, hazardous chemical spills, medical pandemics, and so on) by rendering reliable and timely communications to all stakeholders—anytime, anyplace, anywhere.

Inbound Call Recovery

One company, Telecom Recovery, Inc., provides an interesting package that accomplishes most of the goals of the contingency planner to provide 4Ci. Telecom Recovery provides a backup PBX system designed to answer or redirect inbound calls. The system maintains 4Ci as well as the image of "business as usual" in the eyes of callers and customers. The system can also perform outbound notification, which can be very helpful in getting the word out to responders. A system like the one provided by Telecom Recovery offers a few essential services:

1. It allows the user to redirect inbound calls without having to call the phone company. This is important for obvious reasons. Imagine trying to get Bell South on the phone in the aftermath of Hurricane Katrina.

2. It allows the graceful coordination of hundreds of employees, citizens, or other first responders, by team and function. One cool feature is the ability to actually dial out specific groups of people—even by recovery team—and place them in instant conference with one another. This could be an indispensable means of retaining 4Ci in a disaster because the machine is remembering the phone numbers, not you.

3. This system, unlike hot sites, is used widely for several reasons. First, there is no activation fee. Second, the customer, not the provider, turns it on and off. With more than 500 excavation events every day in the United States alone, telephone cable cuts are far and away the most popular reason this system is used, followed by power failures. Unlike a hot site that a user hopes they will never use, this service is one a user will use. After all, who among us has never experienced a telephone cable cut and outage?

Telecom Recovery states that there is no need to change service providers and the solution is *vendor independent*. That means all local, long-distance, and 800 services stay with your present provider. Since each vendor has different methods for forwarding calls, Telecom Recovery has a proficiency in those technologies as well.

A second provider we have highlighted is called Telecontinuity. This provider offers a similar service but takes a somewhat different approach. Telecontinuity is a standby voice communications network that is offered on a subscription basis. The Telecom Recovery and Telecontinuity examples that follow are not intended to endorse any particular product, but to show what the current state of the art is in the industry. By combining these services with the various command routing and RACF features described in this chapter (and later in greater detail in Chapter 7) essential inbound calls can be transparently restored and critical 4Ci capabilities can be maintained.

Finally, take a look at MessageOne Communications in Austin, Texas (recently purchased by Dell Computers) and Houston-based Cypress Communications. Both offer solutions for recovery of enterprise e-mail exchange services. E-mail is also an essential component of 4Ci. Still more ideas and solutions are discussed later in this chapter.

Telecom Recovery White Paper Highlights the Need for 4Ci and Survivable Communications

The following white paper describes a service by Salt Lake City–based Telecom Recovery as well as some interesting aspects of 4Ci. We thank Telecom Recovery CEO Tim Ruff for his contributions to this section.[1]

---

1. *Best Practices for Communications Recovery for Business, Non-Profit, and Government Organizations* by Tim Ruff, CEO, Telecom Recovery, Inc.

Disasters vs. Typical phone outages

It's a simple fact that the frequency of natural disasters is accelerating worldwide [see Figure 2.2], and disasters most certainly cause significant damage to telecommunications infrastructure when they occur. However, a telecommunications outage is literally thousands of times more likely to be caused by a cable cut, power outage, or equipment failure. In the excavation industry alone there are approximately 185,000 accidents per year that damage telecommunications lines [shown in the D.I.R.T. Report stats; also see Figure 2.3].

Occasionally, outages are caused by critters of all kinds, including bees, squirrels, snakes, eagles, and even bears, and more frequently by copper thieves. Here are a few of the more interesting examples:

January, 2007, Alaska

10,000 Juneau residents lose power after a bald eagle lugging a deer head crashed into transmission lines. (http://www.msnbc.msn.com/id/16875766)

September, 2007, Texas

Bees attack construction worker, who jumps off tractor, hits lever, lowers auger, severs fiber-optic line. The telephone outage lasted 7 hours. (http://abclocal.go.com/ktrk/story?section=news/state&id=5665059)

September, 2008, California

Thieves attempt to steal copper from phone poles outside San Ramon Medical Center. The telephone outage lasted 11 hours. No word about if they were successful ... (Successfully recovered by Telecom Recovery on 9/18/08)

Though cable cuts, equipment failures, power outages and other temporary outages are not quite disasters (unless it disrupts your call center, where any outage

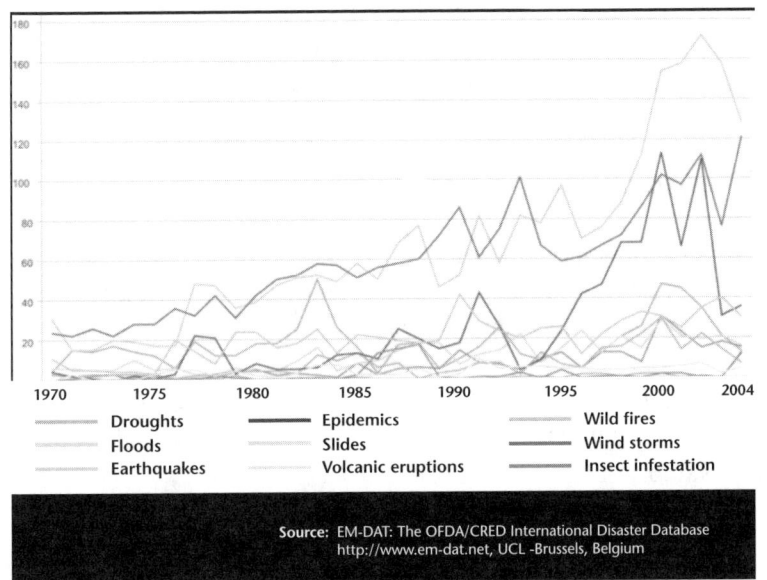

Figure 2.2 Disasters of all types are on the increase. No one knows precisely why.

Figure 2.3   Back hoe.

is a disaster), they can be aggravating, embarrassing, and quite costly nonetheless. It is wise to be prepared for both extremes: the far-more-common short-term outages and rare but potentially longer-term disasters.

The D.I.R.T. Report:

In 2004, an estimated 675,000 excavation accidents in the U.S. damaged underground cables or pipelines. 27.5% (185,000) of the accidents damaged telecommunications facilities. — December, 2005 Common Ground Alliance (CGA) D.I.R.T Report © 2005, Common Ground Alliance, all rights reserved

Inbound Versus Outbound Call Recovery

All of the following items are important tools to help recover communications after an outage, but they all share an important weakness: they can't help callers who are calling inbound into your published and other phone numbers.

- Cell phones;
- Satellite phones;
- Radios;
- Pagers;
- Calling tree;
- Notification system;

- GETS/WPS (Government Emergency Telecommunications Service/Wireless Priority Service).

You can have all the mobile and satellite phones in the world and a top-flight outbound notification system, but it simply cannot help the customer or constituent who's calling you, instead of the other way around.

Several communications methods can be recovered or assist in recovering lost communications. Though not all-encompassing, as communications technology continues to evolve at a rapid pace, this discussion is quite comprehensive and can be used to assemble an effective emergency communications plan.

Inbound Calls

When the local phone numbers are down, the pain can usually be felt immediately by both the callers and the organization. Though unfortunately a common daily occurrence throughout the United States, this type of outage can be the most difficult to effectively recover from.

Almost every local phone number, including direct inward dial (DID) numbers on high-capacity lines, can usually be forwarded to any 10-digit number in the United States by the local phone company. Configuring and testing call forwarding, especially for higher-capacity circuits such as T1s and DS3s, can be intricate and should not be left to the last minute or trusted entirely to the carrier (even though call forwarding is the carrier's product, they're often not able to get complex forwarding right); external expertise in this arena should be sought. Supplying a 10-digit number to forward to that provides sufficient line capacity and manpower to handle the calls, as well as an effective and professional experience for the callers, can also be a dilemma, especially if the best people to answer the calls are still inside the building. Though tricky, recovering calls back to individuals inside a building that's experiencing an outage is more doable than ever, thanks to the ubiquity of phones of all types and the growing number of creative ways available to deliver phone calls (see Figures 2.4 and 2.5).

For a small business, forwarding to a cell phone may be sufficient. For larger volumes of calls, such as call centers or other similarly busy phone numbers, recovery can be as simple as forwarding calls to an alternate location, provided an alternate location exists and has the capacity and people to handle the increased call volume. Where no adequate alternate location exists, forwarding to an off-site backup phone system is preferred, as it can provide the required capacity, transparency, and professionalism for callers, some control to adapt as the situation evolves (the degree of control varies greatly from system to system), and even the ability to route recovered calls right back into the same building on mobile phones, analog/copper lines, VoIP, or any other phone numbers that can still ring. This capability is of great benefit during the vast majority of outages that are usually temporary and require no evacuation of the building, and it can be professional and nearly transparent for the callers as they continue to dial the same phone numbers as before the outage; they may even be completely unaware that an outage has occurred.

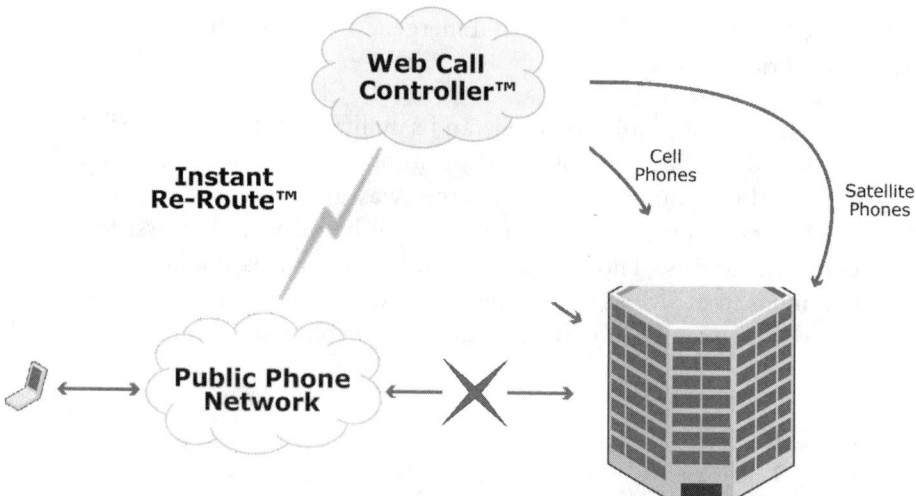

Figure 2.4 Using a commercial recovery solution in combination with the call forwarding solutions by the phone company (see the example of the airline case study at the beginning of this chapter) makes for a superior 4Ci solution—for your organization and its customers.

Figure 2.5 Hear an actual example of a voice telecommunications recovery system now by dialing (888) 843-3285.

The largest obstacle to the installation of an off-site phone system has been the high upfront and ongoing costs of the system, maintenance, hosting fees, and the connected phone lines and Internet service. Where budget is a concern, it is now possible to purchase a telecommunications recovery service in a "shared service" model similar to insurance, where no equipment or software is necessary and it is only activated when needed, after an outage occurs. Also, the power, features, capacity, and control in a shared service are considerably greater than what likely would have been purchased with a wholly owned hosted phone system, and the included around-the-clock support from a shared service can be more than what's available internally for a home-grown solution.

Toll-Free Numbers

Toll-free numbers typically ring onto local numbers and will therefore follow the local number when it is forwarded. Toll-free numbers that ring to dedicated long-distance carrier circuits must be forwarded by the long distance carrier. It is possible to split toll-free traffic—from a single toll-free number—across multiple long distance carriers to add redundancy. Toll-free numbers also have a special resiliency that local numbers do not, as they can still be quickly recovered when the local phone company, or even the long-distance carrier that carries the toll-free calls, completely fails. All organizations should publish at least one toll-free number for this very reason. Toll-free recovery after a carrier failure will take days, rather than minutes, if it is not established well ahead of time by experts in the field.

Voicemail

Voicemail can be a critical tool in handling incoming calls during a telecom outage, especially when incoming calls are spiking. Having different voicemail boxes for different phone numbers is preferred, though a general voicemail box is far better than nothing. Some systems can convert message to .WAV files and deliver them to e-mail, which can be crucial in shortening response times for callers, as Internet outages do not always accompany phone outages.

Faxes

The preferred method of inbound fax recovery is to forward fax numbers to an off-site system that can receive and store faxes for later retrieval. It is also possible to have received faxes converted to .PDF files or other formats and immediately sent to e-mail.

Outbound Communication

Copper/Analog Lines

Copper lines are still often used for fax lines, modem lines, transcription lines, and so on. If the phone lines are down or the power is out, the old-fashioned copper line can often be quite helpful. Copper lines—sometimes called POTS lines, 1FBs, wild lines, or Centrex lines—are line-powered, meaning that if you plug in a traditional corded phone you will hear a dial tone and able to place calls. Ideally, at least one copper line should be installed in each facility.

Mobile Phones

To make an outbound call, the most natural phone to reach for during a phone outage is the one that's most likely in your pocket, your cell phone. Cell phones are obviously quite mobile and are inherently resistant to cable cuts, power outages, and equipment failures; more than 99% of telecom outages fall into one of these categories, so it makes good sense to have mobile phones as part of the recovery plan. They're not disaster-proof, however, as cell phone towers typically have only 12 hours or less of battery backup, so during an extended power outage they can become useless. Cellular towers can be vulnerable to high winds, and cellular networks can be quickly vulnerable to congestion as panicked subscribers flood the air with calls.

SMS/Text

Text messaging, next to a satellite phone, may be the most resilient form of communication in existence; it is the only service that continued to work relatively reliably after 9/11 and Hurricane Katrina. Most common is texting from mobile phone to mobile phone, but it is also possible to text between e-mail and mobile phones, and vice versa. Systems that enable mass text messaging are also available and can be quite reasonably priced if you shop around; some mass-texting features are built into voice mass notification systems.

E-Mail/Instant Messaging

Sending an e-mail or an instant message can be a quick and easy replacement for a phone call when the phones are down. Surprisingly, often when the phones are down, the Internet is still available. It can be a great way to notify individuals or groups the nature of the outage and to carry out business or other tasks until the phones are back up.

VoIP

If a broadband connection is still available while regular phone lines are not, it is possible to use a Vonage, Skype, or similar VoIP service to make outbound voice phone calls to anywhere in the world. (A VoIP phone system, unfortunately, is just as vulnerable as a non-VoIP phone system to cable cuts, power outages, equipment failures, and disasters, so simply having a VoIP system, contrary to some common beliefs, does not provide inherent disaster recovery advantages, at least to the building experiencing the outage.)

Satellite Phones

As the name implies, satellite phones bounce calls off of satellites in orbit, so they can work when nothing else will, even when the entire local telecommunications infrastructure has been compromised. They can be used at the top of Mt. Everest or in the middle of the Pacific Ocean, but they do not work indoors without a special dish and wiring, as they typically require line-of-sight to the southern sky. Satellite phones have dropped in price and improved in quality considerably over the years; a good phone can now be had for less than $700, with plans starting at under $50 per month. Every organization should own at least one satellite phone.

Radios

Two-way and other radio systems can be indispensable for communication, provided whomever you wish to contact also has a radio in their possession. Prices, quality, range, and features vary greatly.

Paging

Still in service mostly within government and healthcare, this communications workhorse also works via satellite and can be a lifesaver when out of range of mobile phones and landlines. Of course, whomever you want to page must have a pager to receive the message.

Mass Notification

Calling Trees

Calling trees, where five people will call five others, who in turn call five others, can work effectively, as long as there's only about, say, five people to contact in total. They are notoriously unreliable when desperately needed, but, not surprisingly, they work quite well when tested during nonemergencies. For any groups larger than a couple dozen, an automated notification system is preferred.

On-Premise Systems

Systems can be purchased that can take a recorded message and quickly blast it out to a group of phone numbers. Phone lines must be supplied, and the number of lines installed will determine the price of the system and the speed of notification. Hosting the system off-site makes the most sense, as a cable cut or power outage could render an on-site system useless at the very time you need it most.

Hosted Systems

Since hosted notification systems or services are hosted offsite, they are inherently resistant to equipment failure, cable cuts, and power outages that affect your building. They typically offer online tools for loading messages, lists, and other settings, and they offer much higher capacity for a speedier notification—a very important advantage. Services can also charge by the minute only when used, making the hosted option more budget-friendly as well.

Important Notification System Features

The following notification features are available on some notification systems and should be sought after:

- Ability to send SMS/text messages to mobile phones;
- E-mail notification;
- Ability to send text-to-speech messages (for when a voice message cannot be recorded);
- Remote activation (dial in, enter secure PIN to activate);
- Polling—the ability to capture touch-tone confirmation and other responses to voice messages;

- Reporting—it is critical to know everything possible after a notification has been sent: when started and completed, which message was sent, who and how many received, who and how many did not receive, and so on.

Toll-Free Emergency Information Hotline

When a mass-notification message is sent, be prepared for an avalanche of incoming calls for hours, days, and even weeks to come. The short, simple message contained in a mass notification is usually sufficient to warn of immediate danger, but not much else. There will be questions of all kinds for some time to come.

To answer those questions, and to ensure that callers receive accurate information, instructions, and reassurance without tying up precious resources such as people, cell phones, and satellite phones, a recorded emergency hotline is ideal for this purpose, and it should be toll free and completely off-premise. The ability to provide multiple messages, rerecord messages remotely, and provide security for certain messages is also helpful.

Web Site

Posting updates on the organization's Web site can be a quick and reliable way to convey critical instructions and other information. Web site posting is also ideal if the information to be disseminated is broad or complex. A process for the rapid posting of information in a conspicuous location on the site should be implemented and tested well ahead of time.

Conference Calls

Whether conducting regular company business during a phone outage or coordinating a recovery with an emergency response team after a disaster, conference calls can be critically important. Premise-based and hosted conference systems abound, with hosted systems having disaster recovery and cost advantages over premise-based systems. Hosted systems may have advanced features that are desirable, such as automatic dial-out to bring in participants, Web-based control of the conference, and more.

Planning, Testing, Reporting

As with all aspects of a disaster recovery plan, communications recovery planning, well ahead of an outage, is a must. People, teams, phone numbers, carrier support numbers, alternate routing plans, and more should be captured and detailed. Once in place, a communications plan should be thoroughly tested at least annually, with test results recorded and utilized for plan improvement.

TeleContinuity White Paper Highlights the Need for 4Ci and Survivable Communications

Another 4Ci service is provided by TeleContinuity. TeleContinuity's patented service allows for 4Ci and inbound call recovery and can allow an organization to stay open for business despite sustaining a disaster. According to Telecontinuity, their system allows key employees to work out of remote locations or their homes, while still maintaining contact with clients, suppliers, production facilities, and staff. And, most importantly, all incoming telephone calls arrive over the previous telephone numbers and extensions—without the need for an expensive secondary PBX or the need to change numbers.

TeleContinuity's service also takes into account the inevitable telephone network congestion that occurs during a time of crisis, when entire telephone networks may be rendered inoperable. This happened on September 11, 2001, and during Hurricane Katrina, for example, and often happens during other area-wide emergencies when cell phone grids become so overloaded that calls can no longer be placed or received over either the public switched telephone network (PSTN) or cell phone networks. With TeleContinuity's service, numbers and telephones stay "connected," eliminating the problem of network and cell phone congestion. As call volume increases during a disaster, TeleContinuity routes around central office and PSTN outages and dynamically allocates sufficient bandwidth to compensate for increased traffic demand.

Essentially TeleContinuity constantly monitors both the PSTN and the Internet for outages, delays, and congestion to route client's incoming call traffic along the most efficient routes that will deliver the highest level of voice quality and call completion. This technology allows calls to be moved easily between the PSTN and the Internet for delivery at any location over any network and on any device (landline phone, cell phone, IP phone, laptop, computer, or PDA). A vendor synopsis of the service provided by TeleContinuity follows. The authors thank TeleContinuity's senior vice president Michael Rosenberg for his thoughtful contribution to this section. TeleContinuity has also contributed a very lucid account of the effects of an avian flu pandemic on telecommunications as potentially thousands of employees stay at home to work. This insightful account can be found in Chapter 7.

Overvew of the TeleContinuity Service

What Is Telecontinuity?

TeleContinuity is a voice communications network that offers continuous communications during telecommunications outages. Used by its subscribers as an emergency standby network, TeleContinuity provides assurance that communications can be delivered via any network, to any talking device, whatever the user's location.

Who Needs It and When Is It Used?

Mission-critical organizations and organizations with critical operations who demand a flexible, dynamic, survivable voice communications solution when faced with a telecommunications outage, natural disaster, earthquake, PBX failure, fiber cut, fire, flood, building evacuation, terror attack, or other catastrophic event need TeleContinuity.

How Is It Offered?

Organizations subscribe on a per-user annual or monthly fee basis independent of current telecom contracts requiring no change of service, purchase of equipment, or software.

What Does It Allow?

TeleContinuity allows subscribers to be in control of their telecom during an outage or evacuation and removes the dependency on the local carrier during the outage.

How Is It Activated?

The user activates the service when they anticipate or are experiencing an outage. The TeleContinuity service is always in a standby mode, waiting for user activation. TeleContinuity can be activated via a Web browser, e-mail, text message, or telephone.

What Benefits Do Subscribers Realize?

Subscribers can be reached at their usual telephone number or through a special toll-free emergency number and can receive and place calls over any working device: home phone, cell phone, hotel phone, IP phone, soft phone, PDA, and so onc.

How Does TeleContinuity Maintain Continuity of My Telecommunications?

TeleContinuity's unique combination of path diversity, network diversity, geographic dispersion, and distributed network architecture effectively reroutes telephone traffic around network congestion and network points of failure.

How Quickly Can Voice Communications Be Restored?

Operations can resume within minutes following a communications disruption, minimizing or even eliminating any negative governmental or economic impact that would have resulted from the interruption.

Can I Keep My Existing Phone Number?

Yes. TeleContinuity's technology, in conjunction with any call forwarding, enables users to be reached at their existing telephone extensions—via any network, any device, and at any location—as though no service disruption had ever occurred.

Do I Need to Preplan Where I Want My Calls to Go?

No. With TeleContinuity, preplanning the telecom recovery sequence and the ongoing work of maintaining up-to-date telephone trees are no longer necessary. TeleContinuity can link together all operating networks and activated users within minutes of system activation.

Requirements for Survivable Telephone Continuity According to TeleContinuity

Every government and private review of the tragedies of 9/11 and Katrina revealed that the most critical need during and following the disaster was a working telephone and a place to work. To maintain continuity of operations, a company or government agency must provide displaced workers with a substitute work environment with access to incoming and outgoing telephone service.

There are five key elements that an effective BCP/COOP (business continuity plan/continuity of operations plan) must provide to assure telecommunications during an emergency or disaster:

1. *Location independence.* Executives and staff must be able to receive calls made to their work numbers regardless of where they are located. Executives and staff must be able to rapidly and without limitation maintain continuous communications while changing location as frequently as necessary as dictated by changing disaster conditions.

2. *Network independence.* To assure continuous telephone service, the emergency voice provider must be able to move calls through all surviving and operating networks, whether PSTN (landline, cell) or IP (softphone, VoIP, laptop, PC, PDA).

3. *Device independence.* The effective solution must allow executives and staff to utilize any communication device available to them during the disaster. The solution must be capable of delivering telephone service via the user's cell phone, laptop, PC, PDA, landline phone, or any other standard voice-capable device.

4. *Survivable telecommunications network.* The emergency voice provider must be capable of delivering calls throughout the disaster event itself, be self-healing, and be relatively immune to the disaster's effects. The service must have geographic diversity and be capable of maintaining service through dynamic rerouting even when service points have been destroyed. The emergency voice provider must have no single point of failure anywhere in their system.

5. *Isolation and independence from the disaster site's local loop.* To increase communications reliability, the solution must reduce or eliminate any dependence on the local loop at the disaster site. The emergency voice provider must provide both forwarding capability and independence from the local loop in case carrier forwarding is lost or is unattainable, so the executives and staff can still be reached.

BCP/COOP for voice solutions that meet these requirements are the only ones that will yield a practical level of preparedness for government agencies and enterprises. These requirements permit executives, managers, and staff to rapidly adapt to changing post-disaster conditions and work around unanticipated changes in the original plan when the backup site are inaccessible, traffic blocks key roads, or there are other unanticipated disruptions.

TeleContinuity's Principles of Call Delivery

TeleContinuity's technology combines three key principles in assuring call delivery:

1. It enables end users and/or telecom managers to interact directly with the TeleContinuity network during a disaster or outage and to instantly reroute their incoming traffic to any network or device and to any location they choose.
2. It reestablishes any directory of extensions and other PBX features such as voicemail and teleconferencing, allowing the client enterprise to appear as if it were operating under normal conditions.
3. It interlinks the PSTN and the Internet to make the call forwarding features of the local telcos survivable and to route traffic around network congestion and failures.

Underlying and Supporting Technology

This includes the following:

1. A network of interconnected points-of-presence spread (POPS), similar to switching centers, from Boston to San Diego;
2. Connections to every major telephone network (PSTN) and to nine Internet backbones, allowing it to route calls around any carrier points of failure or congestion;
3. Rerouting incoming calls to any device and over any operating network to assure call delivery;
4. Important PBX features, such as auto attendant messages, voicemail, and teleconferencing are maintained even during the emergency or disaster. This level of assured survivable services is currently unavailable through any other telephone company, hardware vendor, or service provider.

Underlying Technology in the TeleContinuity Network

At the heart of the TeleContinuity network is a suite of proprietary software developed by the company for the express purpose of delivering survivable disaster-proof telecommunications service. The software is deployed over a network of control points of presence (CPOPs) and transport points of presence (TPOPs) located in colocation facilities throughout the country. The robustness and disaster-proof features of the network flow directly from the tight coupling of its network control software to the physical architecture.

The TeleContinuity network creates any number of subnetworks on demand as customers activate and use the service. Each subnetworks is specially tailored to service the customer on a call-by-call basis. This tailoring is made possible through patent-pending programs that continuously sense and evaluate the existing network links between POPs, from the PSTN to the POP and from the POP to the user for those users that are on IP phones. The company has dubbed this unique capability heterogeneous adaptive dynamic intelligent routing (HADIR).

With HADIR, TeleContinuity's patent-pending software, a user can seamlessly interconnect the PSTN and the Internet, thereby creating the possibility of thousands of call paths—significantly increasing call path diversity. During an emergency, special algorithms read the real-time condition of each network and route calls to maximize completion and quality of service (QoS). TPOPs and CPOPs located throughout the country are connected to strategically selected backbone segments of both the PSTN and the Internet. Each POP hosts specific software agents that enable TeleContinuity to automatically:

- Sense the state of the attached network (PSTN and Internet) continuously;
- Measure and record availability and time to dialtone for every connected carrier to determine carrier congestion;
- Measure inter-TPOP delay, packet loss, and jitter;
- Measure TPOP-to-end-user device delay, packet loss, and jitter;
- Determine the optimum call route through the network;
- Maintain quality of service for VoIP call segments;
- Dynamically reconfigure all or parts of the network in response to any sensed link congestion;
- Automatically route calls around failed POPs;
- Route calls around failed or destroyed PSTN and Internet links or infrastructure;
- Update user profile databases throughout the network to ensure end-user mobility;
- Effect distributed call control throughout the TeleContinuity network.

TPOPs, which are solely focused on voice transport, are controlled via secure virtual private networks (VPNs) by our CPOPs—multiple industrial-grade servers located at other collocation (collo) facilities. These contain routing and user authentication data and will dynamically control routing of calls across our distributed TPOP network.

How Calls Are Delivered to TeleContinuity

Incoming calls are redirected to the TeleContinuity Network by several methods:

- *Toll-free access.* Positioned as the "business card" solution, TeleContinuity provides its customers with an organization-specific toll-free number that is published as the telephone number to dial in the event of a telecommunications outage or emergency. Each toll-free number provided can be treated with custom messaging to mimic the organization's current auto-attendant features. Users calling the toll-free number will be prompted to enter in the extension or number of the person they are trying to reach, and if that person has activated TeleContinuity, the inbound call will be redirected to the subscriber or to a voicemail box in the TeleContinuity network. The major benefit to utilizing a toll-free number is that it is 100% independent of the local carrier or local loop that may have suffered an outage in the disaster. Implementation is quick and does not require the involvement of any technical personnel or engineering staff from your company or from your carrier. A single toll-free number can cover an entire organization, irrespective of their geographic location.

- *Carrier-based redirect.* TeleContinuity can be implemented alongside standard forwarding features that all carriers provide their customers. These central office, tandem, or SS7 (Signaling System 7)-based call-forwarding or switched redirect services can be applied to main numbers, DID (direct inward dial) lines, trunks, or POTS (plain ordinary telephone service) lines. Once activated by a customer's telecommunications manager or by individual end users within that customer's organization with DID lines, the TeleContinuity network will automatically invoke forwarding at the carrier on behalf of the client and redirect those lines treated with forwarding to deliver their calls to the TeleContinuity network. Once delivered to the TeleContinuity network, HADIR will determine the best route to deliver the call to the user.

Both the toll-free and carrier redirect solutions can be utilized together, offering the subscriber extended features and survivability. One implementation method does not exclude the other.

Activating the Service

A subscribing company's or government agency's telecommunications manager or an individual end user with DID lines can activate TeleContinuity by one of four different devices and networks: telephone, Web browser, e-mail, or SMS. Users activate by inputting their ID, their PIN, and the new destination for their incoming calls, and within seconds or minutes the TeleContinuity network has redirected their number and rerouted their calls to their new device.

Users are presented with simple screens to guide them when activating over the Web. An easy to use drop-down list allows users to change the destination number as many times as required. Additional telephone numbers can be quickly inserted into the list and saved in the system for later use. Users can also designate a default destination prior to the disaster if desired or required by their disaster recovery and business continuity plan. However, a default destination is not needed prior to the disaster occurring.

TeleContinuity will automatically switch the original phone lines to its TPOPs using the call forwarding and switched redirect services offered by the local phone company. The TPOP will play the standard PBX auto-attendant messaging features of the subscribing enterprise and reroute the inbound telephone traffic to the user.

Should the user move to a new location, they would call or log back into the system with their new location and device, and their inbound calls are quickly rerouted. By intercepting calls at the highest possible level of the telecommunications infrastructure, TeleContinuity's survivable, fault-tolerant, reliable system will assure telephone-based business continuity immediately following a service.

TeleContinuity's system administration and operating systems provide clients with complete control of their telecommunications capabilities during critical disaster events. To accomplish this, TeleContinuity has created a fully featured Web-based provisioning administration tool that can be customized by each client. This application enables customer-designated personnel, such as the telecommunications manager, to completely control and administer the network for their user base.

Since the application is Web based, it can be instantly and securely accessed from any Internet-connected browser in the world.

A directory services tree structure provides easy navigation for the administrator and clearly displays the information assigned to them by the top-level administrator. A customer with multiple sites can assign a different administrator the responsibility for each site. Site-level administrators would thus have complete purview and control over users located or assigned to their site. This scheme is consistent with current telecommunications site–oriented graphic representations and is easily accepted and understood by corporate telecommunications managers. The application allows administrators to perform the following functions:

- Modify user profiles;
- Change passwords for users;
- Activate and deactivate users accounts;
- Provision users with carrier forwarding services;
- Activate individual users and send calls to default destinations;
- Deactivate for individual users;
- Activate and deactivate en masse any number of PBX users with extensions;
- Determine individual user status and troubleshoot user accounts.

The application also enables administrators to perform enterprise line management (ELM). The ELM module presents administrators with a view of DID numbers and toll-free numbers that terminate at their site in the customer hierarchy. From the ELM screen, the administrator can designate any and all lines to forward to the destination numbers, and, with a single keystroke, the TeleContinuity network will immediately begin the process of communicating with the customer's carrier and begin line forwarding. Administrators can see the status of the forwarding events as they are completed in the same ELM screen. In addition, administrators can have the application kick off e-mails as the line forwarding is completed. The ELM module frees administrators from having to constantly retry the forwarding interactive voice response (IVR) with the carrier to accomplish each forwarding request. The underlying software infrastructure ensures the completion of forwarding through the use of a robust and distributed forwarding request queuing structure. For a summary of TeleContinuity's services, see Figure 2.6.

Summary of TeleContinuity Service

TeleContinuity and VoIP

VoIP implementations are very popular and offer agencies and companies many advantages over the older time division multiplexing (TDM) voice architecture.

- Protection for every telephone number or extension
- Survivable call transport and routing
- Total control of call delivery
- Real-time changes can be made by you or your admin
- Backup voicemail box for every user
- Web-based user and administrative interface
- System activation via toll free, Web, SMS, and e-mail available
- Dedicated toll free emergency service
- Call delivery to any dial tone device—TDM or IP
- Real-time desktop view of the network monitoring
- Comes with downloadable SIP-based softphone
- Configurable with carrier redirect services
- Individual or enterprise line management
- Customizable auto attendant

Real-time view of important TeleContinuity NOC screens

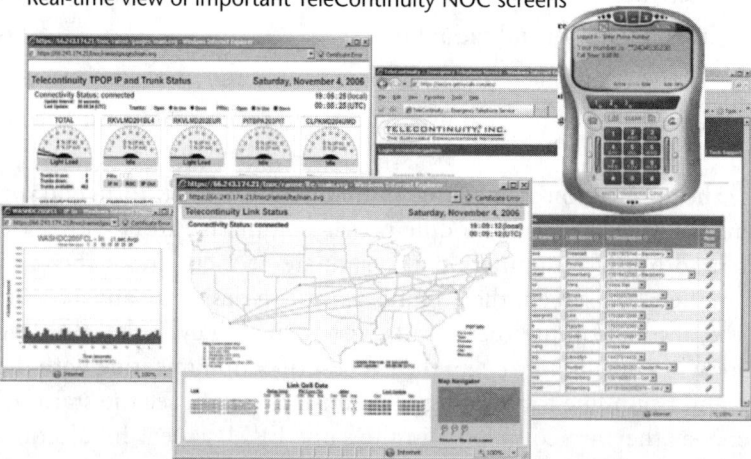

Figure 2.6   Summary of TeleContinuity service.

However, users must be aware that VoIP has serious security issues and is not a sufficiently robust system to assure continuity of operations for voice communications during a disaster, emergency, or evacuation—when continuous voice communication is critical to the agency or company. TeleContinuity's survivable network represents a revolutionary improvement in assuring that VoIP is available during disasters and evacuations. VoIP's vulnerabilities can create outages under normal nondisaster conditions, and they are dramatically exacerbated by any disaster event. For example, during the tragic events of 9/11, several regions of the Internet lost access to domain name service (DNS). Other pieces of the Internet were so bogged down with DNS lookups that the DNS server request queues were overloaded, making it seem like Web sites were not available. Organizations that rely on VoIP/DNS services for telephone access are extremely vulnerable during these types of events. Several architecture, infrastructure, and security vulnerabilities can, individually or collectively, reduce or altogether eliminate the availability of VoIP-based systems during disasters.

TeleContinuity and Enterprise Central Switches

Many large enterprises and government agencies have their own central switches that control incoming and outgoing calls. Enterprises utilizing these central switches usually have no backup switch capable of instantly picking up the telephone traffic in the event of a central switch failure, local loop outage, or evacuation (such as a pandemic) that leaves the switch unattended. The result is a cutoff of all telephone traffic and in many cases the inability to restore traffic until after an extended period.

TeleContinuity ensures that central switch traffic can be almost instantly redirected and rerouted to alternate facilities or have calls individually rerouted to the intended called party with no interruption of service. While central switches are economical and give control of call traffic to the enterprise, they constitute a critical point where telecommunications can break down and leave the enterprise without the ability to continue operations. TeleContinuity is currently being used to back up these critical switches, giving the telecom engineers the ability to maintain telephone communication by switching incoming calls into the TeleContinuity network and rerouting the calls to their destinations—around the failed or abandoned central switch.

SMS-Based Notification Solutions

Just because you are not a common carrier, however, does not mean you cannot avail yourself of other services that are virtually carrier grade. A service worth taking a look at can be found in a company in headquartered in Morrisville, North Carolina: Velleros, Inc. As an emerging industry leader in carrier-grade mass notification solutions, Velleros has developed a product called AlertSlinger, a telecom-grade, service-provider solution that is capable of delivering content-triggered alerts and Web-based applications for emergency notification.

AlertSlinger is based on patent-pending technology that monitors time-critical feeds for Internet-based information. For example, an all-hazards gateway module in the product monitors National Weather Service feeds for all 50 states, Puerto

Rico, and the Caribbean, utilizing National Oceanic & Atmospheric Administration and Department of Commerce information for public messages. Information, developed with the customer, is then verified against an alerting database to determine precisely how the information is to be circulated and what the triggering event is depending on the conditions. Consider, for instance, a 911 center that has experienced a telephone cable cut. Most 911 centers are engineered with diverse routes. Suppose, for example, 24 circuits take one route and 24 take the alternate route. A cable cut that knocks out half the capacity will probably occur on a busy Saturday night, perhaps during a severe weather event, blocking out all 911 calls. In the worst case, a 911 caller will get a recording that states, "All circuits are busy now, please try your call again later." With AlertSlinger, a number of things could happen with these calls. A listing of callers who tried to call 911 but could not get through could actually be *text messaged* to another working device, such as a wireless phone or VoIP phone. The same kind of thing can happen for commercial organizations under the same circumstances. A call to a bank that experienced blockage due to a cable cut could receive a record of who tried to call but was blocked. This record could be delivered to another device in a number of formats, such as e-mail or instant message. That data, in turn, could be bounced up against another database, such as the customer billing database for the carrier or the banking facility to identify the customer. In order to maintain customer goodwill and retention, send a written letter of apology for the outage to the home of the caller. Enclose a discount on future services or a gift certificate. There are many ways to retain an otherwise disgruntled customer through effective use of technology.

There is value in a system like this in a disaster, since AlertSlinger effectively turns phones into automated emergency warning devices. Notification messages are sent to existing landline and cellular phones, handheld data devices (pagers and PDAs), VoIP phones, satellite phones, computers, and so on. Messages can be sent via voice call, short message text, e-mail, voicemail, or data session. Special emergency devices are not a requirement.

With the capability to distribute information over any type of network connection—wireline copper, broadband VoIP, broadband data, cellular, and so on—this AlertSlinger provides an optimal solution. Therefore, additional investment in network infrastructure is not required for a carrier deployment.

AlertSlinger also provides capability for push-type information services (Velleros CAMLOC and STALO products) enabling campus alerts, recorded announcements, and communit-interest broadcasts. Based on a patent-pending technology, authorized personnel can create the information to be communicated via voice or text and activate the alert trigger to send in real time.

AlertSlinger provides a simple, Web-based interface for nontechnical users to act quickly, confidently, and accurately in times of crisis. By using drop-down menus to make simple modifications to delivery method or predefined alert trigger categories, subscribers can manage their accounts on line.

AlertSlinger provides system redundancy by enabling deployment in multiple secure locations and also supports enhanced data encryption techniques for information protection.

Consider Your Responders' 4Ci Preferences (A Parody by Leo and Sharon—or Is It?)

"Baby Boomers"

- Want to talk on the phone;
- Prefer the interaction of the spoke word (seminars, conferences, face to face contact).

"Generation X"

- Want to e-mail;
- Webinars, not seminars.

"Generation Y"

- Just "text" it to me;
- It better be all about me.

"Generation Z"

- Cry long enough, and the bottle or breast will come;
- Who knows what this generation will have for 4Ci in 2030?

Communications Recovery Through Local Number Portability (LNP)

It is arguable to say that voice communication is still the most common method of communication today. Notwithstanding our little parody above, for the moment at least voice communications is still the primary means of responding to disasters. Therefore, the ability for everyone to communicate verbally and immediately after a disaster and, in particular, for emergency responders to communicate with each other are critical requirements of any disaster recovery plan. Businesses, citizens, and public entities like cities, counties, and federal agencies all have this need in common and require real-time verbal information in times of trouble.

We have discussed many ways to maintain 4Ci after a disaster so far in this book. While outbound notification systems, e-mail recovery, inbound call recovery, and other technologies certainly have their place, one technology stands out for voice recovery. That technology is local number portability (LNP). If we can get over a few regulatory issues, LNP has the potential to eclipse many of the technologies discussed heretofore in this book and to become the gold standard in voice recovery.

The PSTN in the United States and other countries remains the central connectivity hub for voice communications. Mobile telephone, land-mobile radio, satellite phone, and VoIP all connect to the PSTN through some form of switch or gateway.

Wireline telephone users are connected to the PSTN in order to communicate with other PSTN-connected users, as well as to other technologies such as wireless and VoIP. As we pointed out earlier in this chapter, recovering voice communications requires at least some level of participation by the local telephone company. Each local telephone company does things a little differently, sometimes on a proprietary basis. This means it falls on the end user to assume responsibility for precisely how their voice communications services will be maintained.

Generally, one of several approaches for recovery has been used to restore communications, as discussed previously. Call forwarding and RACF allow an incoming call to be redirected to a mobile telephone or other telephone number. Call forwarding is completed by programming the call receiving central office switch to receive and forward a call to a designated location or device. This process can be activated locally at the phone set by dialing *72 or other code but can also be activated with the RACF feature previously described in this chapter. In the event that a central office switch is wiped out, however, call forwarding and RACF are not viable means to restore communications. Therefore, in times of emergency, the applicability of this method is extremely limited. Case in point, a flood disaster in 2006 in a Verizon central office in Rochester, New Hampshire, punched a communications black hole into southeastern New Hampshire for several weeks. Ironically, the technology to mitigate this disaster was available but was not used, as described later in this section. This technology was not call forwarding or RACF. Each of these call-forwarding technologies relies on the availability of the serving central office switch. If the central office switch is impacted by the emergency circumstances, then calls processed through that central office switch cannot be completed.

Some companies absorb the expense of setting up separate and identical dialing plans in other central offices, which have sometimes been referred to as *shadow codes*. For example, a telephone number of 214-888-1300 might have a shadow code of 214-821-1300 working out of a separate central office. If the serving central office switch for 214-888 were removed from service for any reason, the calling party would dial the alternate area code and original 7-digit telephone number to complete the call. However, the calling party must be trained to recognize that when their call is not completed, they must dial the shadow code number. While this may work within company dialing plans that serve internal communications, it is not very effective for other kinds of calls, such as from inbound customers. It is, however, sometimes possible to have the local telephone company reroute these calls on an as-needed basis using their own, internal, and often proprietary technology. One such technology is generically referred to as advanced intelligent network (AIN).

AIN can be thought of as a data network that tells the voice network what to do. AIN is a telephone network architecture that separates service logic from switching equipment and allows a service provider to create the capability within their switching fabric to facilitate automatic rerouting of telephone calls. If the receiving central office switch is out of service for any reason, AIN can be used to reroute calls elsewhere. Since AIN is usually a proprietary technology, it is generally capable of rerouting calls only within a single carrier or service provider's network. These are pretty much what have been available (combined with outboard solutions such as those deployed by Telecom Recovery and TeleContinuity) to address the issue of voice recovery. On the horizon, there are systems based on local number portability (LNP), which promise to add a whole new dimension to the science of restoring 4Ci communications.

LNP is a technology used to transfer or "port" telephone numbers from one service provider to another or one technology to another. It is based upon something generically known as the LNP database. Most of us have knowingly or not used the LNP porting process. Many end users, particularly in the Generation X demographic have dumped wireline phones in their home in favor of wireless devices. When they do, they have the option of keeping their old phone number. If you have transferred a wireline phone to wireless, or vice versa, you have used LNP. The process may have taken you a day or up to about 10 days, in some cases. I submit to you that most of this delay is due to regulatory and notice requirement by people like the Federal Communications Commission (FCC) and procedures perpetrated by the carriers themselves. I am not knocking these procedures. We need them in order to keep some semblance of order in the industry. What are needed are extraordinary procedures that can be implemented in times of disaster. First, updating the LNP database can be accomplished by any carrier. Second, as a technology, the change becomes effective in as few as 15 minutes. The benefits to this are obvious: Imagine a caller could dial a telephone number in a time of emergency, and regardless of the status of the wireline central office switch, the call is still completed. You see, the receiving caller's number can technically be ported to a wireless line, satellite, VoIP, or wireless Internet service provider (WISP) almost instantly. Any technology that survives the disaster, even if that disaster is a central office under water, can be enabled through the use of number portability for communications recovery. So what's the issue, and why did people in Raymond, New Hampshire, have to wait weeks? Part of the issue is regulatory in nature, including the need to adopt formal rules. This type of thing has typically been addressed at the FCC in something that is, not surprisingly, referred to as a *rulemaking* docket.

You see, the issue is not a technical issue at all but a regulatory issue. When we owned our phone company, we once updated the LNP database to dial a fictitious "555" number.[2] Within 15 minutes of updating the LNP database, our attorney in Austin, Texas, was able to complete a call via her cell phone to our lab. Obviously, this is not really a legitimate everyday use, but it certainly illustrates the power of LNP as a disaster recovery solution.

So what is the issue? Consider how it looks to the carriers. Let's say a central office floods like one did last year in Raymond, New Hampshire, and how more than 30 did during Hurricane Katrina. In those cases it made economic sense *not* to "port" customers to other providers. Think about it. If your incumbent local exchange carrier (ILEC) had a disaster, and you moved to another provider (Vonage, Time Warner Cable, whomever), you might never go back to the ILEC. The ILEC knows this (and in fairness, so do the others) so don't expect them to volunteer LNP after a disaster, at least under the current rules. If you call them, it may take up to a week, even though as stated earlier, it takes only 15 minutes to update the LNP database.

2. The 555 prefix is the one used on TV. It is not an actual number. That's why TV shows in North America use 555-1212 or Klondyke-5, 1212, etc., when giving out a number. This avoids thousands of people actually calling that number and really creating a problem for the owner of that real number.

So what is the solution? At present, anyone having a disaster can call the FCC at 1-888-CALL-FCC (1-888-225-5322) and coordinate a number move on an emergency basis. We also recommend that you participate in forums at the FCC dedicated to formalizing this process. It is the role of government to mandate the role of carriers in this regard, and the FCC or your state utility commission is the proper forum.

Such efforts geared to fortify the nation's telecommunications infrastructure are ongoing and encompass new and better technologies ever year. At the same time, the rapid evolution of this industry demands constant activity in this regard, according to Barry W. Bishop, senior director, Public Safety for NeuStar. Sharon and I have had the opportunity and honor to meet one on one with Barry and found his insights on the subject worthy of note.

"While strengthening the United States' infrastructure is important, rapid restoration in response to a widespread interruption of service arguably should be the number one priority to ensure real-time accessible communications," according to Bishop.[3]

In the authors' opinion, LNP represents nothing less than a totally new concept for restoring voice communications in times of emergency. LNP utilizes an existing centralized database capable of dynamically rerouting telephone numbers in time of disaster.

There are few systems that could be considered *carrier grade* systems on a par with what a phone company could do—assuming, of course, that your phone company was even inclined to do so.

LNP is not a "mass notification" system per se, but it certainly can help if a very large, Katrina-style disaster strikes. This process uses porting to move a telephone number from a landline to a wireless phone, or vice versa, and it could be a lifesaver under certain circumstances. It allows you to port disrupted numbers elsewhere. (If you are a large enterprise telecommunications user, badger your primary carrier to contact NeuStar, headquartered in Sterling, Virginia.)

LNP is a very sexy technology to government, as well as people like banks and brokers that straddle fault lines or users that lie in coastal areas prone to flooding. In fact, anyone could use LNP . . . and they should.

A little background on Neustar: NeuStar performs numerous services for carriers, such as telephone number administration and U.S. common short codes (short codes, for direct response advertisers as well as content providers, are used to reach mobile subscribers across multiple networks and types of devices to create marketing relationships for brand placement and messaging). Additionally, NeuStar provides various infrastructure services such as technology migrations and network optimization. So, what is NeuStar *best* known for? They manage the LNP database.

As the custodian of the LNP database, NeuStar imparts a beneficial service for disaster recovery and 4Ci. From a carrier perspective, LNP is in our opinion the hottest technology going for maintaining 4Ci after the "big one" hits.

Redirection of telephone service in and out of central offices needs to seamlessly occur when disasters strike these facilities. The dilemma, in this situation, is that tra-

3.   Comments to *Homeland Defense Journal* in a June 2008 article entitled "Communications Recovery Through Number Portability: A Concept for Rapid Recovery of U.S. Voice Communications."

ditional methods of doing this (such as call forwarding or remote call forwarding) are functions of the switch. These functions are eliminated when the switch burns up or goes under water. As a case in point, about 34 central offices in Louisiana went under water during Hurricane Katrina.

NeuStar offers a service called Port DR (for disaster recovery) whereby service providers can point traffic away from a disabled central office switch and toward another switch or technology that is still working. This porting can occur either within the affected service provider's network, or to an affiliated network, or to a competitors network. As we stated earlier, the actual process is largely the same as when you port your landline phone to a wireless phone, or vice versa. NeuStar, using the LNP database, can restore services for anything from a single phone line to an entire area code. During Katrina, some calls to the 504 area code for Louisiana were actually redirected to Dallas, Houston, and Atlanta! Regrettably, this solution is available only to communications carriers. If you are an enterprise user or other consumer of telecom services, about all you can do is nag your local provider to contact NeuStar to implement Port DR. To learn more, see http://www.neustar.biz.

What About 911 Services?

Sometimes when porting outside of a local area, and where the ported number automatic number identification (ANI) is utilized when originating a call to 911, a problem may be created when delivering the originating 911 calls to the correct public safety answering point (PSAP). In all cases, a 911 originated call should default to a designated PSAP. The exception to the rule of course is if the PSAP itself is gone. That was actually the case in New Orleans after Katrina. It is decidedly the opinion of the FCC and emergency responders, however, that 911 calls go to the PSAP they are supposed to go to. At a minimum this avoids some of the confusion noted in our earlier example of the city of Red Oak, Texas. This is why even VoIP carriers like Vonage have to keep records and assure that 911 calls go where they are supposed to go. Emergency impacts on 911 calling varies depending upon the communications method used. (You can take your Vonage phone, unplug it from your office in Dallas, take it to Boston, plug it in, and make and receive the same calls. However, if you dial 911 from Boston, it will be routed to the Dallas PSAP unless you called Vonage in advance and changed it.)

It is likewise important to note that while porting numbers may have some impact on 911 calling, someone's PSAP will always be reached. Generally speaking, one PSAP can transfer calls to another at least under normal conditions. Therefore, the rewards here outweigh the risks.

How Many Numbers Can Be Ported?

As with any approach where a large quantity of numbers is relocated, porting does have potential network implications. However, the porting process is the only real-time approach for moving telephone numbers across various service provider networks and between wireline and wireless communication platforms and technologies connected to the PSTN, such as land mobile radio or VoIP. Barry Bishop should know. He literally was responsible for moving a sizable portion of the 504 area code serving New Orleans to places like Dallas and Houston.

"The events and aftermath of Hurricane Katrina and September 11 punctuated the critical importance and challenges of maintaining real-time communications in times of crisis," according to Bishop (see footnote 3).

Summary

Despite the advances in communications technology, one thing is for sure: the PSTN is a critical part of the national infrastructure for at least the next decade, and steps must be taken to ensure its survivability in the face of manmade or natural disasters.

Bishop has been quoted as saying (see footnote 3):

> While there are a number of technical solutions to provide alternate communications during times of emergency, only the dynamic rerouting of existing call flows using telephone number portability achieves true survivability. The NPAC database is already an integral component of America's communications infrastructure, and it capabilities should be considered as a means of building more resilient, recoverable voice communications. The NPAC is already connected to the requisite technologies and is currently operational in the U. S. and Canada. Additionally, the NPAC can be utilized to re-route to virtually any communications platform and to most service providers connected to the PSTN (limited to areas of portability) without large capital outlays. Finally, the NPAC, as managed by NeuStar, is operated in a manner that incorporates stringent neutrality requirements so that all service providers are assured equitable treatment.
>
> With an existing capability designed to serve the public interest in a low-cost neutral methodology, number portability delivers rapid restoration of voice communications in times of emergency. It should be clear to government and industry that using telephone number portability achieves survivability for our national security and emergency preparedness communications systems.

As for Sharon and I, we could not agree more. We have had the privilege of communing one on one with Barry, under a nondisclosure agreement. We are convinced that NeuStar provides the North American communications industry with an in-place solution as well as the ability to not only manage virtually all telephone area codes and numbers in real time but to also enable the dynamic routing of calls among thousands of competing communications service providers (CSPs) in the United States and Canada in times of disaster. We call on the FCC and state policymakers to formalize these procedures, in addition to offering our own personal support in helping craft such a national policy.

Since 1996, NeuStar has served as the local number portability administrator for North America, operating the Number Portability Administration Center (NPAC). Through this function, performed under the auspices of the U.S. FCC and the Canadian Radio-Television and Telecommunications Commission (CRTC), NeuStar clears all telephone number portability transactions in North America and broadcasts the associated routing updates to enable the dynamic routing of calls among thousands of competing communications service providers. As a carrier-neutral custodian of the LNP database until 2015, NeuStar should be empowered through FCC orders and rulemakings to bring its disaster recovery capability to its full potential.

For Goodness Sake, Let's Implement LNP for Disaster Recovery!

"Given the manner in which telephone numbers are provisioned and maintained in the NPAC, number portability is a proven viable method of restoring incoming calls to customers who have lost service in a disaster," according to Bishop.

As an example, during Hurricane Katrina, millions of customers were out of service, and there was extensive damage to both wireline and wireless central office switching facilities, cell sites, and 911 centers. In response, NeuStar ported approximately 2,000 telephone numbers *across local access transport area (LATA) boundaries* as part of the disaster response. In addition, about 300 blocks of existing numbers were moved across LATA boundaries using number pooling, representing 300,000 more telephone numbers that were successfully moved. No other technology that we know of can move these volumes of numbers, regardless of whether it is within or outside a LATA and despite the nature and extent of the damage in the affected area. While we belabor the Katrina example, be mindful of the fact that Los Angeles and San Francisco are overdue for a major earthquake. A fault line sits right off shore from Seattle that could trigger a catastrophic tsunami, and Mt. St. Helens has not moved. Florida still sits an average of 30 feet above sea level. The bad guys know the locations of the key telecom hubs that serve the financial markets in New York City. We could go on, but you get the message. Regulators and industry need to collaborate on unleashing the power of LNP in responding to disasters. It's a regulatory duty and a challenge that the FCC and state commissions must step up to for the public good.

Telecom Priority

The following government services are available for telecom priority in the event of a disaster.

- Government Emergency Telecommunication services (GETS): priority for selected users, landlines;
- Wireless Priority Service (WPS): wireless priority (next available frequency) for selected users;
- Telecommunications Service Priority (TSP): private line restoral priority, bank wires, and so on.

Note that in Hurricane Katrina, only a small number of eligible users had signed up! Wireless Internet service providers (WISPs) and cable companies can qualify. Has yours?

We recently completed a disaster recovery plan for a Fortune 500 company that contained no telephone numbers whatsoever in the plan for security reasons. Since every employee in that firm had either a Blackberry, laptop, wireless phone, or all three items, standards were endorsed by the company to keep all of the contact information in electronic form. Security standards for the Blackberrys, laptops, and wireless phones, including encryption and passwords, were already in place. So this method was the most useful and secure means of providing contact information. This method also ensured that phone numbers remained up to date, since they were in daily use by each employee. We see this as the future of how plans will be documented as well as an example of how the computers aspect of 4Ci is entering the nonmilitary sector's recovery plans.

CHAPTER 3

What Are We Planning For?

In past books, we have devoted a section like this one to all of the things that can go wrong in an organization and all of the reasons we plan. It is pretty much an accepted fact that disasters break neatly into three categories: natural, human error, and intentional. For completeness, we have often included a fourth: acts of God. With regard to the last cause, sometimes things just happen that defy logic and boggle the imagination. Even so, let's not be too quick to blame God for unusual events since many of them—including the stereotypical lightning bolt from the heavens—most often fit into one of the other three categories. So, let's cover a few.

- Natural: fire, flood, earthquake, tornado, typhoon, volcanic eruption, tsunami, lightning strike, hurricane, winter storm, heavy rain, hail, high straight line winds, wildfire, mud slide, avalanche, fill in your own: _____.

- Human error: telephone cable cut, building collapse, improper equipment maintenance, carelessness and inattention, excavation or construction outage, nontelephone (gas, electricity, and so on), faulty engineering, fill in your own: _____.

- Intentional: terrorism, vandalism, sabotage, theft (such as copper cable or equipment), disgruntled employee, strike or labor stoppage, act of war, fill in your own: _____.

Obviously the reader of this book can think of numerous other categories of calamity. (You can also refer to Table 3.1 for other suggestions.) A quick Google search on the Web will bring you far more ammunition to use with policymakers that addresses the "what can happen" issue than we could possibly present here. Therefore, we have devoted this section to something a little bit different.

We have addressed the philosophical question of whether executives in private sector organizations are responsible, even in part, for natural disasters. We opened these discussions in previous chapters and continue this dialogue here. In that context, we also introduce another concept in this chapter. We deal with the "spillover" effect of disasters on a society or social structure. By spillover we mean this: We have noted that often following a major disaster even private enterprise firms that survive the disaster can still suffer and go out of business. Consider Hurricane Ike, which struck Houston, Texas, on September 14, 2008. Two of our clients, both telephone companies, are based in Houston. Both survived Ike initially. In the weeks that followed, however, it became apparent that these companies' customers, by and large, were not so lucky. The resulting revenue loss (due to the relocation of

Table 3.1 Common Disasters to Business Enterprises—Examples of Natural Causes and Responses

Event	Systems Affected	Impact	Preventative Measures
Fire	All systems—widespread damage	Complete loss of communications, computing, revenue-generating capability up to and including a complete loss of the building	Smoke detectors, fire alarms, fire extinguishers, smoke removal systems, fire stopped cable risers in high rise buildings, sprinklers, infrared scanning, training.
Water damage	Damage to equipment areas	Loss of communications, critical equipment, computer room or telecom hub, partial to total loss of business functionality; electrical hazard	Under floor water detectors, drains with back flow devices, insulation for pipes to prevent freezing, dry-pipe or pre-action sprinklers, inspections of roof, and building plumbing, adequate outside drainage during snow melt and flood season.
Earthquake, tornado, or other widespread disaster	All systems—widespread damage	Partial to complete loss of systems, mobility, power; no personnel (employees will return to families first); recovery centers exhausted; emergency services strained; public telephone network strained or out of service; looting may also be a problem; possible loss of key personnel	Develop plans for dealing with employee families and morale, consider backup power generators, activate computer recovery centers early, develop call out procedure for use in case phones are not available such as a meeting place. Arrange for site security, develop procedures for pick up of material stored off-site which take into account reduced mobility after a widespread disaster.
Lightning strike	Moderate to severe damage to equipment areas	Telephone communications may be out of service; computer equipment of all types down; power may be interrupted; potential for fire	Contact vendors of affected equipment, by public telephone if necessary. Contact building manager for restoral of common systems such as air conditioning and power. Request insurance adjuster on site immediately after any disaster.
Contamination of building	Loss of access to equipment due to building contamination by asbestos, PCBs (found in transformers), chemical spill, or other hazardous material	Public safety officials may shut down the operation in the interest of human safety; building may be totally inaccessible for days, weeks, or months	Arrange for dial-in access to critical communications systems, LANs, or other maintenance functions. Safeguard handling of hazardous materials. Do not store hazardous noxious, or toxic materials anywhere in the vicinity of communications equipment.

Table 3.1 Common Disasters to Business Enterprises—More Than 75% of Telecom Disasters Are Due to Human Errors!

Event	Systems Affected	Impact	Preventative Measures
Programming error	Router, open system, PBX system, intelligent multiplexer, or LAN out of service due to error in programming; ranges from minor service outage to major failure	May support dozens or hundreds of employees; may cause major revenue loss	Do maintenance changes after business hours. Make backup copies of all software, especially before major changes, to allow for a "fallback" if something goes wrong. Back up all PBX class of service indicators and store tapes and disks off site. Maintain a sign off log for all major software changes, and designate responsible individuals.
Improper maintenance	Major PBX or multiplexer failure due to improper performance of routine maintenance	PBX system or multiplexer may support dozens or hundreds of circuits or employees	Train equipment maintenance personnel in proper procedures. Use a "buddy system" for training. (Pair inexperienced personnel with a more seasoned technician) Monitor all subcontractors closely. Schedule all maintenance for off-hours.
Unauthorized personnel	Major PBX or multiplexer system fails due to tampering (intentional or unintentional) by nontechnical personnel	PBX system or multiplexer may support dozens or hundreds of circuits or employees	Lock all equipment rooms and restrict access. Use sign in logs. Keep janitors and custodians out of equipment rooms. Accompany all visitors.
Theft of equipment	Major or minor outage due to theft of telephone, multiplexer, or LAN equipment	PBX system or multiplexer may support dozens or hundreds of circuits or employees	Lock all equipment rooms and restrict access. Use sign in logs. Implement a parcel-pass system. Inspect briefcases and bags. Accompany visitors. Review procedures for reporting theft to internal or external law enforcement authorities.
Sabotage	Major or minor network outage due to intentional destruction of equipment by person internal or external to the company	PBX system or multiplexer may support dozens or hundreds of circuits or employees; worse yet, a disgruntled employee may know right where to strike for maximum damage to the company	Coordinate with human resources department to immediately terminate all passwords, credit card numbers and access codes issued to terminated employees. Collect all keys and badges to sensitive areas. Be especially alert during anylabor unrest or union activity.

Table 3.1 Common Disasters to Business Enterprises—Examples of Intentional Causes and Responses

Event	Systems Affected	Impact	Preventative Measures
Disgruntled employee	Ranges from minor service outage to major failure, since employees know right where to hit you if they are so inclined; one employee with a permanent magnet in the right place can cause more damage than a fire	May affect many employees and systems and cause major revenue loss	Say three "Hail Marys." Although there is a lot in writing about personnel profiling, there is no 100% defense against a deranged or disgruntled employee.
Terrorism	In widespread disaster or terrorist incident, your company's restoral priority may be very much in question	May affect many employees and systems and cause major revenue loss	Train personnel in proper procedures. Use a buddy system for training. (Pair inexperienced personnel with a more seasoned technician) Monitor all visitors and subcontractors closely.
Vandalism	Ranges from minor service outage to major failure.	May affect many employees and systems and cause major revenue loss	Lock all equipment rooms and restrict access. Use sign in logs. Keep janitors and custodians out of equipment rooms. Accompany all visitors. Install security.
Theft	Major or minor outage due to theft of server, telephone, multiplexer, or LAN equipment	May affect many employees and cause major revenue loss; obviously, the value of the equipment is also a consideration	Lock all equipment rooms and restrict access. Use sign in logs. Implement a parcel-pass system. Inspect briefcases and bags. Accompany visitors. Review procedures for reporting theft to internal or external law enforcement authorities. Buy insurance.
Sabotage	Major or minor network outage due to intentional destruction of equipment by person internal or external to the company	Server, mainframe, environmental, PBX system, or multiplexer may support dozens or hundreds of circuits or employees; worse yet, a disgruntled employee may know right where to strike for maximum damage to the company	Coordinate with human resources department to immediately terminate all passwords, credit card numbers and access codes issued to terminated employees. Collect all keys and badges to sensitive areas. Be especially alert during any labor unrest or union activity.

their customers) was enough to cause one phone company to file for bankruptcy and to place the second under severe financial stress. Couple this with the fact that Ike hit the same month financial markets began to melt down worldwide. This meant that the money that would have been available to these small firms to weather the storm (in both the literal and figurative sense) was no longer available. In combination, the results of the two phenomena (one natural, the other manmade) was devastating.

These recent events set the authors to consider just how frail a socioeconomic system is and how susceptible society is to widespread disasters from a general economic perspective. Indeed, when we began looking, we found a number of good pieces in publication. We draw from a few in this section, most notably, a report by Ray Shirkhodai, executive director of the Pacific Disaster Center. Ray provides useful insights on how disasters can erase decades of economic progress in developing nations in a single day. The second is a report by the Public Entity Risk Institute (PERI), www.riskinstitute.org, and notably by Dr. Daniel J. Alesch and Dr. James N. Holly, who discuss these impacts on other social fabrics. The authors hope the readers consider the issues addressed in this chapter in the context of assessing their individual responsibilities to plan for natural disaster—even if they are employed by private sector enterprises. As these facts show, even corporations cannot plan in a vacuum. If disaster strikes the customers of private sector organizations, the impact can be equally devastating and must be considered by the serious recovery planners from both a planning as well as an economic standpoint.

What Are We Planning For?

There is nowhere to hide. Natural and human-induced disasters happen everywhere. One day's sweet-scented breeze can be the next day's tropical cyclone, wrecking havoc in paradise. That happened in Hawaii when Hurricane Iniki devastated the island of Kauai on September 11, 1992. The result of that disaster was the birth of the Pacific Disaster Center as a quasi-federal agency and a new perspective on the causes of disasters and how to mitigate them.

The following section is an excerpt from a number of PDC documents and presentations, including comments by the organization's executive director Ray Shirkhodai and Joseph Bean of the same organization. This section provides the answers to such thought provoking questions as:

- Are disasters really on the rise, or does it just seem that way?
- What can and should communities do in response?
- How can and should such communities address the needs of the poor, handicapped, and disadvantaged?
- What are the responsibilities of public custodians for providing disaster-resistant communities?
- What are the responsibilities of corporate and private sector custodians for protecting the financial assets of corporations? Stated another way, should the stewards of private corporations be held accountable for loss due to natural disasters?

- What is the role of science and technology in these endeavors?
- What is the current state of the art with regard to science and technology in predicting or at least visualizing and describing the effects of natural disasters?
- How do organizations like PDC provide this science and technology?
- How do the resulting visualizations and descriptions benefit both the public and private sector recovery planner?

Natural Disasters Are on the Rise

Is no longer shocking to discover that meteorological natural disasters are happening more frequently and increasing in severity as populations grow and move. Disasters are increasing. Nobody knows precisely why.

Hazards such as severe storms, flooding and drought, high winds, landslides, wildfires, and other weather-related events are causing untold suffering to communities. Modern industrial nations are not exempt. Canada, for example, suffered its longest and most severe drought ever between 1984 and 2002. Similarly, one of the six strongest hurricanes ever recorded, Hurricane Katrina, hit the U.S. Gulf Coast *on August 29, 2005, taking more than 1,800 lives and doing an estimated $81.2 billion in damage.*

Scientists around the globe are attempting to understand why hazardous weather is becoming ever more severe. Those efforts must begin first with demonstrating that the increases in frequency and severity are genuine.

Massachusetts Institute of Technology professor of meteorology Kerry Emanuel, writing in the Nature International Weekly Journal of Science, briefed readers on "an index of the potential destructiveness of hurricanes based on the total dissipation of power integrated over the lifetime of the cyclone." His briefing claimed "that this index is increased markedly since the mid-1970s" and that "this trend is due to both longer storm lifetimes and greater storm intensities." This corresponds with the chart provided as an exhibit by Telecom Recovery.

Over recent decades, there has also been a marked increase in the damage caused by geographical hazards such as volcanic eruptions, earthquakes, and resulting tsunamis. This is not because such events are becoming more frequent. Any apparent increase in the number of geographical events (as well as the sharp rise of related damage estimates) can probably be accounted for at least in part by the fact that there are more people—and not only more people, but also more communities in more places that are known to be at risk. Wildfires cause more problems in densely populated Southern California. Hurricanes do much more damage when they hit seacoasts that had previously been uninhabited but now contain vast new developments. As a result, a higher number of events are reported in detail, more lives are lost, and more human-constructed environments affected or destroyed. Add to this the fact that disasters sell a lot of soap when carried on the evening news programs. That means more disasters happen, they affect more, and public awareness as a result of the media coverage has never been higher.

Human-induced disasters, from terror-related events to oil spills, from industrial accidents to sabotage and the effects of deferred maintenance, are also becoming increasingly common and affecting more people. Similarly the Exxon Valdez disaster and other high-profile cases have contributed to awareness and have also been widely publicized in the media.

Numbers alone, however, cannot tell the story, but they help us to picture the sheer scale of the problem that disaster managers face. In 2006, disasters killed 23,000 people and cost more than $34.5 billion. Now there is a sound bite for you.

Lives are at risk everywhere and every day. As tragic as that sobering fact is, communities, countries, and international regions suffer much more than merely the loss of individual citizens.

When a cyclone, for example, kills dozens or hundreds of people, the impact does not stop there. In addition to death and injury, the incident displaces and dispossesses hundreds of thousands, even millions of people. The storm wipes out or seriously damages economic gains infrastructure development, especially in developing countries, which lack the ability to recover due to resource constraints. It completely tears apart social order and political stability. While a disaster-affected community searches for the missing and mourns the dead, it must still service, supply, and resettle the affected survivors and find ways to reintegrate and reemploy them. Meanwhile, ecomomic production is commonly at a standstill, followed by a protracted period of interrelated social, political, and economic disorder and recovery.

Responding on Behalf of Billions

The disaster-response capacity of communities, nations, and regions is, however, improving. It is not merely keeping pace with disasters; it is actually gaining some ground. Lives are being saved. Considering the fact that larger and larger populations are in at-risk areas (and statistically reduce the effect of this good news), when all loss statistics are examined, things are improving. Whether they are improving quickly enough is a subjective question.

The IRIN News, a publication of the United Nations Office for the Coordination Humanitarian Affairs (UN OCHA), reviewed the impact intent and costs of disasters in 2005. The publication said the following:

> While a number of lives lost has declined in the past 20 years—800,000 people died for natural disasters in the 1990s, compared with 2 million in the 1970s—the number of people affected has risen over the past decade [while] the total number affected by natural disasters has tripled to 2 billion. According to the UAN bureau for crisis prevention and recovery, some 75 percent of the world's population lives in areas that have been affected at least once by an earthquake, a tropical cyclone, flooding or drought between 1980 and 2000.

The IRIN authors added: "Disasters are closely linked to poverty; they can wipe out decades of development in a manner of hours." This is a sad but true fact. Poor communities, regions, and nations are heavily and disproportionately impacted by disasters. Not only do they have fewer resources to assist with recovery, their infrastructure, homes, and other construction is more often and more completely destroyed. The livelihood of their citizens is often linked to natural environments or cultivated lands that are also destroyed or negatively altered.

Industrialized nations, on the other hand, where the dollar values associated with losses can be staggeringly high, themselves struggle with the extraordinary and ever-rising cost of disasters. Hurricane Katrina was the most costly ever in the United States. While the value of the loss is lower because there is less to lose in developing nations, they often pay for this shortcoming in terms of greater loss of life and more extreme economic setbacks.

Regardless of any comparisons in raw dollar amounts, natural disasters cost developing countries from 6% to 15% of their gross domestic product (GDP) on average. Progress is being made on this front, however, at least according to available data. For example, according to the Pacific Disaster Center, Cyclone Val cost 230% of the GDP of Samoa in 1991. Cyclone Heta cost only 9% of Samoa's GDP in 2004.

With human allies and their livelihoods at stake, every item capable of reducing risk and mitigating the impact of natural and human-induced disasters must be used to the fullest. This includes incorporating what is known of potential disasters in the planning of development and infrastructure and taking steps today to lessen the effects of future disasters.

Not if, but when, a disaster occurs, it is vital to have the appropriate science and technology (S&T) in place to support complete, concise, accurate, and actionable communication for early warning within agencies and between organizations right down to the last mile to those potentially affected.

Two experts on these kinds of events, Dr. Daniel J. Alesch,[1] and Dr. James N. Holly,[2] shared some additional insights on this topic in a paper they published through the Public Entity Risk Institute (PERI),[3] http://www.riskinstitute.org, and have graciously added their own reflections on this topic here.

Tight Coupling, Open Systems, and Losses from Extreme Events

Following an earthquake, one watches video images of death and destruction. Within hours, one hears loss estimates, revised continually as the days pass. Public officials guess at the loss of life and structural damage. Insurers report on insured losses. These are the obvious losses, but research indicates that many of the losses from extreme events occur in the weeks, months, and years following. This is an exploratory narrative developed as part of an effort to understand both losses from extreme events and the processes associated with recovering from those losses. The nascent idea came from the Dr. Alesch, Dr. Holly, and PERI's observations that many small businesses in communities that suffer disasters fail financially as a direct consequence of a disaster—even if (and here is the punch line) they suffered absolutely no physical damage from the event.

1. Emeritus Professor and Director, Center for Organizational Studies, University of Wisconsin-Green Bay, e-mail: dalesch@new.rr.com.
2. Center for Organizational Studies, University of Wisconsin-Green Bay.
3. The authors are extremely grateful to the Public Entity Risk Institute (http://www.riskinstitute.org) as well as Dr. Alesch and Dr. Holly for providing support to this section of the book.

The premise is that, while society typically focuses on losses to the built environment, there are less visible, but longer lasting and even more extensive, losses from extreme events. These correlate very closely with the findings of the PDC insofar as the more extensive losses stem from things like disruptions to the relationships between members of the affected community, and between the community and the outside world. These more invisible but equally real adverse impacts reduce individual and collective abilities to adequately perform designated functions. Stated another way, the ability to maintain relationships that are critical to the political, social, and economic well being of the community system also become casualties of the disaster.

According to PERI, urban settlements must be thought of as open systems comprising physical artifacts, people, and patterned interrelationships, both within the system and with other systems that touch its environment. Urban systems interact continually with the larger environment, regionally, nationally, and internationally. The most robust and viable urban systems are those that adapt successfully to random internal variances and to changes in the larger environment. They maintain a dynamic homeostasis in the face of that continual change and, sometimes, considerable discontinuities.

Set in this context, consider the fact that disaster damage is reported in largely actuarial terms, such as $20 billion in damages. Costs associated with extreme events are also usually reported in terms of deaths, injuries, and losses to the built environment. For most of us, these are the phenomena most easily observed and understood, so such reporting makes sense. What is difficult to comprehend, as in the case of the recent earthquake and tsunami tragedy in the Bay of Bengal and the Indian Ocean, is the staggering loss of life and other destruction.

As for losses to the built environment, insurers typically report insured losses to structures. That, after all, is their primary concern. Here, too, lies another flaw. In a conversation with a Pacific Disaster Center executive, it was noted that to an insurance company, the destruction of a $1 billion electric power plant and a $1 billion gaming casino was pretty much the same thing. That's because these organizations think largely in actuarial terms. Such assessments beg the question of who precisely is the best person, company, or public entity to make such calls, since loss of the power plant obviously carries a much higher imputed cost to the well being of the community in general. For engineers, structures and the built environment are the professional focus. Logically, buildings and the built environment would be a central focus of an engineer's interest in extreme events. The same could be said of information technology (IT) companies. Do a Web search on disaster recovery planning and see just how many hits come up regarding protection of data centers, call centers, and the IT environment. Obviously the people who own the computer centers for large organizations (public and private) often believe they own disaster recovery. Losses to life, to data centers, and to the built environment are, of course, important losses. But they are only part of the story. To understand losses to society as a whole from extreme events, it is necessary to look beyond the injuries, damage to data centers, and destruction of structures and the built environment. To be sure, structures have social value to the extent that they facilitate and support the political, economic, and social processes and activities we collectively deem essential to our well being. Just as engineers study the response of structures to the forces imposed on them (earthquakes, high wind, and rushing waters) to learn how to better design structures, we must focus our attention on the adverse effects these same events hold on our social, political, and economic institutions and processes. We must know the nature and proximal causes of the full range of losses before we can truly devise effective strategies to reduce them and before we can devise effective recovery strategies.

The True Impact of Disrupted Relationships

When an urban system experiences an extreme event, not only physical artifacts but also social, economic, and political interrelationships are damaged. The extent of this damage depends on the specific characteristics of the urban system. This includes the robustness, resiliency, and redundancy of its critical components. The extent of damage depends also on the nature of the event in terms of magnitude, duration, and intensity. Longer-term losses to the system, however, depend on the extent to which key elements and relationships were damaged or destroyed, as well as how widespread the damage is in proportion to the total system.

Damage to buildings and infrastructure has an adverse impact on the ability of actors to perform political, economic, and social processes and activities. However, the results of research over the past decade demonstrate that a business need not suffer physical damage from an extreme event in the community in order to suffer business losses and subsequent failure. Many businesses that suffered no physical damage, but whose customers or suppliers suffered losses, were unable to continue business relationships with the undamaged firm.

In such cases, the "undamaged" firms often found themselves without suppliers and/or customers for their products and services. These undamaged firms were victims of ruptured relationships—relationships that were essential to them performing their functions—that were gone by virtue of the disaster. As one example but not as a limitation, consider the fate of two competitive local exchange carriers (CLECs) known to Leo in the Houston, Texas, area. In the aftermath of Hurricane Ike, which struck on September 14, 2008, both companies struggled financially because their customers for local phone lines were also affected. The customers simply stopped paying their bills. This created enormous financial hardships for both CLECs. One weathered the storm, both literally and figuratively. The other filed for Chapter 11 bankruptcy only two months later. It might be more than a coincidence that the one that survived also specializes in disaster recovery.

Relationships among individual components of a system are critical. Following an extreme event, it is not at all uncommon for local businesses to lose customers. The customers move out and may or may not be replaced. If they are replaced, it is often by people with different preferences or buying habits. Even those who stay in the damaged area often spend their available money to repair or replace their homes, not necessarily to buy new sporting equipment or meals in expensive restaurants. Consider the examples of Northridge, California, and Homestead, Florida. Both sustained disasters: the former, to an earthquake; the latter, to a hurricane. In both cases, many residents moved away permanently following the extreme event and were replaced by people with lower incomes, different consumer preferences, and different skills. This changed both communities permanently, resulting in long-term economic, political, and social effects stemming from altered relationships within the community system.

Another reason for losses to undamaged system components stems from a phenomenon that Dr. Alesch, Dr. Holly, and PERI have referred to as *tight coupling*. Tight coupling is a systemic relationship between two or more components in which there is little or no "slack" in the relationship. A simple example is a just-in-time (JIT) relationship between a supplier and a manufacturer. Organizations that rely on JIT for delivery of goods and services from a single supplier are efficient when everything works as it is supposed to work. When an unforeseen event slows or interrupts delivery, however, both parties suffer, along with potential customers who must find a replacement or do without the required services.[4] In an environment with great turbulence and fraught with unexpected perturbations, warehousing might make more sense than JIT delivery because it provides requisite slack. Production costs go up by the cost of warehousing, but they are offset, should there be a large perturbation, by having goods available and being able to continue to do business. Regional, national, and international economic activities have become increasingly tightly coupled.

The attendant reduction in slack means that, when infrastructure or built environment in one area is damaged or destroyed, repercussions flow though to undamaged organizations quickly and without damping. Tight coupling also means that the effects on one component are not attenuated as they pass through to another component because there is little damping of the effect and no "fuse" to reduce the ripple effects. Two examples are useful.

1. A fire in a small two-story building in Berwyn, Illinois, some years ago severely disrupted long-distance telephone service across the United States. It seems that most transcontinental landlines passed through that single building and they were destroyed in the fire.[5]
2. In another example from several years ago, automobile antifreeze prices in the United States tripled for two years and then went back to normal. It seems that only two plants in the United States produced ethylene glycol, and one of them, a small facility in Idaho, burned, reducing the nation's production substantially.

4. See Leo A. Wrobel's previous books, *Writing Disaster Recovery Plans for Telecommunications Networks and LANs* (1993) and *The Definitive Guide to Business Resumption Planning*, Artech House, Norwood, MA, 1997. In those books, Wrobel goes into specific detail on customer "pain thresholds," as well as who defects as a customer, how quickly they defect, and the reasons why. The books describe how to categorize customers' sensitivity to disruptions and disasters (including those using JIT) before a disaster while protective measures can still be taken. Wrobel covers similar concepts in *Business Resumption Planning, Second Edition*, Boca Raton, FL: CRC Press, 2009.
5. See Leo A. Wrobel's previous book, *Disaster Recovery Plans for Telecommunications*, Artech House, Norwood, MA, 1990. On May 8, 1986, fire destroyed the Hinsdale central office in suburban Chicago, disrupting telephone service in varying degrees for up to six weeks. The fire and other examples of telecommunications disasters that have the potential to ripple through an urban system are described in this book. The Hinsdale fire, coincidentally, occurred on Mother's Day—traditionally the highest volume day in the United States in terms of telephone calling.

Another consequence of extreme events has to do with the subsequent redistribution of wealth and income. If one draws a line around a large enough system, it is possible to argue that, in aggregate, the system has not been hurt by the redistribution of wealth and income following an extreme event. The problem, of course, is that aggregate measures mask the consequences of transfers within the aggregate.

Some economists have suggested that the economic activity stimulated by rebuilding (along with transfer payments from insurance and charity) result in areas that benefit somewhat from the disaster. Not true. That proposition would apply only if all of the transfer payments and economic benefits from the rebuilding get to ultimately *stay* in the damaged area. History seems to indicate this is not always the case. Following the Northridge earthquake of 1994, a considerable amount of the money earmarked for the rebuilding of homes and businesses went to contractors and manufacturers from outside the damaged area. Most of the money spent on replacing carpeting and floor coverings, for example, went to large out-of-state manufacturers who were able to bypass local wholesalers, distributors, and retailers, cutting them out of the market almost completely in addition to bankrupting some of them.

Even decisions made outside the affected urban system can sometimes constitute a disaster in a community, by changing relationships between the community and the outside world. At the same time that hurricane Andrew struck Homestead, Florida, the U.S. government closed Homestead Air Force Base. This resulted in significant reductions in community jobs, income, talent, and population. The federal government had also recently adopted the North American Free Trade Agreement (NAFTA), which locals blamed for devastating the local truck farming industry and, for all practical purposes, moving that industry to Latin America. Hurricane Ike hit at the same time world financial markets collapsed, severely impairing the ability for many to recover.

The result of these events shattered almost all the economic and social relationships in the community, resulting in a community that was forever changed while rendering a municipality virtually insolvent a decade after the hurricane.

What Are the Lessons for Us?

Examples of the consequences of disasters and extreme events on social, economic, and political relationships on societies are not limited to the present.

1. Some historians attribute the demise of the magnificent Minoan culture on the Mediterranean island of Crete (circa 2000 B.C.) to a massive volcanic eruption on the nearby island of Thera and the associated earthquakes and tsunami. It is thought that the Minoan fleet was destroyed along with much of its built environment and a significant proportion of the population. The Minoan culture was unable to regain its preeminence in the two centuries that followed. Indeed, the fatally weakened culture declined and eventually fell victim to an invasion from a competing culture in the eastern Mediterranean.

2. When Europeans came to the Western Hemisphere in the fifteenth century, they brought smallpox with them. The introduction of smallpox into the Western Hemisphere constituted an extreme event of extraordinary consequence. Some estimates suggest that smallpox may have reduced the pre-Columbian population of the Americas by as much as 95% in as little as two centuries, ensuring their inability to protect themselves from succeeding waves of European immigration.

In their longitudinal research of more than a dozen communities struck by an extreme event, Dr. Alesch, Dr. Holly, and the PERI found examples of persistent and predictable economic, social, and political costs catalyzed by the event. In some cases, those costs were the consequence of multiple events. These losses included disruption and destruction of economic linkages and locational relationships, diminished production and distribution capabilities, reduced or refocused purchasing power, all of which precipitated reductions in employment. Social losses occurred at the level of the individual, household, family, neighborhood, community, and nation, resulting in demographic dislocation, wealth transfer, and loss, as well as familial disruptions, death, divorce, and mental illness. Political costs manifested themselves as changes in the salience of various issues, the primary focus of political activity, changes in political prominence and fortune, and in how local governments framed issues and problems. Political processes were disrupted, and demographic changes resulted in changed electorates.

Having set the context, what conclusions can we draw about the less visible losses from extreme events and disasters? We submit the following:

1. Urban settlements are complex open systems.
2. A set of elements integral to urban settlements are interrelated in such a way that the whole is more than the sum of the parts.
3. A perturbation to one or more elements has consequences for other elements in the system because of their interrelationships.
4. Urban settlements are largely self-organizing. If we think of a community as a largely self-organizing open system, then it becomes somewhat easier to understand the relationships among elements within the system.
5. Relationships with elements in other systems can be affected adversely even though the physical damage to individual members or segments of the community system may occur in other parts and places of the system.
6. Complex systems (comprising elements capable of individual choice and reacting independently in response to stimuli) adapt continually to disasters and extreme events. The sum of these behaviors can cause the system to morph from one system state to another through time.
7. Disasters and extreme events are major perturbations that require substantial response from a large share of system components very quickly following the event. The extent to which the system maintains its dominant characteristics following such an event presumably depends on the size of the extreme event in relation to the size of the total system, as well as the system's inherent robustness, resiliency, and redundancy.

8. To understand *all* of the effects of a disaster or extreme event, it is important to understand the ability of individual elements of the community system to continue to perform their roles and functions. This understanding must include the interplay of the system or elements with components in the relevant environment.

9. While the system may return to a dynamic equilibrium after a disaster or extreme event, that equilibrium state may never be the same as that which existed before the event. The event fundamentally changes individual elements and the relationships among elements, often spawning new relationships that emerge as new elements, as the system changes behaviors in an attempt to stabilize in the new environment.

10. Depending on the magnitude of the perturbation, individuals, organizations, and institutions will be unable to perform functions and maintain important relationships in the aftermath. When the event is truly extreme, there may be no such thing as a total recovery—if recovery means returning to the status quo ex ante. What happens is that the individual elements of the system attempt to cope in a new setting. Relationships change as some people leave and others move in and as businesses and institutions adjust, more or less, to new circumstances.

11. The post-event viability of an urban settlement depends, too, on the relationships between the system and its environment. Systems struggling to maintain weakened relationships with their environment will have a difficult time maintaining viability in the post-event environment. Smaller settlements whose historical raison d'etre is largely dissipated and that are struggling to develop a reason or being or to reestablish demographic and economic viability sometimes find that an extreme event, such as a flood or conflagration, simply means an accelerated trip into oblivion.

12. The sum of their behaviors can have any of several outcomes. The pre-event community system may survive, but with changes to adapt to the new realities. Alternatively, the pre-event system may not survive, or it may suffer a fatal wound that leads to its longer-term demise, as with the Minoan culture in Crete. The community may simply wither away. Another possible outcome is that a new community system emerges in the place of the old one. Change, however, is not necessarily for the better. The new community may be inferior to the old one in any number of ways.

This all sounds like abstract theorizing, but there are very practical implications. These points have to do with how we collectively frame our understanding of the consequences of extreme events and how we conceptualize and devise recovery strategies. Consider the fact that officials in Grand Forks, North Dakota, rebuilt the central shopping district of their community some years back. It is still largely vacant a decade later. A similar quest to reestablish Princeville, North Carolina, following a devastating flood has not resulted in a better community, despite the millions of federal aid spent there. St. Peter, Minnesota, on the other hand, lost a third of its buildings to a tornado but is doing better than ever. South Dade County, Florida, changed forever; a new community will emerge, but only as Miami continues to expand southward toward the Keys.

Today, governments, institutions, and individuals focus on rebuilding or replacing buildings and structures with the underlying assumption that rebuilding somehow constitutes recovery. By and large, people tend to think in terms of body count or in starkly actuarial terms that neglect to consider the impact to the elements of urban settlements as a whole and the complex interplay of the various elements that make them up. The thinking of the past is not sufficient in this day and age. We have to start looking at facilitating the creation or recreation of the relationships that make communities viable after extreme events or disasters as well.

There are places where such data is available for those inclined to look for it.

1. Insurance companies have forgotten more than we probably know about probabilities of disaster. The downside is that they tend to think strictly in actuarial terms and don't usually concentrate on economic or social consequences.

2. There are a lot of great disaster recovery consultants out there. The problem there is that again they usually work for private firms with the underlying motive of profit. Disaster recovery becomes a "value proposition" to the company. For example, many companies plan for a bird flu pandemic. While we would like to believe that it is out of a sense of caring toward employees, it is more probably a realization that it is difficult to conduct business without a work force.

3. What about the government itself? That might make some sense, given that there are some things that just would not happen if not for government. As a case in point, I like to tune my radio and get only *one* station at a time. I enjoy driving on the interstate highway system, which would not have been built if not for a government mandate. (And even then, it had to be justified as necessary for national defense!) And, of course, my favorite: putting a man on the moon. I remain convinced that 1,000 years from now, that will be the single thing that people really remember about Americans. Neil Armstrong and company. (If you don't believe me, ask yourself this question: Do you remember King Ferdinand and Queen Isabella in the context of Christopher Columbus? Most people do. Who were king and queen of Spain immediately before them? Or immediately after? Get my point?) Getting back to the point, however, government might indeed be part of the answer. Still, there are problems in that regard. Government does not always have cash. Government does not always have the will to do it. Furthermore government does not always understand enough about actuarial issues, in the same manner that private entities don't always understand the public good.

It is with these thoughts in mind that the authors again solicited input from the PDC, which arguably is, like the Osmonds used to say, "A little bit country, and a little bit rock and roll." We find that PDC understands the perspectives of government, by virtue of its working for so many worldwide. The PDC was created with federal funding, which whether we like it or not is one way of a lot of this task is going to get done. Perhaps most importantly, just as there were spinoffs from the U.S. space program, there are spinoffs from the federal program that became the

PDC. For one thing, the place can be referred to as "content central." The PDC receives terrabytes of data as raw material from myriad state, federal, and world-wide organizations. In its years of existence, it has devised tools to draw useful conclusions from that data. What the PDC has not really done yet is apply this data to private sector organizations that seek to go the extra mile in contingency planning. Energy companies come to mind. You have all heard the scare stories. A hurricane hits a major refinery or offshore drilling platforms, and we are all paying $6 for gas. Many energy companies today feel the public pressure to go the extra mile and look into some hard probabilities of this happening and what can be done about it. This is actually a very good sign, as private sector organizations for the first time begin to be concerned about urban settlements and the public welfare as a whole. So where would an energy company go for this kind of experience? We submit to you it might be an organization like the PDC. It beats going hat in hand to the government and has many of the same advantages. In a nutshell, the PDC can be considered a repository of S&T comprised of experience, data, and tools under one roof. The ability to visualize data streams from sensors, satellites, buoys, and other sources is time consuming for the recovery planner but absolutely compelling when completed and presented to policymakers. Consider the six diagrams on the following pages that visualize risks to the Association of Southeast Asian Nations (ASEAN). Pictures are compelling and immediately understandable (see Figure 3.1).

Although S&T by itself cannot prevent disasters from occurring, recent experiences and successes at the PDC and other organizations like them offer real hope for reducing disaster impacts. In the case of PDC, these successes have been achieved principally by using the latest S&T in the process, to increase warning time and accuracy of disaster warnings, by applying risk and vulnerability analysis (RVA) techniques for better planning and mitigation, and by building capacities and partnerships to proactively take advantage of advances in disaster risk reduction. It's that last one, the "partnership" part, that I find most exciting. That's how things get done.[6]

Applying New Science and Technology to Disaster Recovery

Let's put it this way. If disasters are defined as the intersection of phenomena and presence, S&T can now help to limit the overlap of these two principal components, thereby reducing the risks and impacts.

Reducing disaster risks is a multidimensional problem requiring a multidimensional solution. Science and technology alone cannot achieve the ultimate results. The work of culturally sensitive institutions in raising awareness at the community level is also important, to approach the problem from the bottom up.

How can science and technology reduce risks? First, by improving the timeliness and accuracy of disaster warnings, and, second, by improving the speed and clarity of the look of alerts disseminated to decision makers and at-risk populations.

---

6.   See Leo Wrobel's book *Understanding Emerging Network Services, Pricing, and Regulation,* Artech House, Norwood, MA, 1995, where Leo discusses how to form partnerships with regulators and telecom company executives to purvey advanced services.

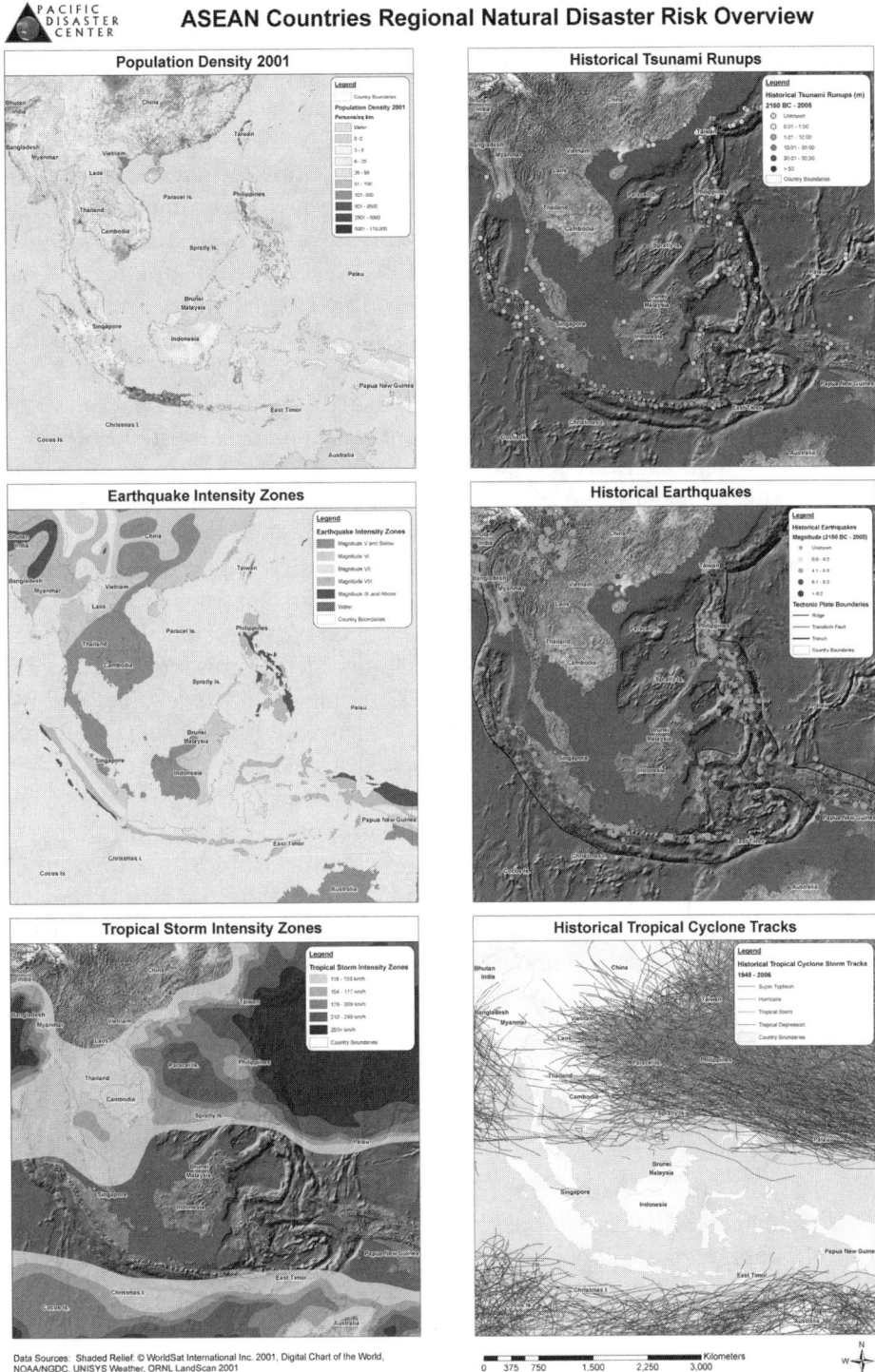

Figure 3.1 ASEAN countries regional natural disaster risk overview. See these diagrams in stunning color at http://www.pdc.org.

As a case in point, since the 2004 Indian Ocean tsunami, considerable efforts have been directed toward establishing disaster early warning systems at regional, national, and local levels. Challenges in this approach involved lack of infrastructure and or interoperability in systems reaching the last mile. However, advances in technology are promising to overcome many of these challenges.

Second, we can best understand risks by using the latest science and technologies. We could be doing a lot more to model events, for example, with the thousands of data sources available today and with things like sophisticated satellite imagery. Even things like the drop in price per CPU cycle of supercomputing and the ubiquity of high-end workstations add immeasurably to capabilities in this regard. Data is not the issue, but making sound conclusions that make sense of the data is.

Statistical and historical disaster data, used in conjunction with indicators that truly and correctly estimate a society's ability to respond and or recover from disasters, are improving disaster management capabilities. These studies not only indicate which communities are most at risk but also exactly how. The results can better support planning mitigation and policy decisions. Urban planners, for example, may use this knowledge to limit the development in an area. Similarly, a study indicating increased risk due to a concentration of population with special needs can lead officials to consider special evacuation measures. Private firms looking to locate costly manufacturing plants and operations centers get to take a look-see at the demographics, topography, and other infrastructure first. Even firms looking to locate call centers, for example, might want to check routing of undersea cables and whether the physical cable or terminating facilities are susceptible to seismic events or sabotage.

Today's S&T can help to better visualize risk, so it stands to reason it can help the contingency planner better communicate risk. Executive management with responsibility for the enterprise as a whole is not against disaster recovery. They will fund a contingency planning effort. Management needs to know, however, that money spent on recovery planning is not money down a rat hole. Today's S&T supports the need for this planning by visualizing what would otherwise be complex and voluminous data. Similarly, various modeling and visualization techniques can be used to convey complex risk information, raising awareness and building knowledge about risks.

PDC's Approach

The PDC and its partner organizations, including national governments, military commands, and transnational collaborators, are a source of proven best practices and subject matter expertise designed to save lives and foster disaster resilient communities. The PDC provides expertise, tools, technologies, and innovative solutions in the area of disaster preparedness mitigation response and recovery for emergency managers and decision-makers. As an applied science, information, and technology center, PDC is chartered to reduce disaster risks and impacts to people's lives and property, and links scientists with risk communities.[7]

---

7. For example, the Pacific disaster centers Asia Pacific Natural Hazards and Vulnerability Atlas have a Web-accessible tool that combines data resources and disaster management tool sets.

The PDC's approach to emergency management shifts the emphasis from reactive (focusing on response and recovery operations) to proactive (with the focus on mitigation and preparedness.)

As we stated earlier, voluminous data is no problem these days, but making meaningful interpretations of it is. The vast streams of data that are now available from sensors, satellites, buoys, observers, and responders are overwhelming. What is needed is the ability to make a clear situational analysis that can be shared in real time or near–real time—preferably in a pictorial format that can be easily understood by decision makers as information products that are instantly understandable and, when possible, fully visualized.

By analyzing, aggregating, and integrating dozens of information resources into a meaningful composite, PDC helps the contingency planner achieve the goal of accurate probabilities gleaned from meaningful data. After these probabilities have been determined, PDC uses concise models rather than charts, full-motion viewers rather than spreadsheets, that are understandable and immediately actionable the moment the information is presented.

While an avalanche of incomprehensible data inspires fear and confusion, a well-designed and shareable data presentation structure becomes an integrated decision support system. Ultimately, the ability to effectively preset and substantiate your figures using hard data gets the plan funded. It also saves lives and helps make communities more resilient.

The overall strategy of PDC is to promote disaster management as an integral part of national-to-local economic and social development.

Fostering Disaster-Resilient Communities

A continually growing array of data streams is available to disaster managers, including seabed sensors, ocean surface buoys, satellites, water-level gauges, seismometers, and other land-based devices. What matters most, however, is the ability to first acquire the right data and then to produce reliable, timely, and easily shareable information on which to base and substantiate executive decisions and actions. To accomplish this, broad collaborations and mutual support are essential. Some of the methods that have been employed by PDC are listed next.

Decision Support Case Studies

PDC worked with the National Disaster Warning Center (NDWC) in Thailand from December 2005 through February 2007, providing technical assistance in order to enhance NDWC's management capabilities, systems, and practices. The focus of the project was on the dissemination of early warnings, especially warnings associated with tsunamis.

PDC, along with Lockheed Martin Information Technology, Sun Microsystems, and the Environmental Systems Research Institute (ERSI), provided NDWC with technical solutions, systems integration, and human resources training to achieve its strategic objective of establishing a scalable and world-class disaster management and emergency communications facility.

The PDC-NDWC collaboration began with an information and communications technology gap analysis and development of a concept of operations in relation to the decision support system. Other steps consisted of a hazard-related data inventory and a proposed system architecture that included cost estimates, a business continuity plan, and a "train the trainer" section to establish and ensure sustainable capacity. The primary product of the project was PDC's integrated decision support system (IDSS) and disaster all hazards warning analysis and risk evaluation (Disaster AWARE).

PDC's first IDSS was deployed for U.S. Southcom in the Caribbean. The more customized Thai version was an evolutionary leap, capable of reconfiguration and expansion to meet future needs. Other versions, tailored to the needs and specifications of specific regions in nations, including Vietnam, are also being developed by the PDC at the time of this writing.

Information Sharing

Realizing that disaster information sharing must be an integrated component of the disaster risk-reduction efforts in the region comprising the member countries of the Association of Southeast Asian Nations (ASEAN); the ASEAN committee on a disaster management (ACDM) and PDC developed a joint framework for enabling a disaster information and communication sharing network or DISCNet. The stated objective of the DISCNet program was to "enhance regional disaster management capability and readiness" by:

- Developing a more effective "information clearinghouse" mechanism to promote regional collaboration and strengthened national capabilities in disaster risk-information dissemination;

- Ensuring disaster management information sharing among ASEAN member countries that will lead to the development of a framework for regional integrated decision support;

- Enhancing disaster management readiness by supporting regional exercises, tabletop or otherwise;

- Strengthening the capabilities of ADCM to integrate its disaster information management system with other entities in the global arena.

The collaboration of PDC and ACDM reached its first major milestone with the 2005 publication of *Information in Communication Technology Assessment for ASEAN DISCNet* and an analysis of the ICT capabilities and capacities of 10 countries. PDC and ADCM have an ongoing partnership, which includes development of the online Southeast Asia Disaster Inventory (OSADI) launched at the tenth meeting of ACDM in October 2007.

Risk Modeling and Mapping

PDC also worked with the state of Hawaii on its dam safety program after a catastrophic dam failure on the island of Kauai in March 2006. After that event, dam owners were mandated to prepare, maintain, and implement emergency preparedness plans for each damn or reservoir. A key element for each plan was a map defining the potential downstream inundation should the dam fail, giving an assessment of the critical infrastructure and population at risk (see Figure 3.2).

PDC was contracted to prepare these analyses on behalf of the dam owners. A critical objective for PDC was modeling potential failures and creating inundation maps of all registered dams for emergency planning purposes. PDC has provided maps and consequence assessment reports for all 135 dams. PDC's data will be used in the creation of evacuation maps and plans.

Capacity Building

A significant element in disaster management is the capacity to plan for, respond to, and recover from disasters, in terms of equipment, facilities, human resources, skill sets, standard practices, and policies and procedures. PDC has worked with many of its beneficiaries and partners to improve their systems, develop information-sharing applications, and bring staff education to the "train the trainer" level, stabilizing the expertise within the organization. PDC conducts extensive analysis of data

Figure 3.2 The same natural phenomena that make life possible, like the floods that produced the fertile valleys of Vietnam, become natural hazards when they intersect with human populations in unmanaged ways.

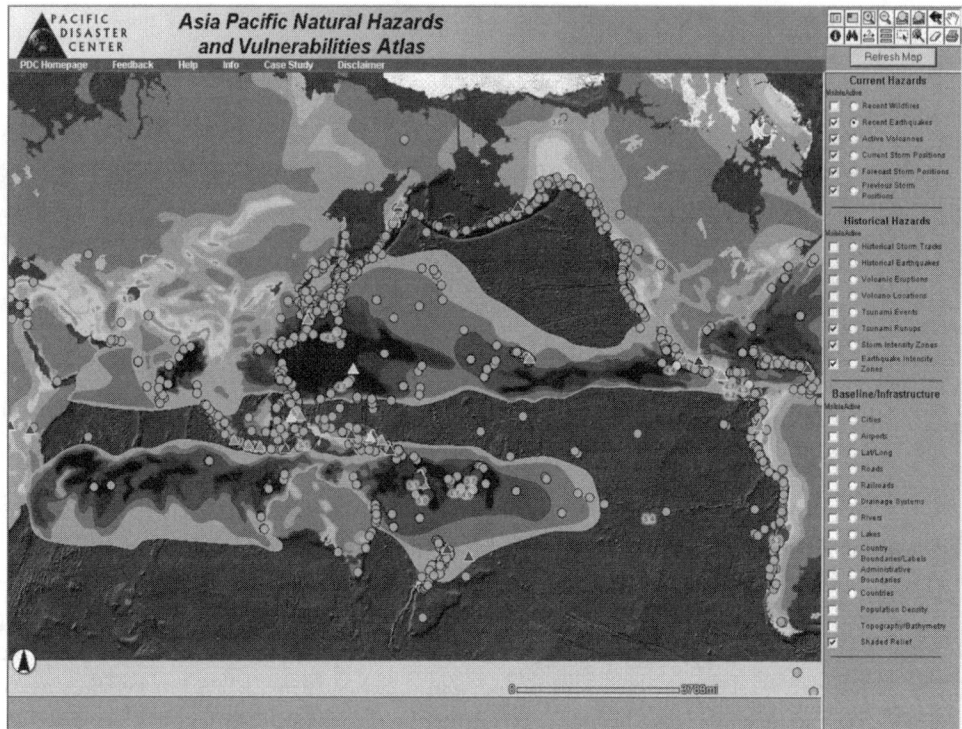

Figure 3.3 Are you ready for some HANDS ON experience with natural disaster data visualization? Visit http://www.pdc.org and click on the Hazards and Vulnerabilities Atlas.

holdings, policies, practices, and procedures to point out gaps and deficiencies in layout plans for maintaining an operating capacity over time (see Figure 3.3).

Summary

Information alone is not the answer to the challenges of disaster management in a world of more than 6 billion people, who are crowding ever more densely into disaster-prone regions, while weather hazards are increasing in number and severity. Nor is S&T alone the answer. Whatever the type of disaster being planned for, analyses, integration, visualizations, and shareable presentation of data are necessary to provide instantly actionable decision support. The necessary information and communications technologies exist, and they are being improved every day. For further information on these topics, contact the authors, the PDC, or the PERI. The authors thank everyone for their thought-provoking contributions in this chapter.

CHAPTER 4

Case Studies and Examples of Quantifying Risk of Natural Disasters to Critical Infrastructure

Weighing the Risks of Natural Disasters

One of the most difficult tasks that face the recovery planner is calculating the odds that a natural disaster will strike. As contingency planners, we consistently try to stack the odds in our favor, in much the same way that a gambler is always on the alert for a "hot tip" from the racetrack before placing his bets. In our business, it is all about information. Information and what you do with it are your most valuable assets in times of disaster. Armed with this information, you are able to make critical decisions on where to build your home or business and what type of materials to build it out of, as well as being able to establish escape or evacuation plans where needed.

Just as an insurance company scrutinizes actuarial tables before underwriting a policy, the same kinds of tools are available to the contingency planner if one knows where to look. But what if a disaster happens where it has never happened before? Some disasters, such as earthquakes and tsunamis, happen so infrequently that not a lot of data is available. Consider the possibility of an earthquake in California or Japan. It is a virtual certainty that one of these will happen in a lifetime. The great Kanto Plains earthquake in Japan was in 1921; the San Francisco earthquake was in 1906—we are quite overdue. It's been so long since one occurred, in fact, that it is guess work at best to determine how modern infrastructure would stand up to a similar event today. Buildings are a lot better, but there are also a lot more people. It changes the whole equation for the contingency planner.

It is also important to remember that disasters, though predictable in some cases, can occur in any form, anywhere, at any time. So where can you get much-needed real-time information? Just because Washington state has not had a tsunami for a while does not mean they do not happen there. Washington had not had a volcano eruption for a while either, but look at what happened in 1980 with Mt. St. Helens.

When planning for these kinds of events, consider sources like the following:

- The National Oceanic and Atmospheric Administration (NOAA), http://www.noaa.gov/, forwards valuable information regarding climate changes on to the general public.
- On a more global scale, there is the Emergency Disasters Data Base (EM-DAT), http://em-dat.net/.

• We have already covered the PDC, http://www.pdc.org/, in Kihei, on the Hawaiian island of Maui, which gathers information about potential disasters from a variety of sources.

In fact, information is only half the equation. For the remainder, please see the following section.

It Is All About How You Present Data and Information

In the last chapter we discussed the socioeconomic impact of disasters on urban societies and how to present complex and voluminous data to decision makers and policymakers through the use of modern visualization techniques. In this chapter, we take a more practical approach. We present several actual case studies by the PDC, which illustrate specific tools, capabilities, and methodologies that can be used to communicate risks associated with many kinds of natural disasters. These techniques and methodologies allow contingency planners to strengthen their case for presentation to policymakers, thereby enabling those who set policy to *proactively* manage and sometimes even prevent catastrophic disasters. By thoroughly understanding the risks, and communicating them effectively to policymakers, the goals of a disaster-resilient community, company, or other entity can be more quickly achieved. For example, mock tabletop exercises and annual disaster drills can be staged based on real-world data. This helps communities and enterprises to work together by being freshly prepared to spring into action when disaster does strike. Drills are scheduled based on real-life scenarios, utilizing storm track maps, weather situation products, and wind damage estimates.

As a quasi-public agency, the PDC is one of the few resources that can not only gauge the effect of disasters in actuarial or monetary terms, it can also help with the more nebulous impact on socioeconomic systems such as were discussed in Chapter 3. Each nation has its own economic capabilities that must be addressed in context when making these assessments, but you get the idea. Of course, each geography, municipality, or nation also faces its own unique threats, which can include such things as were discussed previously in this book, including:

• Drought;
• Earthquake;
• Flood;
• High surf/seas;
• High wind;
• Hurricane;
• Tornado;
• Tsunami;
• Volcanism;
• Wildfire.

Using Public Domain Sources Like PDC to Compute Disaster Probability

Trying to convey the need for disaster recovery planning in a request for funding is a long-time problem, especially in the commercial sector. Disaster recovery planning can be expensive. And like any major capital outlay, the executive presentations and justifications for funding can be like a root canal to the requesting party.

Ironically, in our experience, management is not generally averse to funding a plan, since there are a lot of ways today they can be held accountable if things go wrong. In fact, fundamentally speaking, in order to endorse (and fund) a disaster recovery initiative, responsible management really only needs to know four things:

- What can happen?
- What is the probability it will happen?
- What does it cost when it happens (in terms of lost productivity, sales, market share, and public perception)?
- What does it cost to make the exposure go away?

Sounds easy, right? Well, it's not. If it were that easy, everyone's plans would get funded all the time. Generally speaking, the "What can happen?" part is not always an issue. An executive in Dade County, Florida, knows hurricanes are a possibility. Similarly, the third bullet point, "what does it cost when it happens?" can be accurately quantified if enough consulting horsepower is thrown at the problem. A good constant will not only include productivity loss (which can be estimated by taking the average loaded personnel cost and multiplying by number of employees) but also other issues like lost sales, lost market share, lost customer confidence, legal and contractual liability, and so on. Without question, the requesting party always knows to the dime what they want to spend to make the problem go away, so the fourth bullet point is not generally the issue, either. In our opinion, the item most people fall down on is the second point, probability. A decision-making executive entertaining a funding request may well believe that the probability of the particular disaster or disasters is overstated. The reason for this is pretty much engrained in the human psyche: Everyone has different levels of aversion to risk. Take 10 executives with identical income, assets, family size, demographics, and so on. Every one of them will carry different amounts of life insurance based on their own unique risk perception. Generally speaking, you are not going to convince such an executive to carry more or less life insurance. Why then would you believe you would be able to convince him or her to carry more disaster recovery protection? Yet this is exactly what a contingency planner is tasked to do. The only way to win over skeptical executives, in our opinion, is to use hard, irrefutable facts to make your case. That means finding good sources, and "packaging" the data in a format that management can understand and endorse.

In this section, using data available from public sources, in particular from the PDC, we compute the probability of a disaster in paradise using actual weather, seismic, infrastructure, and demographic data. The first example is a demonstration of a natural hazards and vulnerabilities atlas introduced in the last chapter. This tool is an example of modern modeling techniques that may be relatively unknown to contingency planners, especially those in noninsurance-related corporations. It also might be the next "new thing" in contingency planning, since it bolsters and supports the case of the contingency planner based on real data, a scientific approach to understanding it, and an almost Walt Disney–like method for presenting it. Using the atlas, a repository of raw data drawn from numerous public and private sources can be examined, manipulated, and presented that examines people and infrastructure at risk due to natural hazards present in a given region. The atlas provides decision makers and contingency planners with a resource for understanding the types, frequencies, and severities of hazards that may threaten their communities and thereby provides the horsepower in many cases to "sell" a disaster recovery project to skeptical executives. By compiling voluminous, complex, and often intimidating data into a pictorial and graphical format, this tool can play an important role in raising the awareness of these hazards and their associated risks and of potential mitigation strategies.

Let the Exercise Begin!

Often, a first step is examining the potential for natural hazards that can cause harm to a society's people, infrastructure, or environment. Even for commercial and private sector organizations, damage to sensitive social systems such as that described in Chapter 3 can ripple catastrophically into the organization.

Historical records of these events can be examined using the atlas. In Figure 4.1, the locations of major earthquakes[1] and their intensities are viewed over a geographic area. In this case, it is for the Southeast Asia region, but remember the tools and techniques can be applied anywhere.[2] The earthquake intensity layer, which underlies this diagram, maps the modified Mercalli[3] (MM) intensity that is expected, with a 20% probability, to be exceeded during a 50-year period for a given location. The 50-year period represents the average design life of a building. In Figure 4.1, for example, dark green areas correspond to intensity expectations of V and lower, while orange areas have a 20% probability of experiencing an earthquake intensity of IX or higher over the next 50 years. (Note that these figures are black-and-white diagrams. To see them in color, visit http://www.pdc.org/Artech.) In the most seismically active areas, the symbols representing epicenter locations almost completely obscure the underlying intensity layer at this scale.

---

1. The earthquake epicenter database includes information from NOAA NGDC's Catalog of Significant Earthquakes (2150 BC to present) and Database for Instrumentally Recorded Earthquakes (1962 to present). Detailed information on these and all other hazard atlas layers are included under the "Atlas Info" link (http://atlas.pdc.org/APNHVA/Atlas_Info.pdf) as well as in associated metadata documents.
2. For these examples, the authors used the Asia Pacific region since PDC had the richest repository of existing materials in this area. For other areas, such as the Mediterranean, the same kind of methodology could be used but would require loading revised data. Even in the most obscure areas where electronic data is not available, PDC has become adept at scans and overlays of map data. In this fashion, not only is the subject area portrayed regardless of whether electronic data exists, but after the project of course, electronic data wil exist for future use on other projects.
3. Intensity is a measure of the effect of an earthquake on the Earth's surface. The Modified Mercalli Intensity Scale has 12 levels, ranging from I (not felt) to XII (total damage), and is widely used in the United States by USGS, FEMA, and others. See http://neic.usgs.gov/ neis/general/handouts/mercalli.html.
4. The Help function of the atlas provides details on the usage of these and other tools.

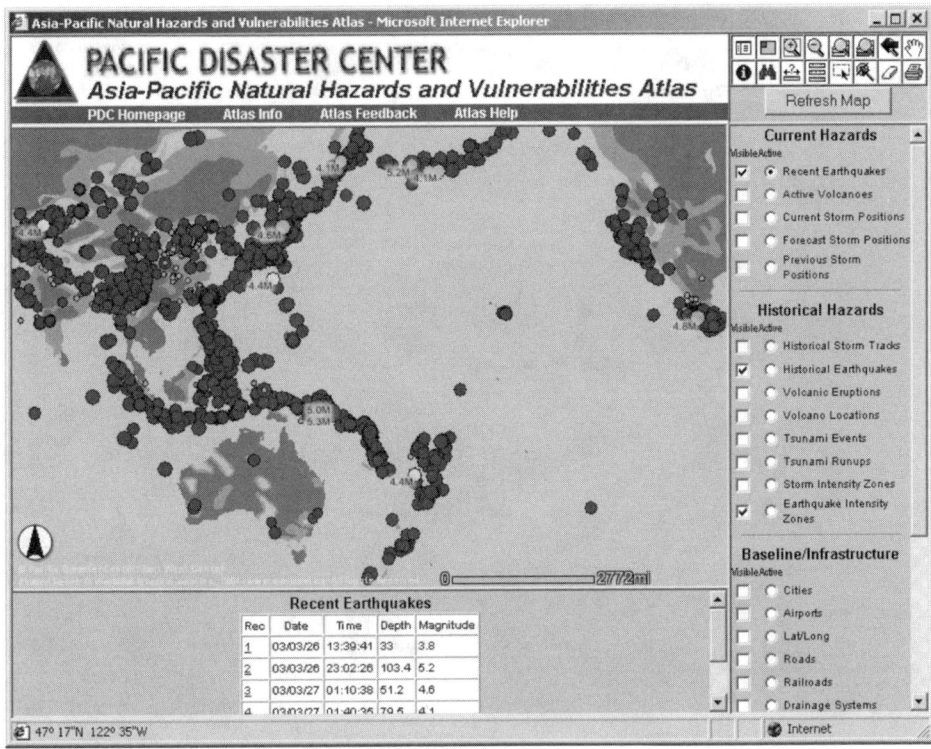

Figure 4.1 Earthquake epicenters clearly outline the "ring of fire" along tectonic plate boundaries. The background shades correspond to earthquake intensity zones and are derived from the earthquake epicenter data. To see in color, visit http://www.pdc.org.

The zoom tool[4] function of the atlas can be used to examine these data at a higher resolution. Additionally, the identify tool and the select tool can be used to view details, including magnitude and date, of recent or historical earthquake events.

Tropical cyclones can be investigated in much the same way using the same tool. Figure 4.2 shows tropical cyclone intensity zones for the same region. These data map areas that have a 10% probability of experiencing a tropical storm of a given intensity during a 10-year period. Darker shades of blue correspond to higher storm intensities. For example, light blue bands show areas that are expected to experience a tropical storm with maximum sustained winds of 118 to 153 km/hr, while the darkest blue areas would expect a tropical storm with maximum sustained winds over 250 km/hr. Additionally, a current tropical storm, Inigo, off the western shore of Australia is shown.

Now let's take this data and compare it to the location of critical infrastructure where we propose to locate our business. Using the PDC's tools, the relationship between these hazards and potentially impacted resources can be observed by displaying the hazards along with population centers, roads, railroads, and airports. In this example, Figure 4.3 shows earthquake risks, along with transportation infrastructure, for a portion of the Philippines including Manila. Figure 4.4 shows tropical storm risks for the same area.

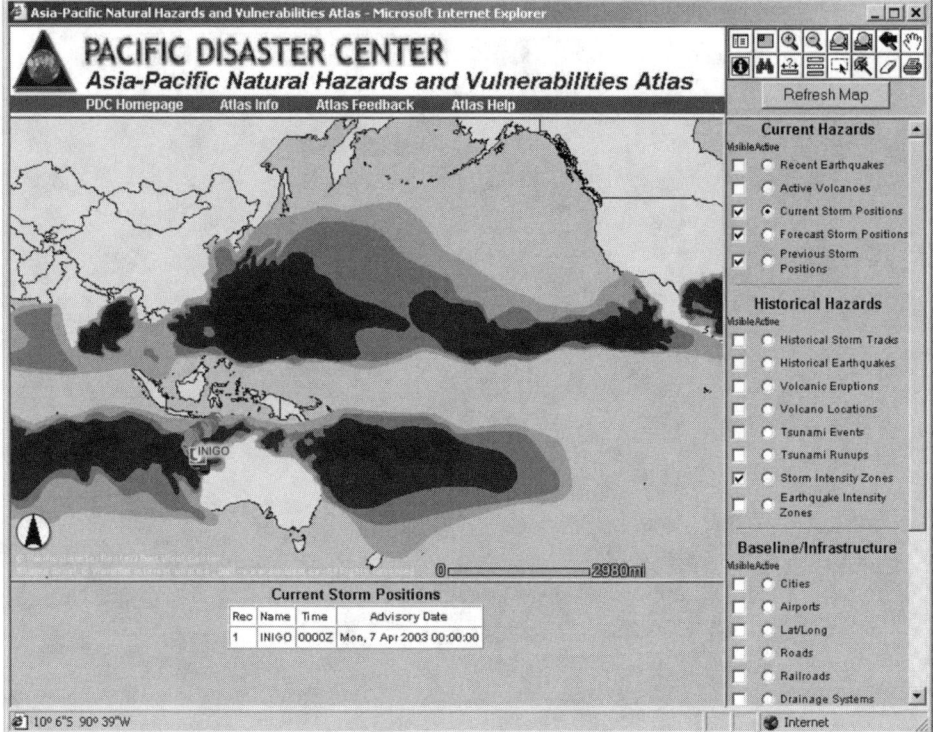

Figure 4.2 Tropical storm intensity zones estimate the most severe storm that a region is expected to experience during a 10-year period based on analysis of historical storm data. Tropical storm Inigo, active at the time this graphic was captured (April 2003), can be seen off the western coast of Australia. To see this figure in color, visit http://www.pdc.org.

As can be seen in Figure 4.3, the frequency and severity of earthquakes in and around Manila places much of the region in the top two categories of earthquake intensity. Most of metropolitan Manila can expect an earthquake of intensity VIII or higher once per 50-year period (20% probability), while the remaining eastern region can expect earthquakes with even higher intensity, IX, during the same time frame.

As Figure 4.4 shows, this same area is also frequented by tropical storms and falls within the top end of the storm intensity zones. By creating a virtual overlay of earthquake and tropical storm risks in Figure 4.5, it is possible to draw some conclusions. Some regions, such as the western portion of Borneo Indonesia, face relatively low risk, while others such as the Philippines face a high risk to both earthquakes and tropical storms, as shown in Figure 4.6.

The next step in the multihazard risk assessment process is to analyze the varying degree of potential exposure of people and infrastructure to the hazards present in a region. The atlas contains both population density and transportation infrastructure (i.e., roads, railroads, and airports) data layers. By combining population density (Figure 4.7) with transportation infrastructure (Figure 4.8), one can examine the relative magnitude (or density) of people and infrastructure exposed to potential harm from future occurrences of earthquakes and tropical storms.

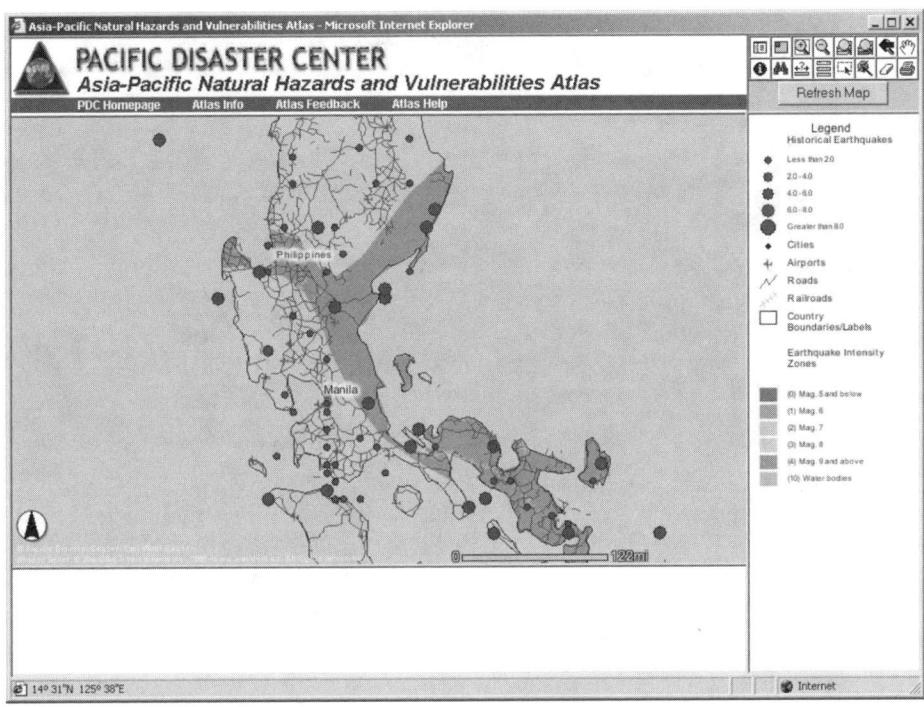

Figure 4.3 Depiction of the frequency and severity of earthquakes in and around Manila. To see this figure in color, visit http://www.pdc.org.

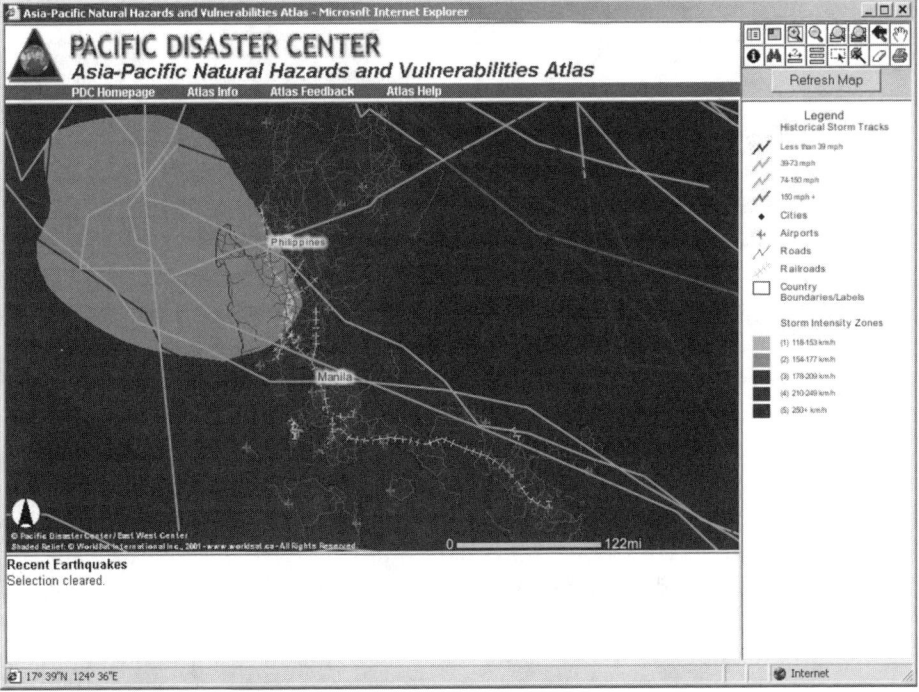

Figure 4.4 Frequented by tropical storms, Manila falls within the top end of the storm intensity zones. To see this figure in color, visit http://www.pdc.org.

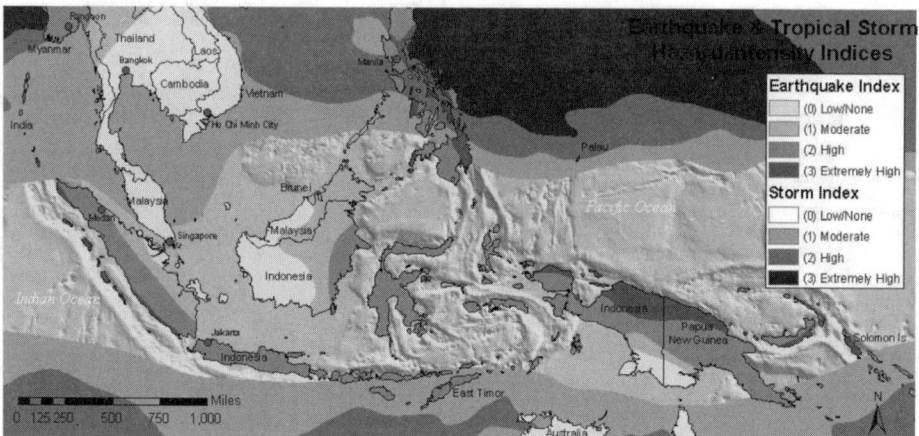

Figure 4.5   Analytically combining both the earthquake hazard index and the tropical storm hazard index into a multihazard index allows the planner to draw broad conclusions about overall risk in an area. To see this figure in color, visit http://www.pdc.org.

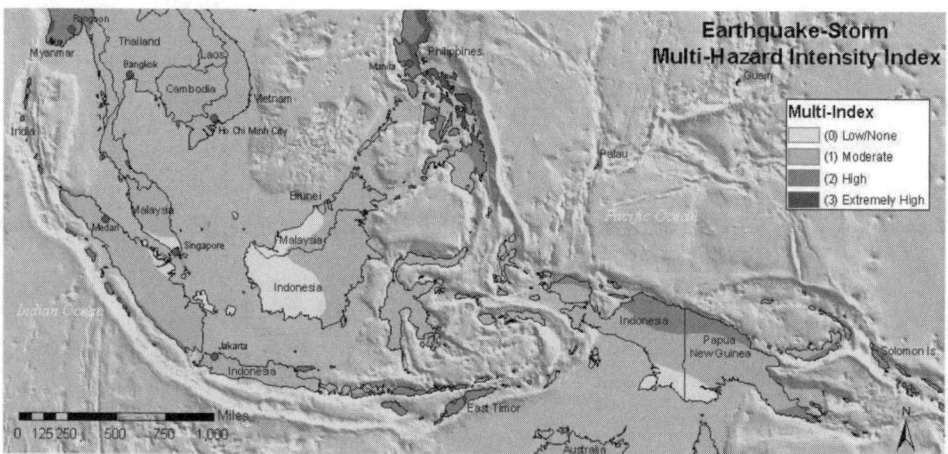

Figure 4.6   The patterns observed here are made even more evident and, more importantly, can be used to assess hazard exposure to various categories of human and natural resources. To see this figure in color, visit http://www.pdc.org.

Finally, the categorized population density data, along with the transportation infrastructure data, are analytically combined with the multihazard index to identify those areas with a high potential for impact from these natural hazards. Specifically, areas with a high hazard index and a high number of people and infrastructure are those that would warrant the most attention from mitigation efforts and be expected to require significant resources during the response/recovery phases of a natural disaster. This result is shown in Figure 4.9.

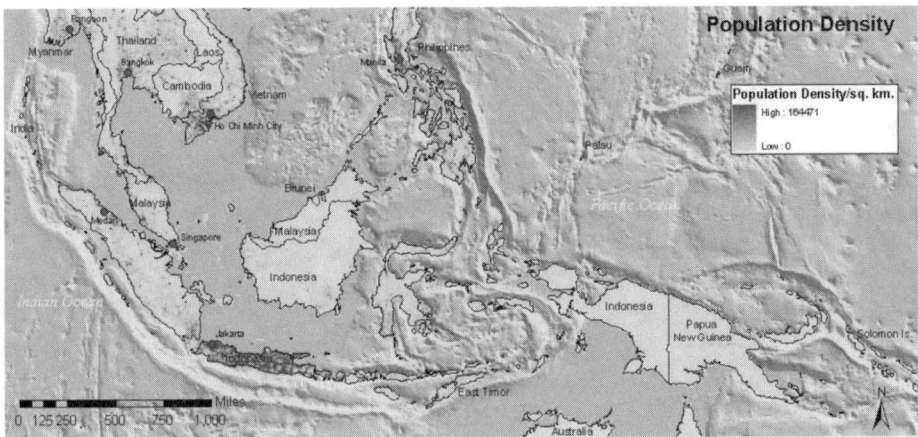

Figure 4.7 Where are the people? Southeast Asian countries have some of the highest popula-tions and population densities in the world. Indonesia, for example, is the fourth most populous country (behind only China, India, and the United States) but has less than one quarter of the land area as the United States. To see this figure in color, visit http://www.pdc.org.

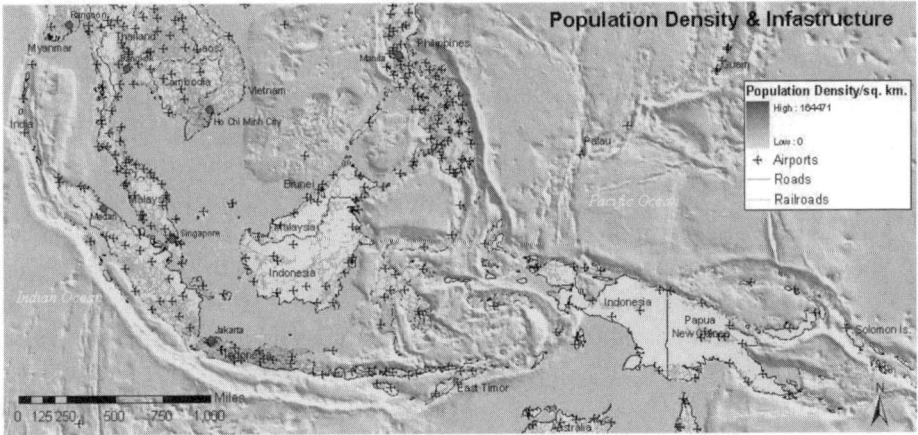

Figure 4.8 Where is the infrastructure? Transportation and infrastructure, including roads, rail-roads, and airports, that are subject to exposure to natural hazards as well as key to a region's recovery following such an event. PDC's atlas supports assessment of the vulnerability of these crit-ical assets for various hazards. Here, it is superimposed over population density data. To see in this figure color, visit http://www.pdc.org.

Section Conclusion

The atlas is a great resource for providing a geospatial context for assessing the risk exposure of human, natural, and societal resources to natural hazards. Try it your-self right now by referring to http://www.pdc.org.

Using this information, it is possible to produce up-to-date information on trop-ical storms, earthquakes, and volcanoes that is dynamically integrated with a com-prehensive set of historical records on natural hazards and disaster events

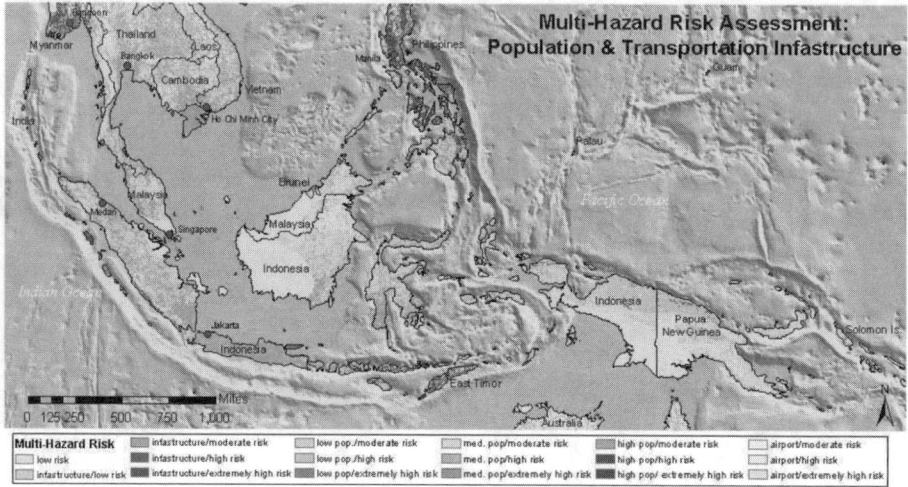

Figure 4.9 Population density data, along with the transportation infrastructure, are analytically combined with the multihazard index to identify those areas with a high potential for impact from these natural hazards.

throughout the region. In this manner, it is possible to characterize hazards, from both a real-time and a probabilistic perspective. Resource data, including people and infrastructure, provides the basis for understanding exposure and vulnerability from these hazards. This data in turn can be used to gauge the vulnerability of anything from a new beachfront hotel complex, to a potential factory location, to even things like vulnerability of undersea telephone cables to seismic and other events.

In summary, the PDC atlas data can be used both within the atlas and also within external analysis tools to develop measures of risk and vulnerability to support decision-making processes by skeptical executive policymakers in private sector organizations. In addition, resource allocation associated with mitigation planning or humanitarian assistance efforts in the public sector can also be prioritized using these techniques. Either way, in the final analysis, the goal of the PDC is not to build atlases that simply record disasters, but rather to present accurate data that can be manipulated by a wide variety of organizations to gauge probabilities and ultimately reduce the loss of lives from these terrible events. The atlas is one of many resources that can guide the development of policies leading to more disaster-resistant business.

Selected Project Profiles and Case Studies

Case Study #1: Flood Surveillance and Early Warning in Phu Tho Province Vietnam by the Pacific Disaster Center

Chris Chiesa[1], Hoang Minh Hien[2], David Askov[1], Todd Bosse[1], Sharon Mielbrecht[1]
[1]Pacific Disaster Center 1305 North Holopono Street, Suite 2, Kihei, Hawaii, 96753 U.S.A.

E-mail: cchiesa@pdc.org, daskov@pdc.org, tbosse@pdc.org, smielbrecht@pdc.org
[2] Department for Dyke Management, Flood and Storm Control Ministry of Agriculture and
Rural Development, 2 Ngoc Ha, Ba Dinh, Ha Noi, Vietnam, E-mail: hmh@netnam.vn

Abstract

This paper is about a collaboration between Pacific Disaster Center (US) and Vietnam's Central Committee for Flood and Storm Control (CCFSC) and Disaster
Management Center (DMC). As part of an overall program for capacity development within Vietnam's disaster management community, especially at the national
and provincial level, PDC, CCFSC, and the Ministry of Agriculture and Rural
Development's DMC have collaborated to better understand and enhance the current state of disaster management in Vietnam in terms of

- Monitoring and observation systems/networks;
- Analysis and decision support system (DSS) capabilities;
- Warning and notification dissemination.

Toward these goals, a pilot project was undertaken to automate flood-related
information feeds and incorporate them into a flexible, easy-to-implement decision
support application. Additionally, flood inundation model outputs and data sets
detailing critical assets and infrastructure were prepared and included in the
Web-based application.

More specifically, a map-based user interface for the decision support application based on PDC's natural hazards and vulnerabilities atlas and on the DSS
deployed to Thailand's National Disaster Warning Center was developed and
deployed using open source map server technology. The flood-prone Phu Tho Province, northeast and upstream of Hanoi, along the Red River, was selected as representative of both the need for such disaster management tools and the capacity,
especially in terms of information and communications technology (ICT), to field
such tools.

The technical scope of the pilot project in Phu Tho included

- Assessing ICT capabilities and needs;
- Review and enhancement of GIS hazard data;
- Development of hazard products applications;
- Sharing results and findings in a seminar.

The project was initiated in December 2006–January 2007 with an ICT survey
of DMC and of the Phu Tho Province Dyke Management and Flood and Storm
Control Office. Subsequent activities included the open source map viewer and map
server development, GIS database development focusing on flood hazards and
at-risk infrastructure and populations in Phu Tho Province, and flood modeling and
consequence assessment for various dyke breach scenarios along the Red River.

The project concluded with an international seminar on "Best Practices" in
Disaster Management: Disaster Risk Assessment and Early Warning Systems in
November 2007, which featured panel sessions and skill development seminars in

risk assessment, flood modeling, and Web-based crisis management software. Experts from Vietnam and at least six other countries made presentations contributing to the best practices theme. Following the international workshop and skills development event, representatives from Phu Tho Province participated in a day-long training session on the use of the map viewer.

This paper describes the recently concluded pilot project and discusses next steps toward implementation of a nationwide system in all of Vietnam.

Introduction

Pacific Disaster Center, in collaboration with Vietnam's Central Committee Flood and Storm Control (CCFSC), Department of Dyke Management and Flood and Storm Control (DDMFSC), and its Disaster Management Center (DMC), has undertaken a pilot project aimed at better understanding, illustrating, and enhancing the current state of disaster management in Vietnam in terms of monitoring and observation systems and networks, analysis and decision support system (DSS) capabilities, and warning and notification dissemination and computer-based decision support systems.

This project grew out of a then three-year engagement with the government of Vietnam. The initial phase, the pilot project, was designed to be limited to riverine flood threats in Phu Tho Province, with the development of a national, all-hazards Vietnam Disaster Center (VDC) as the longer-term goal.

The objectives of this first set of activities included raising awareness of the need for coordinated disaster management in Vietnam, clarifying to stakeholders the benefits of coordinated disaster management, validating the concept of an all-hazards approach to disaster management, and generating exemplar products to provide insight into issues ranging from disaster management to data sharing and access. This project is broken into five main tasks. Each is described in a following section.

Task 1: Project Definition and Pilot Area Selection

Although VDC is envisioned to be a multi-hazard facility, representatives from DDMFSC and CCFSC expressed their desire to limit the scope of the initial collaborative effort to flooding, one of the most prevalent and destructive hazards in Vietnam, to help ensure that the project could be successfully completed within a 12-month timeframe. To further focus this effort, PDC and DDMFSC staff agreed to select a representative "pilot" site to refine and validate assessment methodologies and applications. Through various discussions between PDC and Vietnam stakeholders, a pilot project focusing on flooding in Phu Tho Province, located in the hazard-prone northern region of Vietnam, was defined. Phu Tho Province is located within the floodplain of the Red River, approximately 50 miles upstream from Hanoi. This floodplain plays a critical "safety valve" role in the flood mitigation strategy for protecting Hanoi from severe flooding, although at the potential expense of homes, crop land, and livestock within the affected retention basins. These retention basins would be intentionally flooded by breaching levees along the Red River in cases of extreme flooding. A better understanding of the resultant flooding, the impact on people and infrastructure, and considerations of appropriate warning mechanisms thus became the focus of a pilot project to assess capacities and gaps.

Following an initial kickoff meeting in Hanoi (in December 2006), PDC worked with DMC staff to identify other stakeholders in Vietnam's disaster management community, including the Hydro-Meteorological (HydroMet) Service and affiliated universities including the Hanoi-based Water Resources University. These stakeholders helped PDC throughout the course of the pilot project to identify and secure data resources that would be used for subsequent flood modeling and impact assessment tasks. Additionally, PDC and DMC staff traveled to Phu Tho Province to meet with the provincial DDMFSC representatives, review their capabilities and procedures, and baseline the ICT systems. The provincial DDMFSC representatives were very excited by and encouraging of the PDC/DMC project. They provided an overview of their capabilities, which primarily included facsimile-based communication with the central DDMFSC (in Hanoi) and central and regional offices of the HydroMet service. Most urgent warnings from the provincial office are hand delivered to district and village offices and officials.

Task 2: Preliminary Data Gathering and Monitoring and Warning Capacity Survey

Once the Phu Tho study area was established, PDC worked with DDMFSC to conduct the ICT survey and to clarify the availability of data. PDC project members traveled to Vietnam in the spring of 2007 for this purpose and to inventory and collect data as well as to document flood monitoring and warning capabilities at both the national level and within the selected study site.

Assessment of ICT

PDC conducted an assessment of the information and communication technologies in order to characterize the current state of ICT equipment and personnel, and to understand any potential shortfalls. The assessment was conducted by using a questionnaire to interview DDMFSC's ICT personnel in Hanoi as well as their colleagues at the provincial office in Phu Tho. Generally, the DDMFSC IT environment was found to have the basic components (servers, networks, Internet connectivity, PDC) necessary for early warning and decision support. However, there were limited provisions for redundancy (i.e., spare servers and components) and the computing environment was not robust (i.e., server room was not on separate electrical and air conditioning services, limited backup and off-site data storage, and so on). Recommendations were made to DDMFSC to meet the minimum requirements for "enterprise-class, high-availability computing."

Data Inventory and Gathering

During this visit, PDC staff also conducted an inventory of data resources at DDMFSC. The team learned that geographic information systems (GIS) data were available from DDMFSC, including baseline infrastructure/boundaries, elevation, population, critical infrastructure, communication and transport networks, land cover, and data on river systems, including historical flooding, dykes and other flood control devices, textual historical data on flood frequency and severity, and real-time data from hydro-met and river gauge stations.

Additionally, PDC project staff visited the Phu Tho Province office in June, 2007, and collected the priority datasets, including critical facilities such as medical clinics, schools, government buildings, communications, and transport facilities, for further analysis in the flood impact study.

Survey Flood Monitoring and Warning Capabilities

PDC and DDMFSC surveyed the availability and suitability of meteorological monitoring equipment, hazard assessment tools, warning dissemination systems, and practices serving the study site and at the national level. PDC found that the Hydro-Meteorological Service of Vietnam (HMSV) has two networks of real-time hydrological and metrological monitoring equipment for Vietnam: one for measuring rainfall, and another composed of gauging stations along several large rivers in Vietnam, including the Red River. The monitoring equipment collects and transmits vital information such as rainfall amount and river elevation data. These data are made accessible by HMSV to various agencies in Vietnam, such as DDMFSC/DMC. Combined with predefined thresholds for flooding, these real-time data are used to create rudimentary map products that visualize potential flood areas for DDMFSC staff and serve as a basis for warning dissemination. They are not well integrated with other geospatial data, nor does it appear that automated warnings are generated based on these data (i.e., any warning requires manual review and assessment of these data).

Task 3: Hazard Assessment Product Definition

In the Phu Tho study area, PDC and DMC focused on a flood hazard area between the Red River and the Black River. This hazard area is the designated site for intentional flooding by forced breaching of levees at predetermined sites along the rivers, which takes place when water levels exceed an official threshold. The forced breaching is intended to safeguard Hanoi from flooding, albeit at the expense of the local community.

PDC staff worked with DMC staff to document the breach locations and the estimated flood volume and flow rates for the purpose of creating a modeled simulation of the flood using statistical modeling software and GIS analytical tools. This simulation, shown in Figure 4.10, was shared at the final workshop.

PDC and DMC staff next collaborated to develop prototype flood-hazard warning products depicting critical facilities in this area that would be susceptible to flooding based on parameters supplied by DMC. These products, including maps and tables, were used to evaluate potential impacts caused by releasing water into the flood retention area by computing which facilities would be impacted, by how much water, and at what time. Table 4.1 shows these data for several selected facilities. This information is valuable for evacuation planning and other mitigation efforts, including relocation of critical services to nonflood-prone areas. These

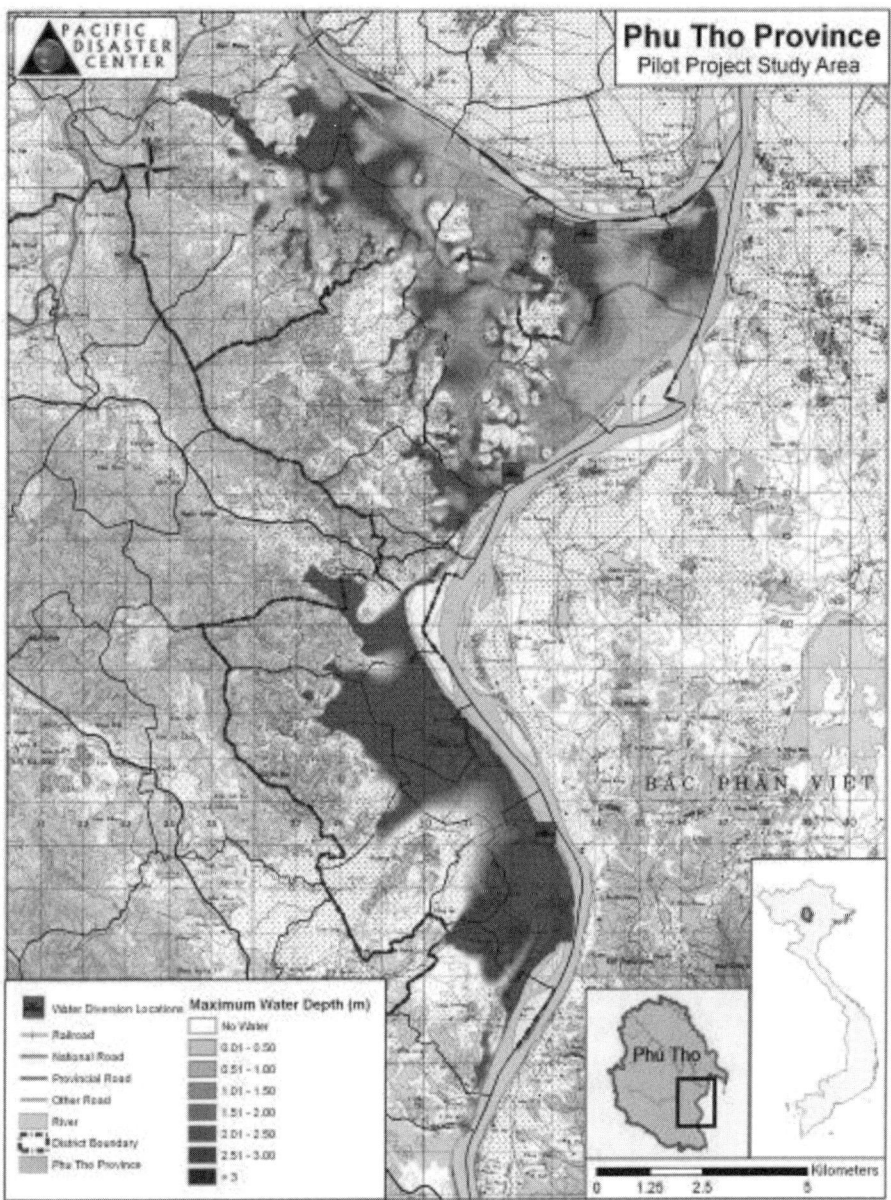

Figure 4.10 Flood extent and depth model output for Phu Tho Province study area. To see in this figure color, visit http://www.pdc.org.

products are also useful for increasing public awareness of flood risk, an important step in the risk-reduction process.

Task 4: Monitoring and Warning Capability Enhancement

Working with DMC staff, PDC designed and implemented a process and associated software applications to automatically capture, parse, and database these data streams, and to further automate the process of creating alerts/warnings based on their values and predetermined alerting levels.

Table 4.1 An Analysis of Flood Depths and Time to First Arrival for Several Facilities Within the Phu Tho Province Study Area

Infrastructure Type	Commune	Distance to Nearest Breach Location (km)	Time to First Arrival (hours)	Name of the Nearest Breach Location	Maximum Water Depth (m)	Latitude	Longitude
Pumping Station	X. DËu D¬ng	0.87	2.57	Thuong Nong	3.44	21.2476	106.3134
School	X. DËu D¬ng	2.11	8.98	Thuong Nong	1.02	21.2453	106.3008
School	X. DËu D¬ng	1.90	8.34	Thuong Nong	1.00	21.2415	106.3031
Clinic	X. De NËu	5.79	46.20	Thuong Nong	0.11	21.2425	105.2653
Commune Office with Police	X. De NËu	5.89	42.35	Thuong Nong	0.54	21.2409	105.2644
Post Office	X. De NËu	5.78	42.99	Thuong Nong	1.83	21.2413	105.2655

Specifically, PDC established a secure file transfer protocol (SFTP) feed to gather, from the Vietnam Hydro-Met stations, data on rainfall and observations of river water levels. These data are automatically retrieved, processed, and ingested into PDC's Enterprise Geospatial Database (EGDb) at regular intervals. The water level values are compared to predefined alert levels in order to determine the appropriate alert level for each station. An interactive multisource map viewer application, also known as the "Viet Nam Hazards and Vulnerability Atlas," which uses open communications standards for GIS data and map requests, is dynamically updated to show the alert level at each station. The Viet Nam Hazards and Vulnerability Atlas is shown in Figure 4.11.

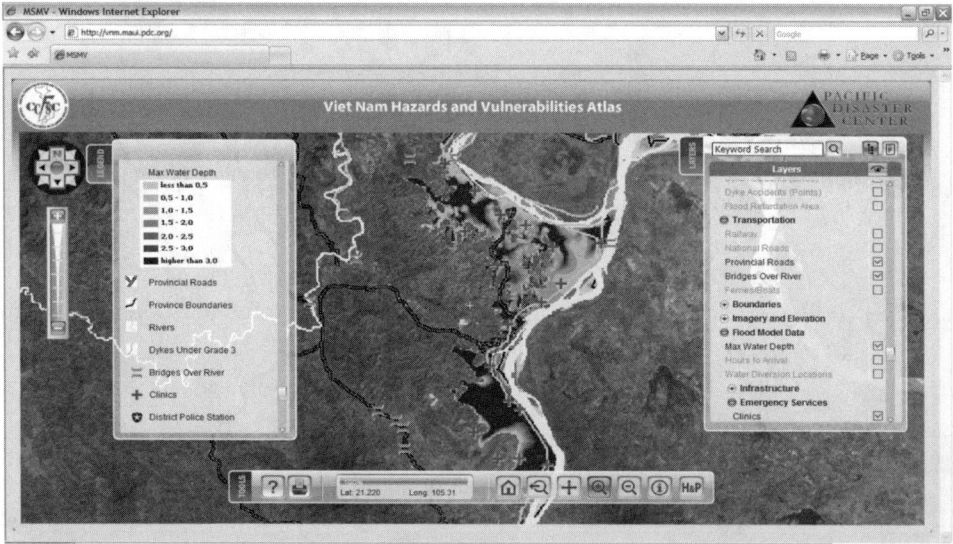

Figure 4.11   Viet Nam Hazards and Vulnerability Atlas.

Task 5: International Workshop on Best Practices

At the conclusion of the pilot project, PDC and DMC organized and participated in an international seminar on "Best Practices" in Disaster Management: Disaster Risk Assessment and Early Warning Systems in Hanoi on November 6–7. Vice Minister Nguyen Ngoc Thuat of the Ministry of Agriculture and Rural Development greeted the conferees with opening remarks. Other presenters included City Planning and Development Officer Thomas Aguilar from Marikina City, Philippines; Director of Disaster Prevention and Civil Defense Division Chul-Do Kim from Busan City, Republic of Korea; Dr. Smith Dharmasaroja, chairman of the Committee of the National Disaster Warning Administration, Thailand; Dr. Wei Sen Li, deputy executive secretary, National Science and Technology Center for Disaster Reduction, Taiwan; Mr. Chuck Dolejs, et al., representing ESi911; and Mr. Christopher Nielsen representing Danish Hydraulic Institute (DHI). PDC Chief Information Officer Chris Chiesa and Hazard Mitigation Specialist Sharon Mielbrecht provided presentations, and the Vietnamese speakers and presenters included top officials of the Central Committee on Flood and Storm Control (CCFSC), Ministry of Agriculture and Rural Development (MARD), Department of Dyke Management and Flood and Storm Control (DDMFSC), Disaster Management Committee (DMC), and Hydro-Meteorological Service of Vietnam (HMSV).

The second day of the conference featured skills development sessions in flood modeling, and risk and vulnerability assessment methodologies, led by DHI and PDC, respectively. Finally, representatives from CCFSC and Phu Tho Province participated in a day-long training session on the use of the map viewer on the third day of this event. The training marked the practical launch of the Vietnam hazards atlas in the open-source/multisource map viewer.

Next Steps Toward All-Hazard Disaster Management

The U.S. Trade and Development Agency (USTDA) has expressed interest in "next step" activities that PDC has conceptualized in cooperation with CCFSC/DDMFSC/DMC. After a series of discussions with USTDA, PDC has prepared a proposal outlining an expansion of the pilot project to include flood monitoring in central Vietnam and development of a national architecture for all-hazards disaster management.

PDC looks forward to a long and productive engagement with Vietnam, working in partnership to strengthen the country's disaster management capacity.

Case Study #2: Assessing and Reducing the Impacts of Disasters in the Asia Pacific Region

Stanley Goosby[1], Chris Chiesa[2], Sharon Mielbrecht[3], and Todd Bosse[4]

[1] Chief Scientist, Pacific Disaster Center*, Kihei, Hawaii. E-mail: sgoosby@pdc.org
[2] Senior Manager, Pacific Disaster Center, Kihei, Hawaii. E-mail: cchiesa@pdc.org
[3] Hazard Mitigation Specialist, Pacific Disaster Center, Kihei, Hawaii. E-mail: smielbrecht@pdc.org
[4] Project Assistant, Pacific Disaster Center, Kihei, Hawaii. E-mail: tbosse@pdc.org

Abstract

As part of its Risk and Vulnerability Assessment Program, the PDC has developed and implemented a strategy to assess and reduce the impacts of natural hazards within the Asia Pacific region. This strategy supports the PDC's goal of building "safe and sustainable communities" within the region through effective disaster risk reduction. It offers two multiscale and multihazard risk-reduction methodologies, and, where appropriate, utilizes *Internet map viewers* to assist decision makers and communities to better understand their vulnerability to multiple hazards.

For multihazard assessments at *regional and national scales*, a vulnerability-exposure-sensitivity-resilience (VESR) methodology is used to assess the spatial variability of vulnerability to natural hazards, and to assess their impacts on populations and infrastructure. Most recently, this methodology has been applied to the Lower Mekong River Basin to assess vulnerability to flooding.

For multihazard assessments at *municipal or community scales*, a risk-reduction framework methodology integrates three components that allow for more detailed multihazard risk reduction: a *multihazard risk and vulnerability assessment*, the *development of mitigation countermeasures*, and recommendations for the implementation of an overall *risk-reduction strategy*. Most recently, this methodology has been applied to the municipality of Marikina City, Philippines, to assess earthquake and flooding risks to existing infrastructure and planned development activities.

Finally, PDC has also developed and deployed Internet map viewers, which allow communities and decision makers to visualize and better understand hazard information and other critical data. Where appropriate, these map viewers constitute an important piece of the risk and vulnerability program's strategy and provide an effective education and outreach tool for disaster managers, for decision makers including legislators and land-use planners, and for students and teachers.

VESR Methodology: Assessing Vulnerability to Hazards at National or Regional Scales

VESR is an advanced, multihazard methodology that portrays and assesses the spatial variability of vulnerability to natural hazards (as well as the impacts of natural hazards on the people and infrastructure) within individual countries or a region. The VESR methodology addresses the questions "Which locales or cities in a region are most vulnerable to hazards?" "How does vulnerability vary across the region?"

and, more importantly, "What can be done to further explore the reduction of vulnerability and its underlying components?" Accordingly, this methodology can provide a first-level assessment, which can potentially assist decision makers who are allocating resources, considering infrastructure improvement investments, or might potentially pursue a more detailed assessment offered by PDC's risk-reduction framework methodology.

The underlying premise of VESR is that vulnerability (V) to natural hazards is related to the risk (frequency, severity) of exposure (E) as well as the presence (P) of populations sensitive (S) to that exposure. Further, a population's resilience (R), or ability to endure and/or overcome hazard impacts it may experience, serves to mitigate its vulnerability to hazard exposure. Each of these components is modeled as a geospatial layer or "surface," usually by combining various indicators or surface subcomponents.

Most recently VESR methodology has been applied to the Lower Mekong River Basin to assess vulnerability to flooding. In this particular case, these factors included:

- E = exposure to floods over time as a function of frequency and severity;
- P = presence (and number) of people exposed to flooding;
- S = their sensitivity, or susceptibility, to negative impacts of flooding as computed from measures of awareness, fragility, remoteness, and access to high ground;
- R = resilience of subjects, or their ability to mitigate and rebound from effects of flood exposure, computed as a function of per capita GDP, food and water security, and governance measures.

As a first principle, *vulnerability* requires *exposure* to a hazard by a population of concern. This study specifically addressed the vulnerability of people to flooding. Accordingly, there is no vulnerability to flooding where there is no flood risk. Similarly, there is no vulnerability in a location, even if it experiences extreme flood events, if there are no people. Additionally, not all flood events produce undesirable effects. In the Basin (as in many parts of the world), annual flooding contributes significantly and positively to agricultural practices. Floodwater recession marks the beginning of the rice-planting season. In fact, an absence of typical annual flooding in the Basin is likely to have a significant and negative impact on crop production.

Computing flood exposure, therefore, is more than simply combining flood depth and flood duration for a location. For this activity, typical flooding was used as a baseline against which less frequent (5-year, 20-year) events were assessed to develop an exposure surface.

Similarly, once exposure and *presence* (reflecting the number of people, as measured by population density) are combined, the concepts of *sensitivity* and *resilience* must be explored. Not all people subjected to the same level of flooding are impacted the same way—some experience little or no loss, while others experience significant loss of property, livelihood, or even life. In other cases, two populations may experience similar levels of initial impact, but may have different recovery experiences.

The question of why one population is more resilient to disasters than another was explored in this study through the construction of sensitivity and resilience surfaces that incorporated socioeconomic variables, including infant mortality, access to electricity, access to safe drinking water, GDP, as well as measures of physical remoteness of a community.

By combining the exposure, presence, sensitivity, and resilience surfaces, a spatial pattern of vulnerability emerges. This concept is represented in Figure 4.12, which depicts vulnerability to flooding in the Lower Mekong Basin region of Southeast Asia.

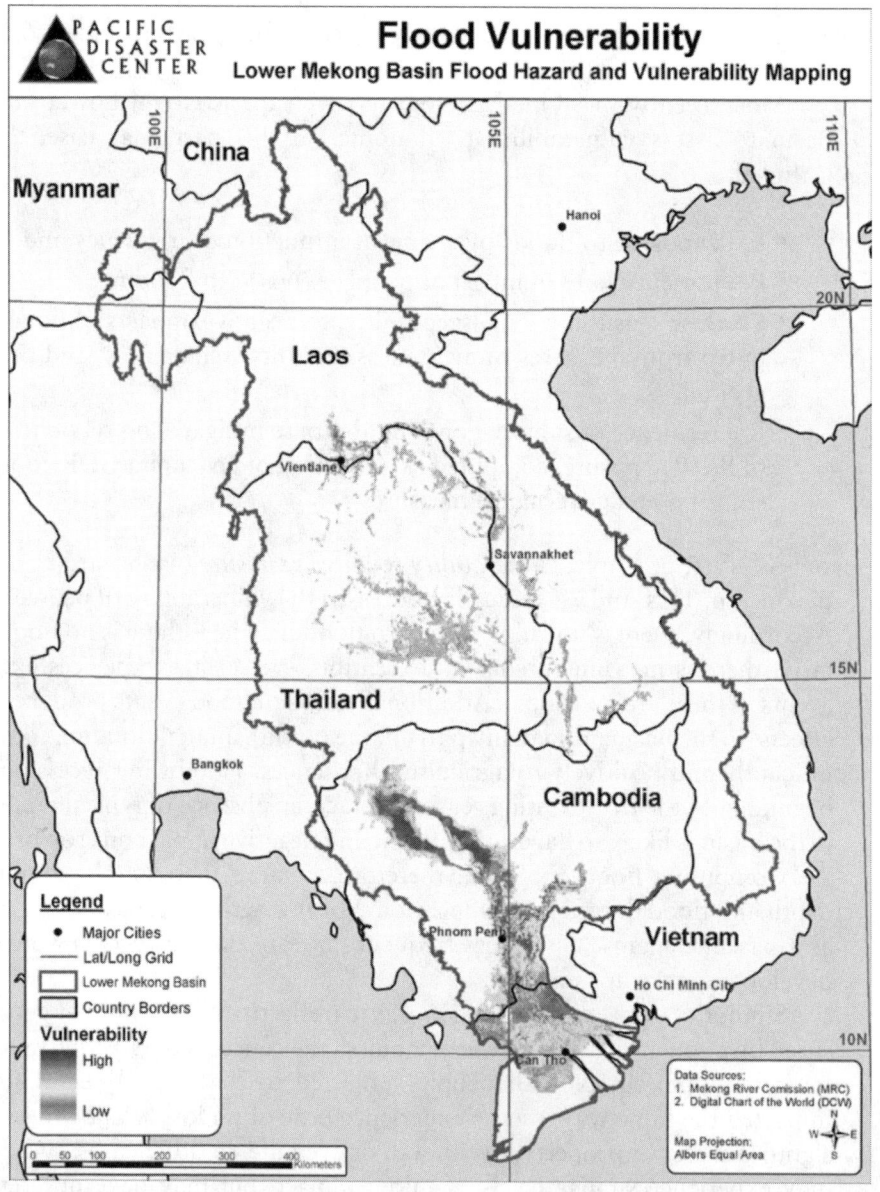

Figure 4.12   Flood vulnerability surface for the Lower Mekong Basin flood hazard and vulnerability mapping project. To see this figure in color, visit http://www.pdc.org.

Areas exhibiting the highest level of vulnerability can potentially be targeted for further investigation through risk-reduction framework methodology described next.

Risk-Reduction Framework Methodology: Assessing Vulnerability to Hazards on a Local Scale

PDC's three-part risk-reduction planning framework methodology has been developed and implemented to support disaster reduction and risk management at the local level in both urban and island (in this case, the Pacific Islands) environments. The first component of the framework, the *risk and vulnerability assessment* (RVA), identifies recurring hazards in a given location and assesses vulnerability to these hazards through analysis of the potential impacts upon the economy, society, and environment, as well as critical facilities and infrastructure. Examining the root causes of vulnerability across these sectors is essential to the development of a sound and viable strategy for risk reduction.

Based on the results of this multihazard assessment, the second component of the framework methodology, *development of mitigation countermeasures*, outlines a process for defining and prioritizing mitigation actions aimed at reducing a community's vulnerability to identified hazards. The third component, *development of a risk-reduction strategy*, builds on the integration of the first two and ultimately incorporates hazard mitigation into the decision-making processes that guide community sustainability and promote disaster resilience.

Risk and Vulnerability Assessment

Local-level RVAs form the basis of mitigation planning processes by identifying and raising awareness about the kinds of hazards that can affect a community and their potential impacts to physical assets and infrastructures, as well as social, environmental, and economic sectors. In addition, these assessments identify specific areas most vulnerable to damage from natural hazards, estimate the possible costs of damages, and assist in determining where losses might be reduced or avoided through mitigation activities. RVAs benefit planners, decision makers, and emergency management personnel by supporting disaster preparedness activities and by guiding future development policies.

The PDC and the municipality of Marikina City, Philippines, recently conducted a multihazard RVA for the city to gain awareness of earthquake and flood risk and to determine the vulnerability to social and economic sectors and critical facilities. Marikina City has also taken steps toward incorporating risk-reduction principles into its long-term planning initiatives. The city's goal is to build a safer and more sustainable community by becoming more resilient and resistant to disasters.

The initial step in the RVA process was to identify and profile the hazards affecting the study area. It was determined that the earthquake hazard facing Marikina City is not one that occurs with frequency; however, its potential for extensive damage is significant. On the other hand, flooding has proved to be a persistent and ongoing hazard, and has been the focus of successful mitigation efforts over the past decade. Hazard profiles had been previously established by various experts, who cited historical events of each hazard type to determine the frequency of occurrence, probability of future occurrence, potential magnitude or intensity, geographic extent, and conditions that increase or decrease vulnerability.

Three additional steps completed the RVA for Marikina City: (1) identification of vulnerable areas; (2) estimation of damages; and (3) assessment of potential losses. PDC's RVA methodology incorporates "best practices" from both the Federal Emergency Management Agency (FEMA) and comprehensive hazard and risk management (CHARM) methodologies.

To *identify vulnerable areas*, PDC used a geographic information system (GIS) to develop earthquake and flood hazard maps, which depicted various levels of risk to each hazard from existing data sets. Criteria were developed for each hazard to characterize the conditions that determined low, moderate, or high degrees of risk. These criteria were based upon existing maps and studies. Hazard maps were then overlaid with other GIS data sets, such as the city's building layer, municipal structures, schools, markets, and emergency services facilities to develop a visualization of the vulnerability of these structures. A number of facilities were determined to be of significant social or economic importance, and were therefore deemed "critical" to the well being of the city. Combined hazard layers provided yet another level of analysis, identifying those areas vulnerable to *both* earthquake and flood hazards.

An *estimation of damages* was possible once the hazard risk maps and asset layers were combined, and potential impacts across economic and social sectors and to critical facilities could be estimated. A significant facility-specific data collection effort was also initiated during the course of this project to further develop the city's asset database and to be able to quantify potential losses.

Understanding the level of risk to individual and multiple hazards, the vulnerability of assets, and the value of these assets to the community allowed for an *assessment of potential losses*. More detailed loss assessments were possible where critical facility and business information such as location, ownership, number of employees, estimated replacement costs, and estimated value of contents were known.

Development of Mitigation Countermeasures

Rarely are enough resources available to mitigate all known vulnerabilities in a community to every possible hazard. A process of mitigation project identification and prioritization is necessary. This ensures that appropriate mitigative actions can be carried out and that the costs and benefits associated with these actions are well understood. Careful consideration of how various mitigation strategies and countermeasures contribute toward loss reduction, risk reduction, and community sustainability is best accomplished by a dedicated group of professionals, experts, and community leaders who represent government, business, and community interests.

Mitigation countermeasures can be divided into three types of actions—physical, informational, and strategic. Among physical countermeasures are such actions as upgrading infrastructures, strengthening design and safety standards for new buildings, and retrofitting existing buildings. Informational countermeasures are those that increase public awareness of hazards and the preventative actions that can be taken to prepare and respond through outreach and educational activities.

Strategic countermeasures are key to the development and implementation of an overall risk-reduction strategy. These countermeasures establish laws, policies, and guidelines that build capacity and assist in the implementation and enforcement of physical and informational actions.

Development of a Risk-Reduction Strategy

The formulation of an overall *risk-reduction strategy* synthesizes outputs of the risk and vulnerability assessment and mitigation countermeasure development activities. This third step of the framework methodology aligns proposed risk-reduction activities with policies aimed at accomplishing the study area's long-term initiatives. This ultimately promotes disaster resilience and sustainable development. Risk reduction cannot eliminate disasters, but it can significantly reduce their *impacts* through effective prevention methods, planning, and preparedness. When implemented, these measures can increase resistance to hazard effects and improve community resilience. Such strategies could include measures such as the update of building codes, improvement of building practices, changes in land use planning, and environmental management policies.

Communicating Risk: The Internet Map Viewer

As a component of PDC's overall risk and vulnerability assessment program, PDC has developed and deployed Internet map viewers to effectively and efficiently disseminate and view hazard maps and GIS data within a standard Internet browser. Where appropriate, these map viewers constitute an important piece of the risk and vulnerability program's strategy, and provide an effective education and outreach tool for disaster managers, for decision makers including legislators and land-use planners, and for students and teachers.

For example, as part of the Marikina City RVA, PDC implemented a prototype Internet map viewer, which assisted city planners, public safety officials, and educators in the city to better understand the spatial context in which multiple hazards impact urban environments. The *Marikina City Internet map viewer* in Figure 4.13 supports decision-making processes, allowing visualization and assessment of earthquake and flood hazard risk zones to assist with land use, zoning, development, and mitigation policy decision making (Figure 4.14).

Beyond city officials, the general public of Marikina City will also find many uses for the map viewer. Educators will use it to teach students and community members about hazards and to promote community preparedness. It will also increase awareness about the location of public safety buildings, such as police and fire stations, as well as hospitals, transportation networks, and water systems information, and will point out areas that are vulnerable to natural hazards.

As a final step in the risk-reduction program for Marikina City, PDC is working with city officials and staff to transfer the operational capacity of the Marikina City Internet map viewer to the municipality itself. PDC has prepared a map viewer implementation plan outlining the necessary steps for local deployment of the map viewer, including hardware and software acquisition and configuration, data management, and, most importantly, human resources skills development.

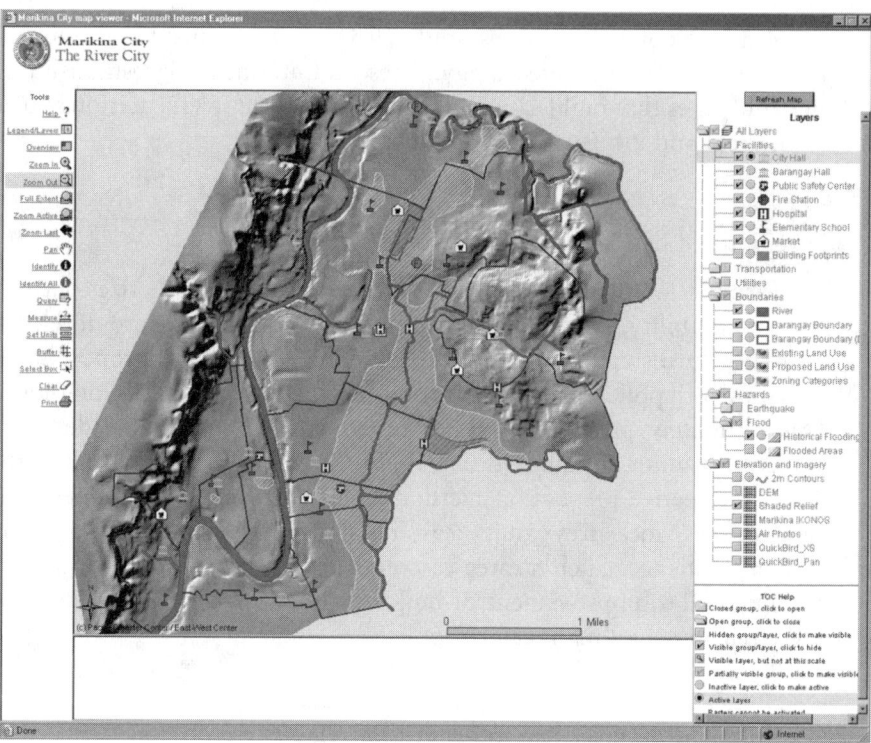

Figure 4.13 Marikina City Internet map viewer showing critical facilities and barangay boundaries, overlaid with flood-prone areas and a shaded relief image of Marikina City.

Conclusions

Enhancements to standards of living through economic, social, and environmental improvements are often hampered, halted, or reversed due to natural disasters. It is therefore prudent for communities, however large or small, to familiarize themselves with the impacts of recurring hazard phenomena, outline vulnerabilities, and manage the risks posed by these hazards. PDC's risk and vulnerability assessment program strategy promotes awareness of the impacts of multiple hazards through the implementation of methodologies by which hazards and vulnerabilities are identified at varying scales. Both the VESR and the risk-reduction framework methodologies provide an understanding of hazard impacts. The risk-reduction framework methodology further offers guidelines to help develop and prioritize appropriate mitigation countermeasures that ultimately contribute toward the development of an overall risk-reduction strategy.

In addition, communicating hazard risk and conveying vulnerability is important throughout this development process. The PDC's Internet map viewers utilize the power of GIS to visualize, query, and manipulate hazard and infrastructure data. The broad accessibility of these map viewers through the Internet provides the means to effectively communicate, educate, and promote risk reduction.

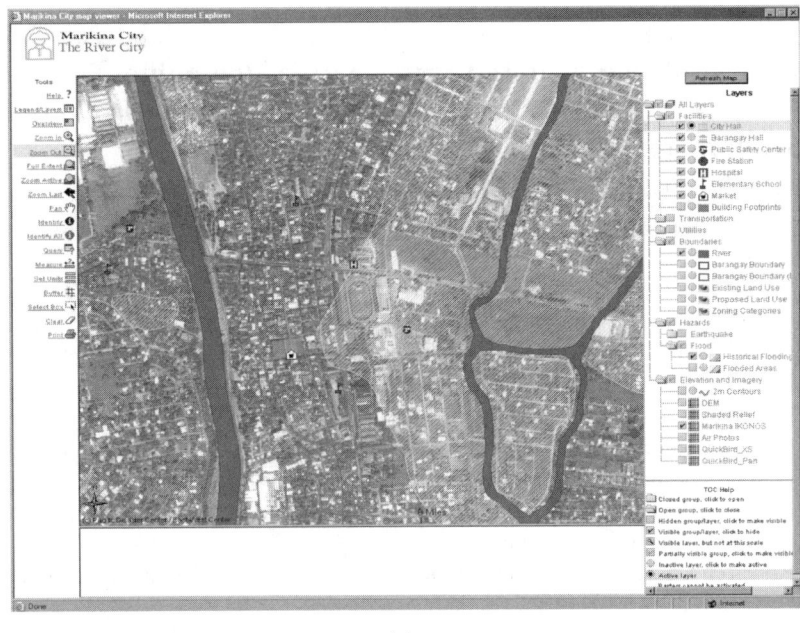

(a)

(b)

Figure 4.14   (a) Marikina City Internet map viewer showing critical facilities near city hall against an IKONOS image background and overlaid with flood-prone areas. (b) The same location is shown overlaid with peak ground acceleration.

Case Study #3: A United States–Japan–Philippines Collaborative Planning Process to Implement a Multihazard, Urban Risk Reduction Strategy for Marikina City, Philippines

James Buika[1], Stanley Goosby[2], Sharon Mielbrecht[3], Dr. Allen Clark[4], Julie Borje[5], Tomas Aguilar, Jr.[6], Dr. Haruo Hayashi[7], Dr. Norio Maki[8], Dr. Machiko Banba[9], and Kenneth Topping[10]

[1]Senior Manager, Pacific Disaster Center, 590 Lipoa Pkwy, Kihei, HI 96753, jbuika@pdc.org

[2]Chief Scientist and Senior Manager, Pacific Disaster Center, sgoosby@pdc.org;

[3]Hazard Mitigation Specialist, Pacific Disaster Center, smeilbrecht@pdc.org;

[4]Executive Director, Pacific Disaster Center, aclark@pdc.org

[5]Director, Center for Excellence, 2/F Marikina City Hall, Marikina City, Philippines, julie.borje@marikina.gov.ph

[6]Director, Marikina City Development Authority, Marikina City, Philippines, jun.aguilar@marikina.gov.ph

[7]Research Center for Disaster Reduction Systems, DPRI, Uji, Kyoto, Japan, Hayashi@drs.dpri.kyoto-u.ac.jp

[8]Earthquake Disaster Mitigation Research Center, NIED, 1-5-2 Kaigan-dori, Wakinohama, Chuo-ku, Kobe 651-0073, Japan, maki@edm.bosai.go.jp

[9]Earthquake Disaster Mitigation Research Center, NIED, Japan, michibanba@yahoo.co.jp

[10]President, Topping Associates International, 504 Warwick Street, Cambria, CA 93428 Kentopping@aol.com

Abstract

From 2002–2004, two research application organizations, the U.S.-based PDC and the Japan-based Earthquake Disaster Mitigation Research Center (EDM) collaborated with the mayor of Marikina City, Philippines, and her planning staff to undertake a sustained and multilateral, collaborative planning process involving a broad coalition of community-based stakeholders, including local subject matter experts. Marikina City is a municipality within Manila with a population of 437,000. The result of this planning effort has been to create a multihazard (flooding and earthquake) urban risk assessment, an accompanying Marikina City Internet map viewer and training program, and a city-adopted Marikina City safety program, which outlines strategies and actions for short-term and long-term disaster risk reduction.

Eight key components are identified as part of the successful multilateral, planning process and institutional capacity development program. For project planning purposes, this paper provides other research application organizations that are beginning community-based disaster management projects of their own with a realistic road map for evaluating the potential for project success and for facilitating the disaster risk-reduction planning process at the community level.

Introduction

The purpose of this paper is to present key components of the two-year, multilateral, community-based, planning process that have promoted the development and implementation of a multihazard, urban risk assessment and strategic plan for

Marikina City, Philippines. Each of these components contributes to local institutional capacity development within a community-based planning framework. These components are (1) financial resources for sustained involvement by international experts; (2) advocacy by progressive and proactive political and planning leaders; (3) a risk-reduction planning framework that makes sense to local stakeholders; (4) sustained feedback from city decision makers; (5) a proactive stakeholder advisory committee; (6) involvement of local subject matter experts; (7) an informed citizenry through project participation and training; and (8) awareness and lessons learned from repeated disaster and emergency events. Combining these complimentary components has created an effective researcher-practitioner-stakeholder coalition, which is undertaking risk-reduction planning at the community level (Buika and Comfort, 2004).

A Two-Year and Multilateral Collaborative Planning Process

During 2003 and 2004, PDC and EDM partnered with Marikina City, Philippines, to conduct eight city-sponsored, community-based planning workshops and three training sessions. EDM had established the partnership earlier, but helped the city create the *Marikina Safety Program: Comprehensive Earthquake Risk Reduction Strategy and Action Plan* (EDM, 2004) primarily during 2003. PDC transitioned the project with EDM between July 2003 and March 2004, to create the *Multi-hazard Urban Risk Assessment for Marikina City, Philippines* in 2004 (Pacific Disaster Center, 2004). EDM planners participated in each of the 2004 PDC workshops. Both of these complimentary and overlapping planning processes have improved comprehensive disaster management and sustainable development initiatives underway for the municipality. The result has been the City's formal adoption of the *Marikina City Safety Program* (2004).

Marikina City's contribution to the project has been significant. The Marikina City Center of Excellence provided continual access to city personnel, resources, and data. The center also provided training facilities, organized eight major meetings, and organized a city exhibition, fieldwork, and field trips. From EDM and PDC guidelines, city personnel provided data sets, collected additional data, and arranged for interviews with city personnel. The National Defense College of the Philippines also provided training facilities. Several hundred stakeholders representing approximately 50 local government agencies, businesses, and regional and national organizations participated in the overall planning process.

Background

About Marikina City, Philippines

Marikina City is one of 17 cities in metropolitan Manila, comprised of 14 barangays covering 21.50 square kilometers, with a population that is estimated at approximately 437,000. Although established in 1630, it was not formally incorporated as a city until December 8, 1996. Marikina City, like the rest of metro Manila, is characterized by rapid growth due to high fertility and immigration. It is important to note that the city is internationally recognized for its strong advocacy around disaster risk reduction.

The active West Valley Fault creates the western boundary of the city. There is evidence of four dated, strike-slip earthquake occurrences in last 1,400 years, including a scenario earthquake of magnitude 7.2 (personal communications, Dr. Renato Solidum, director, Philippines Institute of Volcanology and Seismology, 2004) [see Figure 4.15]. The fault system has created the Graben Valley in which the Marikina-Pasig River meanders through the city. Cyclone-driven rains can cause overflow of the river banks, posing a repeated flooding risk.

Marikina City Planning Initiatives for Sustainable Development

The policy focus of Marikina City is on economic development in order to improve the environment and quality of life of its citizens. Marikina City's has prioritized incorporating disaster risk management as a critical component of its economic development policy in order to obtain overall economic sustainability. The city recognizes that reducing disaster risk and improving economic resilience to disasters creates an incentive that attracts outside business interests and investors. This visionary planning approach will contribute to Marikina City's goal of becoming a disaster-resistant community (Topping, et al., 2004). Marikina City is extremely proactive in terms of implementing policies, programs, and initiatives to achieve this goal.

Five plans and programs guide Marikina City development strategies. These are as follows:

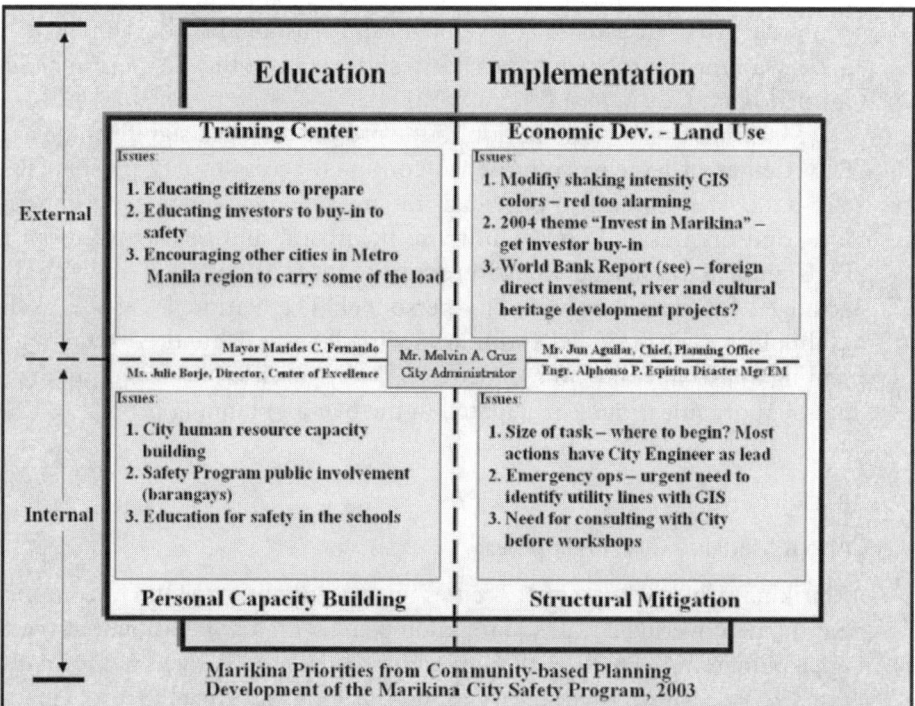

Figure 4.15 Marikina City, Philippines, as viewed in the PDC Marikina City Internet map viewer (Pacific Disaster Center, 2004). The map viewer can be launched at www.pdc.org/marikina. The Marikina-Pasig River, West Valley Fault, critical facilities, barangay political boundaries, and flood zones are shown overlaid on a hill-shaded relief map.

1. *Marikina Safety Program: Comprehensive Earthquake Disaster Risk Reduction Program and Action Plan* (March 5, 2004);
2. *Marikina City Comprehensive Land Use Plan* (2003);
3. *Marikina City Long-Term Master Plan* (2003);
4. *Invest in Marikina City Program* (2004);
5. *Flood Mitigation Program* (2004).

These initiatives include priorities in the social, environmental, and economic sectors. In the social sector, Marikina City has policies that include poverty alleviation and expanding Internet access. In the environmental sector, the municipality has programs that address traffic management, pollution control, "green space," and enhanced flood control measures. In the economic sector, the city has conducted a study to promote initiatives that would facilitate future development modeled after those of Singapore. The city has also recently zoned "areas for priority development" as part of its revised Comprehensive Land Use Plan.

The Earthquake Disaster Mitigation Research Center's Planning Process with Marikina City

International experts from the Earthquake Disaster Mitigation Research Center facilitated five workshops with Marikina City officials in 2003 and developed the community-based *Marikina Safety Program: Comprehensive Earthquake Disaster Risk Reduction Program and Action Plan (2004)*. This has been formally adopted as the *Marikina Safety Program*. It has identified 10 strategic directions and activities to reduce earthquake hazards in both the short term (1–2 years) and the long term (3–5 years).

The five Marikina City planning workshops were as follows:

1. *Problem Identification Workshop: Understanding Marikina City and Its Earthquake Threat*, January 2003 (EDM, 2003a);
2. *Risk Assessment and Goal Setting Workshop: Understanding Marikina's Earthquake Risk and Setting Marikina's Future Vision*, May 2003 (EDM, 2003b);
3. *Planning Workshop: Prepare Conceptual Framework*, July 2003 (EDM, 2003c);
4. *Implementation Workshop: Develop a List of Tentative Programs/Projects for Further Evaluation at the November Workshop*, October 2003 (EDM, 2003d);
5. *Planning Workshop: Stakeholder Resource Assessment and Priority Evaluation Workshop*, November 2003 (EDM, 2003e).

Pacific Disaster Center's Planning Process with Marikina City

PDC's effort with Marikina City officials continued the implementation of the Comprehensive Earthquake Disaster Reduction Program and Action Plan developed in 2003. As part of the project transition, PDC and EDM conducted the

PDC-Japan Community-Based Risk Planning Workshop in March to understand city priorities, plans, and available data.

PDC completed tasks on behalf of the Marikina City government including an earthquake and flood mapping and assessment, data collection of infrastructure and hazard information and its integration into a GIS, development of a Marikina City Internet map viewer, as well as education, outreach, and training for city stakeholders through three workshops and three training sessions.

The outcome of the 2004 PDC Marikina City Multi-Hazard Urban Risk Assessment Project has been to provide city planners and policy makers with specific multihazard disaster management information, an integrated risk assessment methodology, and associated tools for analysis to

1. Provide city planners with the capability to estimate disaster losses impacts;
2. Provide policy makers with a basis for developing regulations and policies that can reduce the impacts of earthquakes and floods on their communities;
3. Strengthen the city's strategic plan for economic and sustainable development;
4. Attract future financial and business opportunities to the city;
5. Engage stakeholder development of multihazard risk-reduction strategies as part of the ongoing Marikina City Safety Program.

Discussion

Key Components of the Planning Process

For establishment of a sustained, multilateral planning process at the community level, eight key components have been identified. These components have promoted a multihazard risk-reduction planning process, resulting in the Marikina Safety Program, which guides local disaster management activities supporting Marikina City's long-term goals of economic sustainability, poverty alleviation, and its recognition as a world-class city.

1. *Financial resources for sustained involvement by international experts.* Marikina City secured resources, funds, and in-kind support for international expertise in 2003 and 2004 to shepherd an extensive planning process for a multihazard risk and vulnerability assessment. The Japanese and American teams funded the majority of the project. At the same time, Marikina City has dedicated a substantial amount of in-kind planning and personnel resources. In fact, Marikina City's mayor actively solicited international community support for disaster risk-reduction activities at the onset of the project.

2. *Advocacy by progressive and proactive political leaders.* The mayor; city administrator and directors of planning, engineering, and public works; and Center of Excellence all embrace and advocate for a strengthened hazard mitigation planning process for Marikina City. This strong leadership has brought the city the World Bank's City Develop Strategy Program, which resulted in a Marikina City long-term master plan for sustainable development, a revised and promulgated comprehensive land use plan

designating areas for priority development, and the Invest in Marikina City Program focusing on business development. The Marikina Safety Program was developed and enhanced during the 2003 and 2004 planning efforts.

3. *Risk-reduction planning framework that makes sense to local stakeholders.* PDC provided Marikina City with guidelines for a three-step, multihazard, risk-reduction planning framework that was simple to follow and made sense within the context of Marikina City's goals of sustainable economic development. The *3-Step, Risk Reduction Planning Framework* (Pacific Disaster Center, 2004) guided project activities, resulting in a risk and vulnerability assessment (step 1), development and review of mitigation countermeasures for flood and earthquake from the Marikina City Safety Program (step 2), leading to recommendations for risk-reduction policy implementation (step 3). In step 1, both flood and earthquake risk-assessment products were tailored to meet Marikina City's requirements to support decision-making processes. The framework and risk products were repeatedly presented for stakeholders' review, feedback, acceptance, and corroboration in each of the three planning workshops. For step 2, EDM conducted three planning workshops of their own to collaboratively identify, prioritize, and assign resources to further develop suggested mitigation countermeasures. Countermeasures were divided into three types of actions—physical, informational, and strategic. Step 3 synthesized the previous two activities (steps 1 and 2) by aligning proposed planning strategies and risk-reduction strategies with effective planning and development policies aimed at accomplishing the city's initiatives outlined in Marikina's Long-Term Master Plan and the Comprehensive Land Use Plan, thereby promoting sustainable development.

4. *Sustained feedback from city decision makers.* EDM's collaboration with Marikina City to understand and document the city's issues, priorities, and directions assisted PDC in developing a common project direction, a vision, a goal, and objectives in 2004. The vision is to "build a safe and sustainable community." The project goal is to "incorporate disaster risk-reduction strategies as part of the Marikina City planning process by involving the community." During each of the planning workshops, Marikina City personnel cooperated by completing written feedback forms requested by the international teams. [Figure 4.16 identifies the city administrators' priorities as well as advocacy activities for internal and external training and mitigation activities as expressed during the development of the Marikina City Safety Program.]

5. *Proactive stakeholder advisory committee.* The project was organized to effectively integrate EDM and PDC research teams into a community-based Marikina City disaster management organization [Figure 4.17]. This organizational structure helped to organize resources, follow city protocol, and create efficiencies in the research, assessment, and planning process. Together, this built a coalition of researchers, practioners, and stakeholders. This coalition has been vital in developing local institutional capacity that will continue the disaster reduction planning process beyond the life of the project.

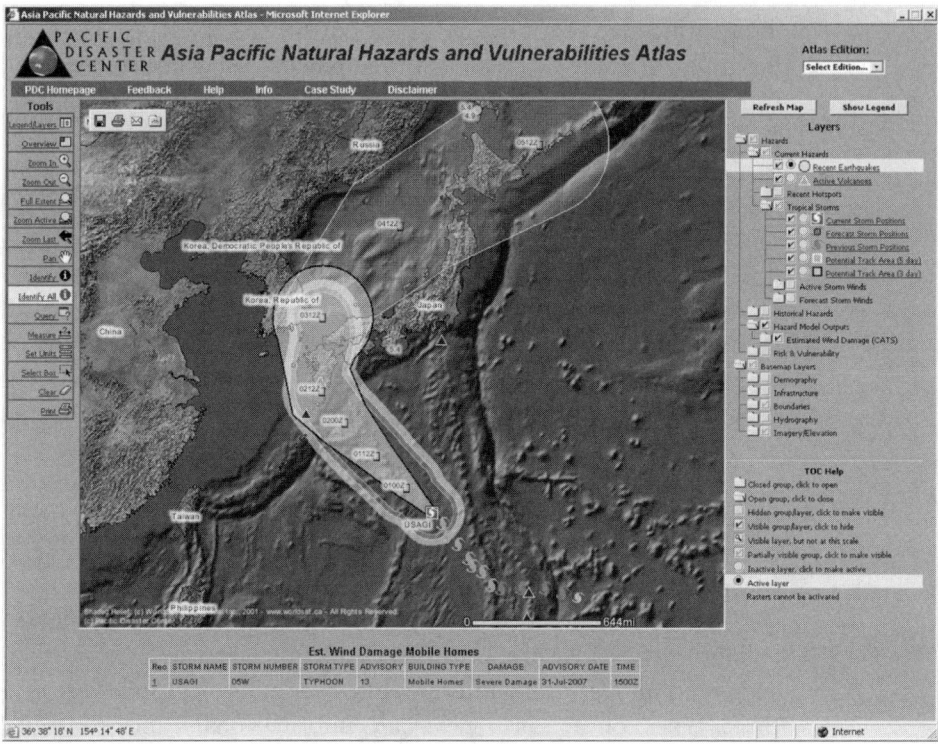

Figure 4.16   Marikina City priorities derived from community-based planning workshops. To see this figure in color, visit http://www.pdc.org.

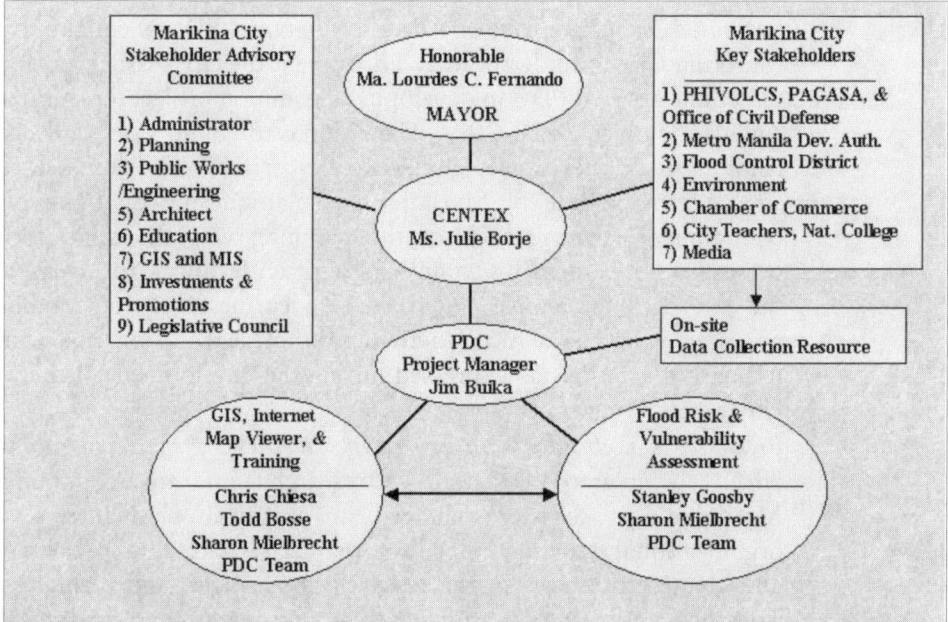

Figure 4.17   PDC project organization and stakeholder partnerships. Nine city representatives advised the PDC and Marikina City. A larger group of Marikina City key stakeholders provided local, national, and institutional expertise. Two PDC research teams worked with both stakeholder groups.

6. *Involvement of local subject matter experts.* Local experts from science and engineering organizations, training colleges, and the construction industry were introduced to Marikina City administrators as invited guest speakers and participants in the stakeholder advisory workshops and to support training sessions. Through the formation of this researcher–practioner–stakeholder coalition, these experts provide an understanding of the planning process underway and are available to continue to collect and provide data as well as to advise, promote, design, and complete mitigation projects. Local subject matter experts have supplemented the international expert teams and now create the basis for Markina City technical capacity development in the areas of earth science, meteorology, engineering, as well as local construction practices.

7. *Informed citizenry through project participation and training.* Through a series of workshops, five in 2003 and three in 2004, stakeholders participated and have begun a sustained educational exposure to hazards, risks, and the mitigation planning process. PDC has developed the Marikina City Internet map viewer (http://www.pdc.org/marikina) as an interactive, decision-support tool to assist city planners, public safety officials, and educators to better understand the relative spatial context between known earthquake and flood hazards and the infrastructure, economy, society, and environment (Pacific Disaster Center, 2004). A train-the-trainer program on the Marikina City Internet map viewer is continuing the hazard mitigation planning process by educating many stakeholders. It is now ready to be applied to barangay leaders, communities, schools, businesses, and the general public.

8. Awareness and lessons learned from disaster and emergency events. Repeated disasters and emergency events provide critical educational and motivational windows of opportunity for policymakers and the public to better understand and reinforce the importance of long-term mitigation planning and to take mitigation actions (Buika, et al., 2003). In 2004, the Philippines experienced recurring natural disasters, including a magnitude 6.9 earthquake, four active volcanoes, seven tropical cyclones, and major floods and landslides—collectively causing hundreds of deaths and widespread damage and internal displacement. These calamities and their associated human and financial consequences, when combined, have heightened national awareness and have prompted a call for action.

Conclusions

Development of disaster risk-reduction strategies in a multilateral environment is a complex undertaking that has been demonstrated to succeed in this project through the building of an effective and dedicated researcher–practitioner–stakeholder coalition, created, organized, and sustained through a comprehensive planning process involving workshops and training sessions. Multilateral, collaborative project teams are encouraged to review and incorporate key components identified as part of the Marikina City, Philippines, planning project.

Acknowlegments

The PDC is a public/private partnership sponsored by the PDC Program Office (ASD/NII). The content of the information does not necessarily reflect the position or policy of the U.S. government and no official government endorsement should be inferred. Since 2001, the East-West Center has been the managing partner of the PDC.

References

Buika, J., S. Goosby, and S. Mielbrecht, 2003, *Natural Hazard Risk and Vulnerability Assessment and Mitigation Plan for the Territory of American Samoa*, Proceedings, International Symposium for Remote Sensing of the Environment, Honolulu.

Buika, J., and L. Comfort, 2004, *Building Researcher and Practitioner Coalitions: Safeguarding Our Future Against Disasters*, Natural Hazards Observer, Volume XXIV Number 2, pp. 1–2.

EDM, 2003a, Earthquake Disaster Mitigation Research Center, Japan, City of Marikina, Philippines, and Philippine Institute of Volcanology and Seismology, 2003, *Problem Identification Workshop: Understanding Marikina City and Its Earthquake Threat*, EqTAP Marikina City Workshop, CD 1, January 27, 2003.

EDM, 2003b, Earthquake Disaster Mitigation Research Center, Japan, City of Marikina, Philippines, and Philippine Institute of Volcanology and Seismology, 2003, *Risk Assessment and Goal Setting Workshop: Understanding Marikina's Earthquake Risk and Setting Marikina's Future Vision*, EqTAP Marikina City Workshop, CD 2, May 7, 2003.

EDM, 2003c, Earthquake Disaster Mitigation Research Center, Japan, City of Marikina, Philippines, and Philippine Institute of Volcanology and Seismology, 2003, *Planning Workshop: Prepare Conceptual Framework*, EqTAP Marikina City Workshop, CD 3, July 28, 2003.

EDM, 2003d, Earthquake Disaster Mitigation Research Center, Japan, City of Marikina, Philippines, and Philippine Institute of Volcanology and Seismology, 2003, *Implementation Workshop: Develop a List of Tentative Programs/Projects for Further Evaluation at the November Workshop*, EqTAP Marikina City Workshop, CD 4, October 2003.

EDM, 2003e, Earthquake Disaster Mitigation Research Center, Japan, City of Marikina, Philippines, and Philippine Institute of Volcanology and Seismology, 2003, *Planning Workshop: Stakeholder Resource Assessment and Priority Evaluation Workshop*, EqTAP Marikina City Workshop, CD 5, November 2003.

EDM, 2004, Earthquake Disaster Mitigation Research Center, Japan, *Marikina City Comprehensive Earthquake Risk Reduction Strategy and Action Plan*.

Marikina City, 2003, *Appraisal of the Marikina City Long-Term Master Plan*, prepared by CPG Consultants Pte. Ltd.

Marikina City, 2004, *Invest in Marikina Program*.

Marikina City, Philippines, 2003, *Marikina City Comprehensive Land Use Plan*.

Marikina City, Philippines, 2004, *Marikina Safety Program: Comprehensive Earthquake Disaster Reduction Program and Action Plan*.

Marikina City, 2004, *Flood Mitigation Program*, Paper presented at the *Pacific Disaster Center-Marikina City Stakeholder Advisory Workshop for Community Disaster Risk Reduction Planning, Marikina City*, Marikina City, August 20, 2004.

Pacific Disaster Center, 2004a, *Multi-hazard Urban Risk Assessment for Marikina City, Philippines and Guidelines for Development of Multi-Hazard Risk Reduction Planning Framework for an Urban Environment*.

Pacific Disaster Center, 2004b, *Marikina City Internet Map Viewer*, www.pdc.org/marikina.

Topping K., H. Hayashi, S. Tatsuki, N. Maki, S. Tanaka, M. Banba, T. Kondo, K. Tamura, K. Horie, K. Hasegawa, M. Fukasawa, Y. Karatani, 2004, *Strengthening Economic Development Through Disaster Reduction Strategic Planning in the Asia-Pacific Region*, Asia Conference on Earthquake Engineering, March, 3–5, Manila, Philippines.

Case Study #4: Risk and Vulnerability Mapping and Assessment Supporting Disaster Risk Reduction

Pacific Disaster Center, 1305 N. Holopono Street, Kihei, Hawaii, USA:
Christopher Chiesa, Chief Information Officer, cchiesa@pdc.org
Stanley Goosby, Chief Scientist, sgoosby@pdc.org
Sharon Mielbrecht, Hazard Mitigation Specialist, smielbrecht@pdc.org

Abstract

Pacific Disaster Center in Hawaii has been activity engaged in disaster risk-reduction activities in the Asia Pacific region for more than a decade. During this time, PDC has developed and applied a four-component risk-reduction framework to assess, communicate, and address risks from a variety of natural and man-made hazards. A key component to this framework is the risk and vulnerability assessment (RVA) process. The RVA process, or method, relies heavily on geospatial data and GIS analysis tools to map hazards and the people and structures to which they pose risks. GIS applications are also used to communicate the results of the RVA to decision makers and the general public. Case studies illustrate the RVA process and results at three very different scales: a large metropolitan area in the Philippines, a small island nation in the South Pacific, and a multicountry watershed in Southeast Asia. Finally, guidelines for implementing an RVA, including the collection of required geospatial information, are presented.

Disaster Risk Management

Every year around the world, natural disasters affect millions of people and cause extensive damage and economic losses. The United Nations International Strategy for Disaster Reduction's (ISDR) *Living with Risk: A Global Review of Disaster Risk Management Initiatives* (2004) estimates that approximately 100,000 lives are lost due to natural hazards yearly and that the global cost of natural disasters will exceed $300 billion a year by 2050. This money is necessarily diverted from other national and municipal investments in environmental, social, educational, and infrastructure sectors, any of which produce better returns toward the goal of fostering sustainable and resilient communities.

The result is a negative effect on the overall quality of life—hampering, halting, or reversing economic, social, and development initiatives.

RVA is one key program area in which the PDC assists decision makers and communities to better understand their risk and vulnerability to a wide variety of hazards in order to develop and implement appropriate risk-reduction strategies. To this end, PDc has developed an integrated risk-reduction planning framework as shown in the following two figures. Composed of four components, the framework is the product of PDC's applied research, based on various long-term risk-assessment and mitigation-planning projects, including (1) Multi-hazard Urban Risk Assessment for Marikina City, Philippines; (2) American Samoa Hazard Mitigation Plan, 2003; and (3) Lower Mekong Basin Flood Vulnerability Assessment.

Risk and Vulnerability Assessment Goals, Methods, and Data Sources

PDC's RVA program has developed a risk-reduction framework that outlines a four-step process to assess and address risk from natural and human-induced hazards. PDC has successfully implemented this process, working closely with decision makers and planners, to achieve goals of disaster resilience and sustainable development. Advocacy building among policymakers and engagement of stakeholders and community members, in fact, are components of an important first step that guides and supports the entire risk-reduction process, namely risk acknowledgment.

The framework also includes a risk and vulnerability assessment component that assists communities in understanding and quantifying hazards and their potential impacts. The framework further outlines effective ways of communicating risk to the various stakeholders, including decision makers, elected officials, and the general public. The final step in the process identifies and prioritizes mitigation countermeasures that target high-risk areas for implementation.

The flow charts in Figure 4.18 outline the components of PDC's integrated risk-reduction planning framework. Within these, place- and culture-sensitive processes and methodologies are used to gather required data, perform analyses, communicate results, and implement solutions.

PDC has successfully applied the risk-reduction framework in a range of urban, rural, and island environments, and at varying scales, throughout the Asia Pacific region, demonstrating its adaptability to meet individual community needs.

Starting with the overarching goals of saving lives and creating sustainable and disaster-resilient communities, PDC helps all stakeholders develop a common understanding of the risks; assesses the risks by collecting and reviewing hazard and impact data from appropriate sources; organizes opportunities to communicate the risks through live events, reports, maps, and Web-based applications; and develops proposals for addressing the risks.

Applications of the RVA Process

The following case studies are examples of the application of the PDC RVA process. Each is quite unique because the needs and circumstances of the study areas differ.

Marikina City, Philippines

The purpose of the study that resulted in the publication of *Multi-hazard Urban Risk Assessment for Marikina City: Philippines and Guidelines for Implementing Multi-hazard Risk Reduction Strategies for an Urban Environment* was to provide Marikina City officials with an integrated, multihazard framework for assessing risk and mitigating the impacts from riverine and urban flooding and earthquakes on Marikina City, including its critical facilities, businesses, and people.

The final products of the Marikina City project were (1) a multihazard risk and vulnerability assessment, and (2) a series of guidelines and representative examples for continuing the risk-reduction and mitigation planning processes. Separately, the project also provided a customized map viewer for Marikina, an implementation plan for the map viewer, and a comprehensive training manual. Figure 4.19 illus-

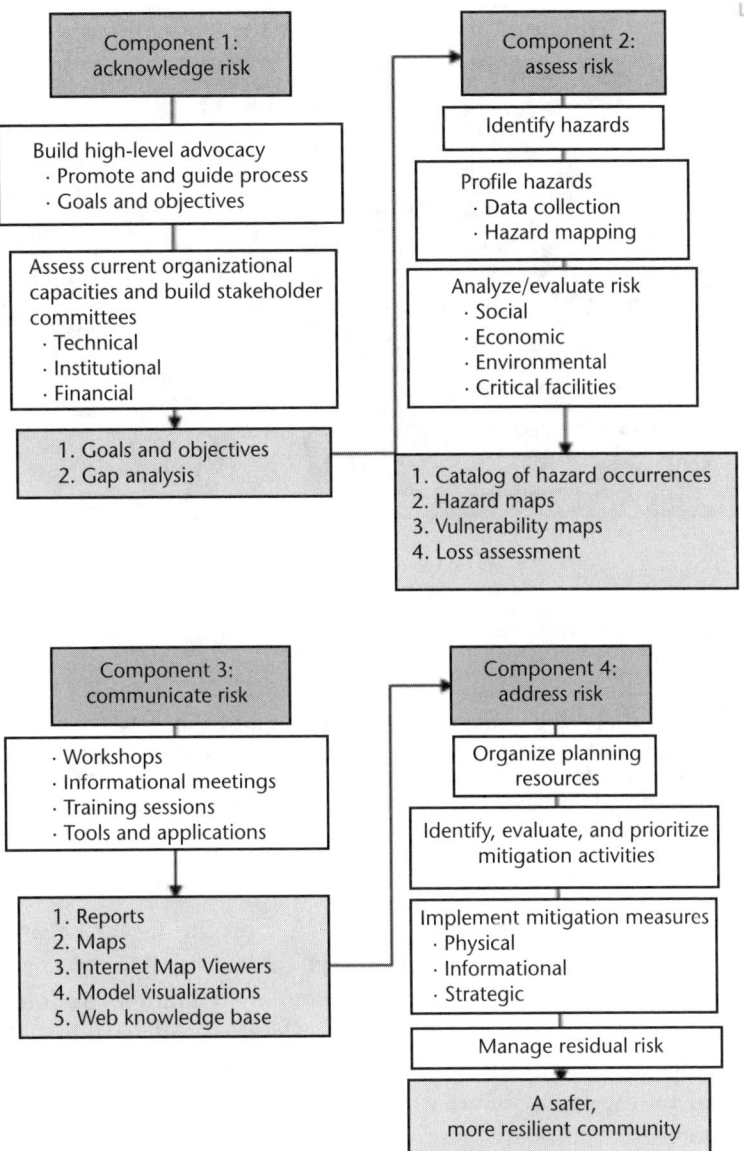

Figure 4.18 These charts outline the components of PDC's integrated risk-reduction planning framework. Within these, place- and culture-sensitive processes and methodologies are used to gather required data, perform analyses, communicate results, and implement solutions.

trates an example risk assessment product for one of several candidate sites for a development project.

To accomplish all the objectives outlined by officials and stakeholders for this project, PDC worked with a very large stakeholder community representing more than 30 national, metropolitan, and local organizations from government and the private sector, and collaborated with affiliated international experts as well.

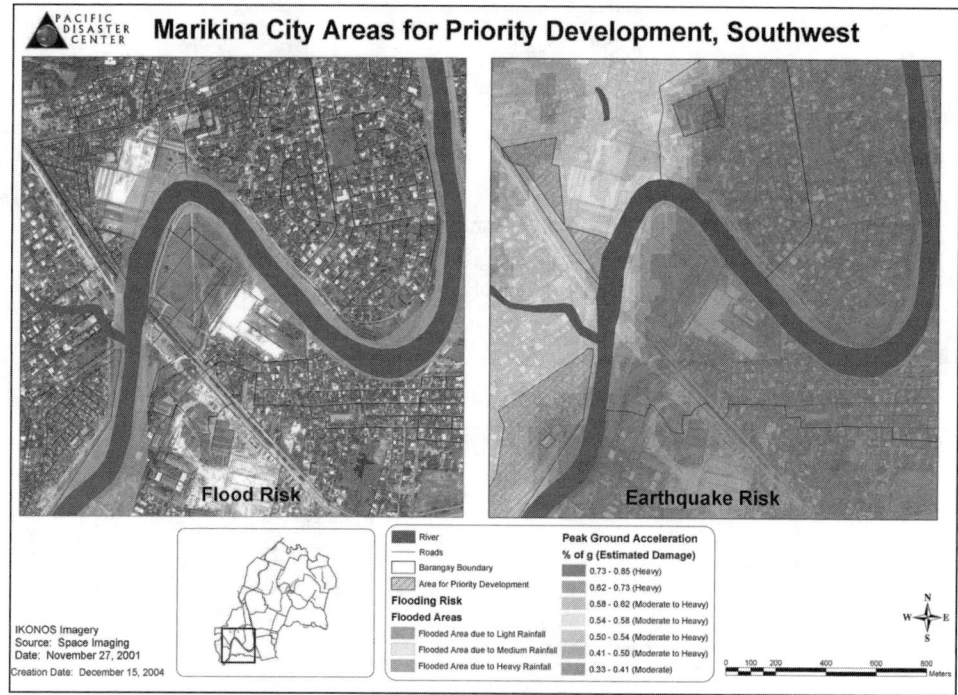

Figure 4.19   Flood and earthquake risk maps are superimposed with high-resolution satellite imagery and street maps for an area containing one of the proposed sites for potential future development (vacant land in the center of the graphic) within Marikina City, helping planners to better understand the hazards against which they needed to mitigate.

Ordinarily, when conducting an RVA, a profile of each hazard is developed, citing historical events of each hazard type to determine the frequency of occurrence, probability of future occurrence, potential magnitude or intensity, geographic extent, and conditions that increase or decrease vulnerability. Previous efforts in Marikina City by a team led by Professor Haruo Hayashi, Kyoto University, and funded by the Japan International Cooperation Agency (JICA) had developed a comprehensive database on seismic hazards. Therefore, only flood hazards required extensive research by the PDC team.

Once the hazard data were established, PDC was able to move quickly into the work of identifying vulnerable areas and developing data layers of hazard areas and assets for the Marikina City map viewer. As these data were prepared for this use, guidelines and templates were developed to expand the critical facilities building inventory as well as that of the business sector.

Enhanced data inventory allowed a more comprehensive assessment of economic, social, and critical facilities damage estimates and potential loss calculations. Other guidelines were developed to facilitate the identification and prioritization of mitigation countermeasures to advance Marikina's previous flood mitigation successes. In addition, a hazard analysis of three areas of priority development was performed to inform the city's long-term planning process.

All the outcomes of the assessment done for Marikina City were incorporated into a Web-accessible map viewer that will continue to reflect real-time circumstances for the use of policy makers, planners, decision makers, and emergency managers.

American Samoa

The hazard mitigation process in American Samoa followed the requirements and guidance provided by the Federal Emergency Management Agency (FEMA) of the U. S. Department of Homeland Security. The guidance standardizes the overall process but allows flexibility in determining how the planning process is best adapted to each jurisdiction. In American Samoa, traditional leaders and chiefs retain authority and respect along with the territorial government. Any planning process must respect the Samoan culture or *fa'asamoa*, the Samoan way of life.

The hazard-mitigation planning process for American Samoa, therefore, was guided by federal requirements and by the people and government of the territory. The methods used in the hazard-mitigation planning process were drawn from several sources. The primary references were FEMA's state and local mitigation planning how-to guides "Getting Started: Building Support for Mitigation Planning" (FEMA 386-1), "Understanding Your Risks: Identifying Hazards and Estimating Losses" (FEMA 386-2), and "Developing the Mitigation Plan: Identifying Mitigation Actions and Implementation Strategies" (FEMA 386-3).

The American Samoa Mitigation Plan addressed the full range of natural hazards threatening American Samoa: tropical cyclones (including storm surge), floods, earthquakes, tsunamis, landslides, and drought. Risk maps for landslide and flood maps are shown in Figure 4.20.

The development of a comprehensive natural hazard risk and vulnerability assessment was necessary to gain an understanding of the risks of natural disasters to the people of American Samoa. The PDC and UH Social Science Research Institute (SSRI) team, in collaboration with American Samoa government representatives, examined the vulnerability of critical infrastructure to various natural hazards. The assessment provided a compilation of information and dataset requirements to officials of the government of American Samoa for comprehensive planning purposes to save lives and reduce property losses in future disasters.

The assessment was formatted to meet the FEMA Interim Final Rule guidance document, profiling each hazard event to assess vulnerability and estimate potential losses by jurisdiction and to assess vulnerability and estimate potential losses to critical facilities. FEMA realizes that data are not always available to create a complete risk assessment, so the assessment indicated where data were available and where information gaps existed.

Using data compiled on historical natural hazard events between 1960 and 2003, the assessment examined the six natural hazards, with storm surge treated as an associated hazard to tropical cyclones. In many cases, historical data were sparse or conflicting, with the result that some details, which had minimal impact on the study outcomes, had to be left for later resolution. Numerical models were not used in this assessment.

Meetings were held with government officials, academics, the American Samoa GIS Users Group, the American Samoa Power Authority (ASPA), and other stakeholders and partners to assess the availability of data for the risk and vulnerability

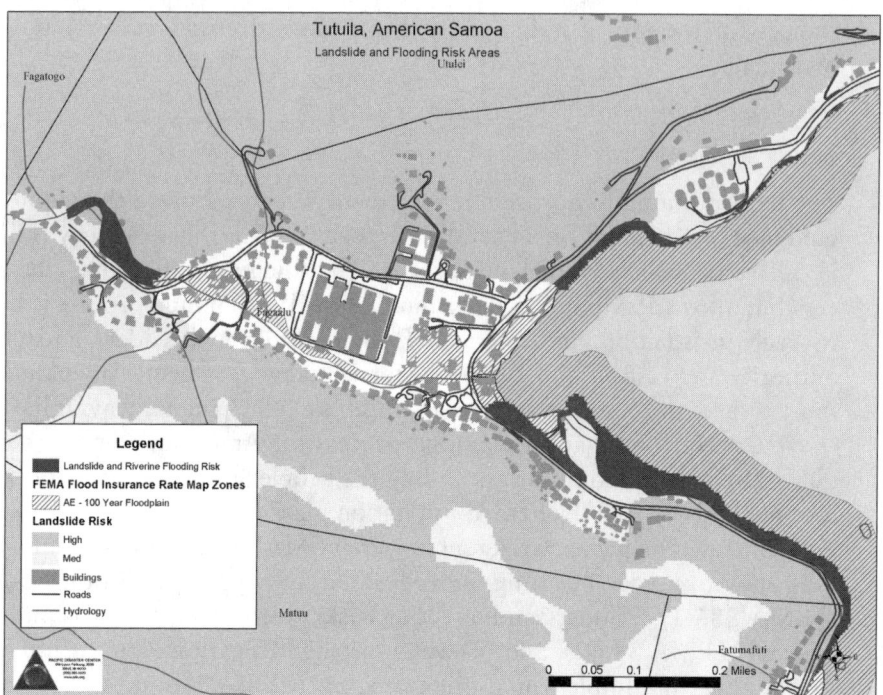

Figure 4.20 A multihazard risk map shows threats to critical infrastructure along the coast in American Samoa from landslides and flooding (including storm surge). To see this figure in color, visit http://www.pdc.org.

assessment. The GIS Users Group provided digital copies of existing data layers. ASPA and the GIS Users Group offered to compile additional information for the risk and vulnerability assessment, recognizing that this effort could improve the conditioning of data and increase their data holdings. In reciprocity for data, the PDC team agreed to the return of all processed and newly created data for use by the GIS Users Group.

The PDC team conducted follow-up meetings and intensive data collection sessions and developed the hazard layers required for the risk and vulnerability assessment. They used national and international databases on climate and extreme weather events, as well as on geologic hazards. Formats for asset/infrastructure layers and hazard layers were established, and data collection began in earnest.

For some of the hazard layers, only printed maps existed. The project team digitized the flood insurance rate maps (FIRMS) and the base flood elevations from paper maps provided by the American Samoa government. Landslide risk maps, as well as landslide occurrences, were also digitized. The National Oceanic and Atmospheric Administration's Pacific Services Center (PSC) helped the American Samoa government acquire IKONOS imagery, which provided a base layer for adjusting detailed maps and information.

Information was compiled on the impacts of tropical cyclones Tusi, Ofa, and Val from FEMA and reports of the Territorial Emergency Management Coordination Office.

PDC and the Social Science Research Institute at the University of Hawaii at Manoa obtained disaster frequency information from the Centre for Research on the Epidemiology of Disasters (CRED)/U.S. Office of Foreign Disaster Assistance (OFDA) database and received potential flood loss data from FEMA.

As data resources were collected, each hazard type was profiled. Vulnerability reports were compiled by jurisdiction, and an estimate of potential losses of critical infrastructure was developed.

The American Samoa Mitigation Council had adopted a specific goal: "Reduce the risks of all identified hazards to the territory, thus alleviating loss of life and property from tropical cyclones (including storm surge), floods, landslides, tsunamis, earthquakes, and droughts and ensure the overall well being of the people of American Samoa." In collaboration with its partners and local stakeholders, PDC developed a hazard mitigation plan in keeping with that goal, one that envisioned changed building codes, improved land use management and regulation, and better regulations for floodplain management as well as specific mitigation projects.

Lower Mekong Basin

The study that resulted in the publication of *The Lower Mekong Basin Flood Vulnerability Atlas* (Chiesa, et al., 2005), was undertaken in response to the question, "How does vulnerability to natural hazards vary across a region?" and the related questions, "What contributes to the vulnerability and its spatial variation?" and, more important, "What can be done to reduce vulnerability and its underlying components?"

The geospatial analysis methodology that was applied in this study helped answer these questions. Additionally, it can support policy development and decision making to reduce the factors that contribute to natural hazard vulnerability.

Preliminary work addressing food security in Africa, especially as influenced by drought, floods, and other natural hazards (Cicone, et al., 2003), based on a conceptual framework developed by Turner, et al. (2003), was adapted to investigate vulnerability to flooding within the Lower Mekong Basin, using geospatial information technologies including GIS software and GIS-based analytical models. The resulting approach explored vulnerability (V) as a function of exposure (E), presence (P), sensitivity (S), and resilience (R). It generally used physical and environmental databases available at a 1-km spatial resolution and socioeconomic databases at a provincial and district level.

The Mekong River watershed is subject to periodic flooding events that place life, property, and livelihoods at risk. The impact of these flood events on populations varies as a function of physical factors, such as weather patterns and topography, as well as social factors that determine the populations' preparedness to cope with floods and their ability to recover from them. Hence vulnerability to flooding is determined from assessing a combination of physical and social factors. This study applies an analysis strategy to examine vulnerability to extreme flood events as a function of both sets of factors.

The method employed is premised on a conceptual framework referred to in scientific literature as vulnerability, exposure, sensitivity, and resilience (VESR). The underlying concept is that vulnerability (V) to a natural hazards event is related to the risk of exposure (E) to the hazard by the presence (P) of populations that are sensitive (S) to that exposure. Resilience (R), or ability to endure or overcome impacts, may lower the overall vulnerability of a population (see Figure 4.21).

Specifically, these terms are geospatially computed and then combined as follows:

$$V = P \times E \times S \times (1-R)^{1/4}$$

The availability of regional area data on physical and social conditions in the Lower Mekong Basin, at scales on the order of 1-km resolution, provided the opportunity to create a model of vulnerability that could be expressed as a map. This scale of analysis proves useful as a practical solution to synoptically observe regional conditions for such a large area.

Physical and social data about the Lower Mekong River Basin were used to create quantitative indicators of exposure, presence, resilience, and sensitivity. The indicators were then combined and visualized to aid in communication about the highly complex underlying physical and social processes involved in flood disasters.

Among other findings, it could be seen how areas of overall similar levels of flood vulnerability required different mitigation strategies ranging from improved early warning capabilities for some regions and investments in the establishment of local reserves of emergency relief supplies in others. Furthermore, direct and indirect impact of potential investment projects on overall vulnerability in the region, both positive and negative, could be modeled via various proxy measures developed under this assessment project.

Considerations for Undertaking an RVA for a Large Metropolitan Area

Frequently PDC is asked by its partners to undertake an RVA study such as those illustrated here. To most effectively accomplish this task, PDC suggests the following guidelines:

1. Consider starting with a pilot project to gain a better understanding of the risks, consequences, and range of stakeholders.
2. Consider limiting the pilot project hazard assessment to one or two key hazards. For example, for a coastal city such as Busan, consider typhoon risks (and associated flood and landslide hazards) as these likely to be the cause of most hazard-related losses.
3. Consider starting with a smaller study area, which provides a representative mix of the risks and vulnerabilities experienced by the entire city. This allows the team and stakeholders to gain familiarity with the overall RVA process and to refine data collection and data analysis steps to better reflect desired project outcomes.
4. Start the process with a kick-off meeting and stakeholder workshop to outline and communicate project goals, scope, participants, timelines, and anticipated outputs and outcomes, as well as to provide stakeholders with a detailed understanding of the project methodology, and provide the project team with insights into interagency communications needs and processes.

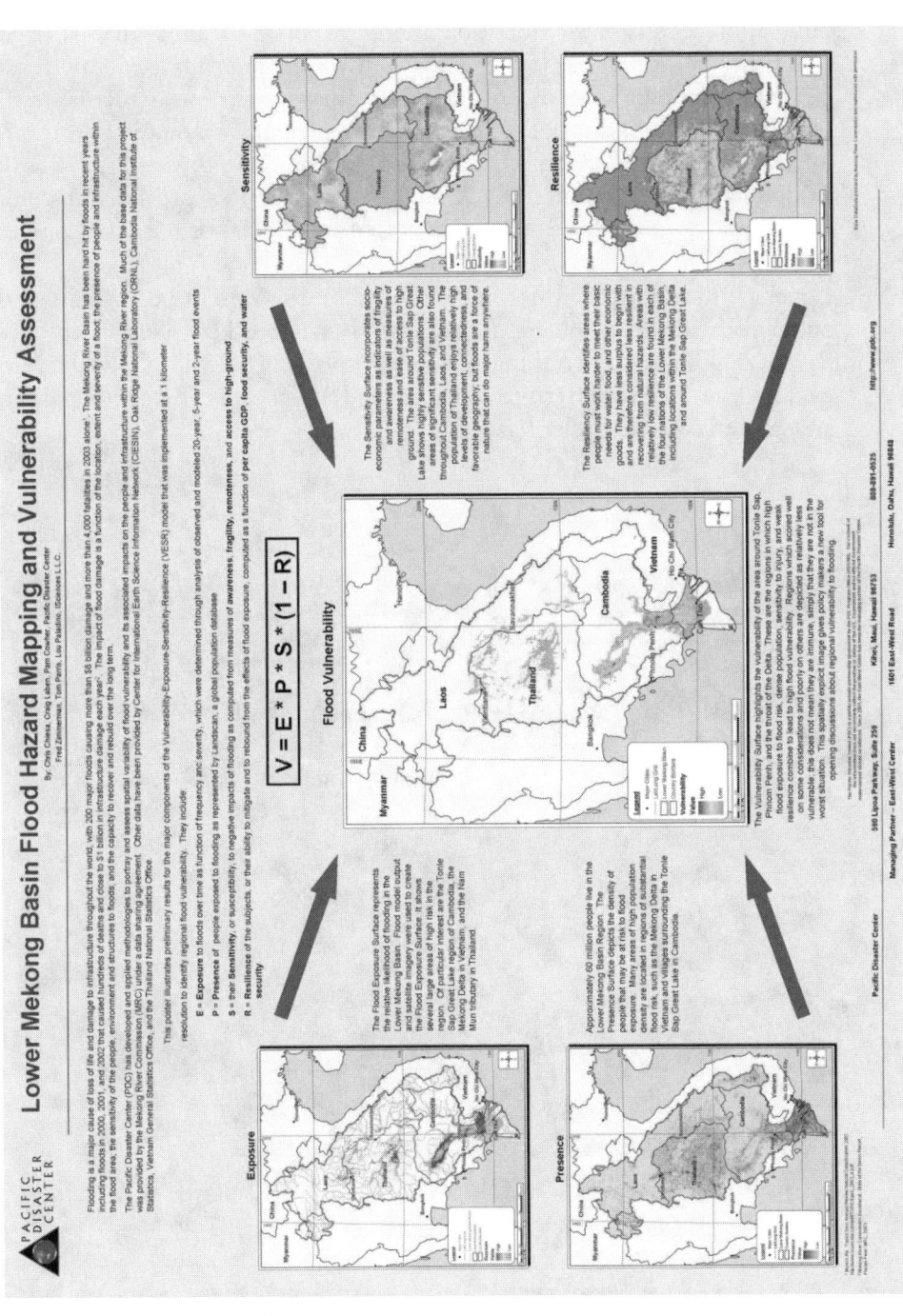

Figure 4.21 This graphic illustrates the application of the concept of vulnerability (V) in relation to exposure (E) and presence (P) of populations that are sensitive (S) to that exposure, and how resilience (R) may lower the overall vulnerability of a population.

5. Develop a data template and conduct a series of interviews with stakeholder agencies to begin the process of identifying, collecting, and managing required data and information (i.e., GIS, imagery, maps, and so on), especially population data (e.g., census data, demographic data, vulnerable population centers or facilities); data on transportation infrastructure (e.g., roads, bridges, tunnels, ports), water and waste water infrastructure (e.g., pipelines, pumping stations, treatment facilities), power infrastructure (e.g., transmission lines, substations, generation plants), and communications infrastructure (e.g., radio, television, telephone towers, switches). Additionally, data related to known hazard zones, previous disaster events, and so on will have to be collected from local or national sources (Figure 4.22).

6. Conclude the project with an RVA workshop to share project results and outcomes with stakeholders as well as to facilitate key agencies to gain technical skills in undertaking the RVA process. This will allow them to explore how to most effectively extend the RVA to the entire metropolitan area and to consider other hazards.

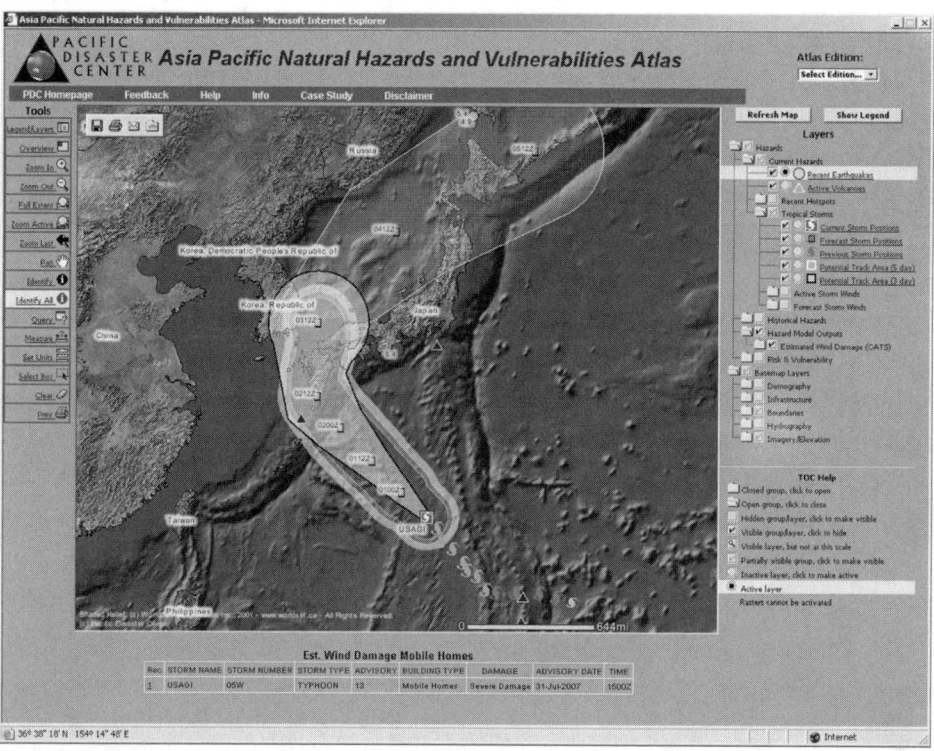

Figure 4.22 Risk and vulnerability assessments are one key component to an overall risk-reduction program. Other components include early warning systems as illustrated here for typhoons is the East China Sea.

References

Chiesa, C.; Cowher, P., et al. (2005) *Lower Mekong Basin Flood Vulnerabilities Atlas*. East-West Center/Pacific Disaster Center, Honolulu.

Cicone, R.; Chiesa, C.; Parris, T.; and Way, D. (2003) Geospatial Modeling to Identify Populations Vulnerable to Natural Hazards, *Proceedings, 30th International Symposium on Remote Sensing of Environment 2003 in Honolulu, Hawaii.*

Goosby, S.; Buika, J.; Mielbrecht, S., et al. (2004) *Multi-hazard Urban Risk Assessment for Marikina City, Philippines*. East-West Center/Pacific Disaster Center.

ISDR (2004) "Living with Risk: A Global Review of Disaster Reduction Initiatives." Report. United Nations, Davos, Switzerland.

Mielbrecht, S., et al. (2003) "American Samoa Hazard Mitigation Plan." Report. Pacific Disaster Center, Kihei, Hawaii. {

For readers interested in more information on this section, please refer to http://www.pdc.org/ArtechHouse. There you will find insightful instructions, color diagrams, an overview of available mitigation data and tools, and other useful materials. The Wrobels and Artech House thank the Pacific Disaster Center for their generous contribution to this book and to the overall well-being of the contingency planning profession.

CHAPTER 5

Satellite Communications for 4Ci

Advantages of Satellite Communications for 4Ci

I, Leo Wrobel, am probably what you would call a self-professed satellite bigot. I suppose this tendency was engendered at a relatively young age when I worked extensively with the technology in the U.S. military. When all hell breaks loose (literally or figuratively), there is nothing quite like a satellite link. The benefits of satellite technology both in terms of restoring telecommunications services and 4Ci cannot be understated. For that reason, we concentrate rather heavily on these services in this section.

The military is a big user of satellite communications, and for good reason. Just like the military, satellite offers myriad benefits to the contingency planner, since after a major disaster (earthquake, hurricane, tsunami, and so on) it might be your only connection to a twenty-first century telecommunications infrastructure. With that thought in mind, let me digress for a moment.

In addition to being a satellite bigot, I am also an admitted science fiction nut. Both passions came together recently in a manner that underscores some of the points I would like to make in this book about satellite. Please humor me for just a moment while I make my point.

I recently finished a terrific science fiction series entitled "Weapons of Choice" by an author named John Birmingham. The story begins in 2021, with a U.S.-led aircraft carrier task force in the Pacific that is instantly transported to June 1942 through a botched scientific experiment. They arrive near Midway Island on June 6, 1942, with crews unconscious. The 1942 Navy was very much awake and quite alarmed by the ships that appeared among them in the middle of the night, one of them flying a Japanese flag. They opened fire on the twenty-first century time travelers. However, they had no way of knowing that by 2021, ships had become capable of fighting autonomously without a crew through a computerized *combat intelligence* (CI) function. The CI (which interestingly enough used the voice of a 1980s female pop star) defended the fleet with twenty-first century technology, such as hypersonic cruse missiles, metal storm ammunition, and subfusion plasma yield warheads. It was a very bad day for the 1942 Navy, which was steaming toward the Battle of Midway and what would have been the biggest U.S. victory of World War II. Not only was this victory in the Pacific null and void, so was most of the subsequent history in the book. While things that go bang are fun for me to read about, equally riveting was the technological, political, and social chaos that ensues in the book. Some 60% of the 10,000 or so twenty-first century crew members were women and minorities. You can imagine how that went over in 1942 America. But you will have to buy the book and read all about this yourself. (I suggest that if you do buy the book, beware that the following paragraphs may contain spoilers.)

Anyhow, the book got me thinking about just how important communications is to all of us and how much we all take it for granted until it is gone. For one thing, it was astounding to see how dependent the twenty-first century people (military and otherwise) had become on satellite communications. Everything the "uptimers" did as far as voice communications, broadband, video links, email, weather forecasting, reconnaissance, global positioning, and a host of other applications suddenly turned into a blue screen that had no signal. The world of instantaneous communications to which the twenty-first century inhabitants had long since become accustomed and the secure feeling of being constantly bathed in wireless broadband on demand was suddenly gone. A good portion of this book is devoted to how the uptimers coped with the loss of everything from GPS-guided weapons and smart bombs to 600 channels of bad TV to watch.

There was some innovative contingency planning that happened in the book. For example, the time travelers restored "21C" telecommunications applications relatively quickly by implementing one of their own disaster recovery solutions—including one that traveled along with them to 1942. The solution was to bounce radio signals off the troposphere using a system called *tropospheric scatter*. This is the same technology I worked on in the Air Force 30 years ago before satellite became the mainstay. It is still used as backup, since any potential adversary will shoot out the eyes and ears (meaning the satellites) first. Through a relay arrangement from Hawaii to an aircraft carrier, then from the carrier to an AWACS plane, then from the AWACS plane to a stealth cruiser, the 21C crews were able to video conference from Hawaii to the mainland United States (see Figure 5.1). Thus, having operated with the technology myself, I knew it would work. In an uncanny coincidence, about the time I was reading the book I received a call from the Defense Information Systems Agency (DISA) with some routine questions about disaster recovery. While I had them on the line, I asked them if the DoD still used "Tropo" to back up satellite communications. They sure do. The book was technologically correct on that little tidbit of backup communications. In another part of the book,

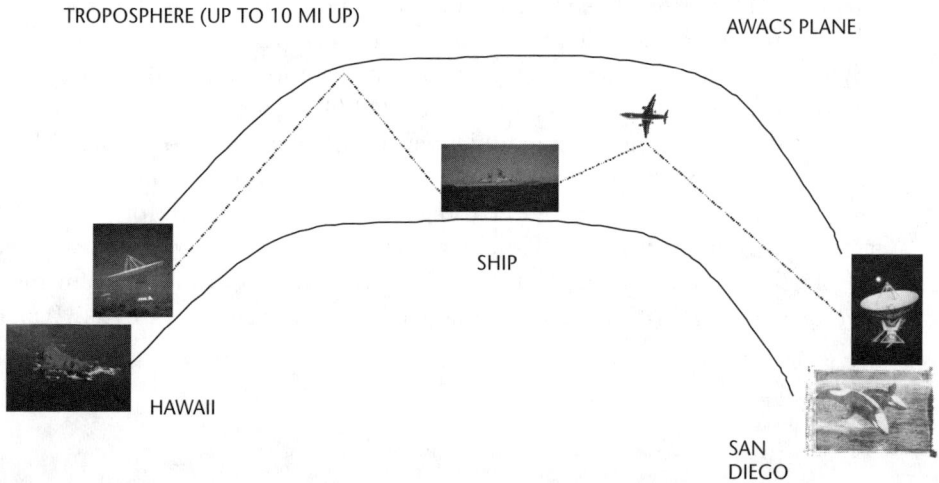

Figure 5.1 Use of tropospheric scatter systems to back up satellite.

the 21C guys hijack part of the 1942 cable plant in Washington, D.C., and add coaxial cable and other technology available at the time to build their own high-speed DSL networks! (Can you imagine a DSL modem using vacuum tubes? You could probably do it. It would be the size of a refrigerator and would probably heat a small home.) Throughout the book, the twenty-first century transplants lamented about what they would pay for "one lousy satellite" because none were in existence in 1942.

Besides, my point here is that there is nothing like viewing a particular technology from a perspective 70 or 80 years in the past to have an appreciation for what we have today and how easily we take such things for granted. That school of thought plays directly into disaster recovery. When today's companies lose critical communications links, they feel as lost as the hapless twenty-first century uptimers, ripped from their secure world of instant communications and tossed into oblivion. It happens all the time.

Therefore, today's companies and organizations must realize that when terrestrial communications is damaged due to a disaster, satellites provide a critical lifeline. As far as getting communications to remote areas far from traditional landline communications, satellites may be the only link available to an "uptime" twenty-first century communications infrastructure. For that reason, let's look at some of the characteristics of satellite from a 4Ci perspective.

Satellite Phones for Command and Control

In any recovery effort, you need to be concerned with two types of communications. One type would be the links that connect your customers to you. The second type is what you use to speak to one another internally in order to coordinate the recovery. Customer communications is no longer limited to the phone, and neither is 4Ci communications. Customers use incoming phone lines, access to the Web, or a variety of other modes of operation. While customer lines would seem to be of primary importance since they represent your organization's cash register, they are in reality number 2. Your main priority is to reestablish essential command and control. Stated another way, your organization may have the finest disaster recovery plan in the world, but it is essentially useless if you can't get in contact—now—with the people you will need to recover.

How will you call people back to work if, for example, your serving telephone central office is down? Don't think that does not happen. On April 16, 2007, a one-story Verizon central office in Raymond, New Hampshire, flooded due to swiftly rising water from a Nor'easter storm. The flood destroyed the electronic switching equipment, which caused 6,000 telephone lines to be completely down while damaging another 12,000 telephone lines. Days later with the central office still down, Verizon distributed free mobile phones to police, fire, EMS, other essential services, as well as to any area residents in need. The cost in terms of lost productivity and customer confidence was enormous.

Also note that under some circumstances, you may not be able to count on your wireless phone either. A mobile telephone serving office (MTSO) at some point usually has to terminate back into the wireline network. Sometimes this is through an *end office*, and occasionally it occurs in a more important central office called a *tandem*. Either of these types of central offices can have problems that affect both the wireline and the wireless networks. Tandems not only switch traffic for the immediate area, they also switch traffic for other end offices. This, among other things, makes the disaster more widespread as well as making tandems very tempting targets for terrorists.

In situations like these, a handheld satellite phone can be a godsend. Units available today typically are pretty rugged insofar as being waterproof and shock or dust resistant for use in harsh environments. They offer many of the same features as regular or wireless phones, such as the ability to store many numbers as well as to speed dial them after a disaster. (Where is *your* callout list at this moment—in the building that just burned down?) Many are also headset or Bluetooth compatible, allowing both hands to remain free. Remember, you could be writing, installing equipment, tending the wounded, driving in wicked weather, or doing a number of other tasks that will require use of both hands in a disaster. Many handheld satellite phones also include features like voicemail or call forwarding as well, which can be very useful.

A typical satellite handset provides 3 or 4 hours of "talk" time and about a day of standby, so don't forget extra batteries. It might be some time before the lights come back on again. You will need a clear line of sight to use a satellite phone. That could be a problem if the storm is still going on when you need to make a call. Some of these phones require that you dial an international access code; others, 00, then the number, and so on. This is because, technically speaking, the calling area of a satellite is the planet with the possible exception of the north and south poles. In any event, you should know how to dial and what access codes to use in advance to avoid wasting valuable time in a disaster.

Satellite phones are still somewhat expensive compared with other wireless options. Phones often cost a few hundred dollars and rate plans range from $1.00 per minute up. However, they are worth it.

Even before the military began using satellite, it invented the phrase *command, control, and communications* (CCC, or 3C). These were the three most essential elements in order for the military to accomplish a given mission, such as getting our missiles out of the ground before the Russians did. In order to do that, the military quickly maneuvered a large number of people. That's the command part. Targets had to be acquired and the situation accurately assessed—ergo, the control part of the three Cs. Finally, orders had to be dispatched as to exactly what to do, how to respond, and who was responsible to do it. That's the communications part.

More recently, computers were added, making the acronym command, control, communications, and computers (4C). The military was one of the first to do that, too, since decisions had to be made faster than people could make them. In recognition of the fact that decisions are now made in minutes or split seconds, computers have become indispensable to the command and control process in view of their rapid response time.

The current term of art with the military, as best I can tell these days, is 4Ci. (That's why we focused the entire book on it.) This term is the same as 4C, but the i is added for intelligence. That, in so many words, means using the four Cs combined with an accurate assessment of the disaster situation—hence the addition of intelligence. (Hey, by the way, remember my time traveler example? Ci was also used as an acronym for *combat intelligence*. In that example, the focus was communications.)

So where do satellite phones play into this process and why are they so valuable to the recovery process? It's because they preserve and aid in the restoration of everything contained in 4Ci.

It is not a coincidence that the Defense Information Systems Agency (DISA) is the largest supplier of satellite services in the world. Satellites assure that command, control and communications are preserved. Satellite technology can also recover data through myriad technologies, including mobile broadband area networks—both voice and data. Finally, and perhaps most importantly, a handheld satellite device can provide on-the-spot situational analysis under virtually any conditions and in spite of any damage to the existing telecommunications infrastructure.

In the critical first hours after a disaster, preservation of 4Ci is worth far more than a dollar a minute. In fact, I submit to you that today's business is ready for this level of precision. I have even coined a new phrase—b4Ci—in my latest book and named one of my companies after that term. The *b* is for business.

In summary, don't think about satellite phones as backup for customer communications, at least initially. Think of them in terms of b4Ci because when it comes to recovery planning, it's all about 4Ci.

Satellite Technology Overview

Satellite communications have grown by leaps and bounds over the last few years. Satellite is essentially microwave radio aimed upward—it uses essentially the same frequencies as microwave radio. The technology has gone from elaborate teleports and 16-ft dishes in years past to pizza pan dishes that fit on the side of a building. In fact, in the case of GPS and freight tracking technologies, the units literally fit in your hand.

Notwithstanding timing delays (it takes a fraction of a second for the signal to go from the Earth, up 22,300 miles to a geosynchronous satellite and the same distance back), satellite is a clean, reliable, and cost-effective disaster recovery solution. Even so, it is interesting to note that some corporate enterprise networks are virtually 100% satellite already. Network television is a prime example. So are a whole bunch of point-of-sale systems from rental cars to retail checkouts. Ever see a whole checkout line come to a standstill during a very heavy rain? I have! It's costly to the retailer if people get frustrated and leave. It is probably prudent, therefore, in these cases to explore another technology for backup. Network television providers do the opposite of what your organization probably does. Your organization may back up terrestrial landlines with satellite. TV providers who already use satellite may back up the satellite with fiber and landlines. Granted, it is probably not practical for your company to license and install a Tropo system, but, even so, there are ways to back up satellite with other technology.

Is Satellite Right for All Applications?

As I stated earlier, a geosynchronous satellite orbits 23,500 miles above the Earth, or about a 50,000-mile round trip for a conversation. Light travels at 186,000 miles per second. On paper, at least, that equates to a little over a quarter of a second of delay. That's enough to become noticeable before one even considers other things that can slow a circuit down, such as latency in Internet protocol and other protocols as well as capacitive and reactive components in circuits. How much it affects data that previously used a landline should be a topic of discussion between you and the satellite vendor. If your equipment sends out short "blocks" of data before expecting an acknowledgment from the far end, this will have a greater impact on performance over a satellite than long blocks will. Absent these kinds of concerns, which can be addressed by the satellite vendor, satellite can be a remarkable disaster recovery vehicle. Here's why:

- *Ubiquitous coverage*: A group of satellites can cover virtually the entire surface of the Earth.
- *Instant infrastructure*: Satellite service can be quickly provisioned into areas where there is no terrestrial infrastructure.
- *Independent of terrestrial infrastructure*: Satellite service can provide additional bandwidth—on a diverse path—to provide redundancy as well as overflow capability during peak usage periods.
- *Temporary networks*: As stated earlier, the military, Department of Homeland Security, and others find satellite to be the only practical, short-term solution for getting critical information into and out of an area.
- *Rapid provisioning of services*: Satellite services can be deployed quickly anywhere within the *footprint* of a satellite.
- *Disaster recovery*: In times of widespread disaster, such as hurricanes, earthquakes, or tsunamis, solutions provided via satellite may be readily available and more reliable than land-based connections.

If you are looking for a great reference and easy read on this topic, download or order the "First Responder's Guide to Satellite Communications," published by the Satellite Industry Association (SIA). Their Web site is http://www.sia.org.

Why Use Satellite for 4Ci?

Satellite is highly survivable in and of itself, both in terms of physical equipment as well as robustness of the technology itself. As a wireless technology, it is not susceptible to the notorious *backhoe fade* that often plagues today's communications-dependent organizations. Satellite is also independent of the terrestrial infrastructure. This could be significant, since the communications systems in many third-world countries may leave much to be desired. The path diversity or physical redundancy of this medium pays dividends during normal operation, particularly in areas prone to frequent circuit outages. In addition, the equipment is becoming increasingly compact, with truck-mounted transmitters, or *uplinks*, becoming commonplace. Just as it was for the hapless time travelers I mentioned earlier, such a link might become your only connection to a twenty-first century communications infrastructure in the event of a widespread event. Finally, some organizations use satellite for "surge capacity" when network conditions warrant. With Internet protocol infrastructures becoming commonplace, it is easy to cycle up satellite links as an additional path when extra capacity is needed. This has the added benefit of testing or exercising your disaster recovery link on a regular basis. If the satellite link is used from time to time for overflow capability, you can be pretty much assured that it will work if a disaster occurs and it suddenly must become the primary path.

If you are in an area prone to widespread disasters, (hurricanes, tsunamis, earthquakes, and so on), satellites are worth a serious look. A couple of things to look out for include the issue of mobility. A few of my clients have learned some serious lessons in this regard. I won't name the clients but can convey the lesson. One client was a large insurance company that had to respond to a series of devastating hurricanes in Florida. I never realized until then how important physical landmarks are to human beings in being able to find one's way around. Imagine several cities completely flattened and devoid of buildings, trees, road signs, and the like. Florida does not even have many hills. In this case the topography was a veritable moonscape, and it was many weeks before claims could be processed for some of the affected populace. So what became the ideal solution for the next series of hurricanes? GPSs, which are based on satellite technology. It's a funny thing but that "lady in the box" that navigates us in the rental car will still know where she is after a hurricane. Her directions are based on vertical and horizontal coordinates, not physical landmarks. This means you will always be able to tell where your business used to be—even if it's not there any more. (Myself, I have *two* ladies in the car telling me where to go: the GPS and Sharon. I'm sure many of you understand!)

While we are talking about mobility, consider what it would be like trying to navigate downtown Los Angeles with a foot of glass and deep crevices in the streets. Or getting around the Gulf Coast when all the bridges are out, particularly to hundreds of miles of barrier islands. Before you buy into a satellite solution, no matter how good it is, ask the provider how he or she is going to transport the equipment to its critical destination, because mobility *will* be an issue.

Who, and What Services, Can Use Satellite for Recovery?

Satellites interoperate in numerous ways. These include fixed location to fixed location, fixed location to mobile locations, point-to-multipoint locations, and mobile-to-mobile locations. The important consideration is that the communicating locations be within the coverage area, or *footprint*, of the satellite (see Figure 5.2).

Types of Satellite Service

Satellite service comes in many sizes and shapes. The optimal configuration is determined by the dynamics of your business and what you are backing up or recovering. *Fixed-to-fixed* configurations can be used to include command centers and "first

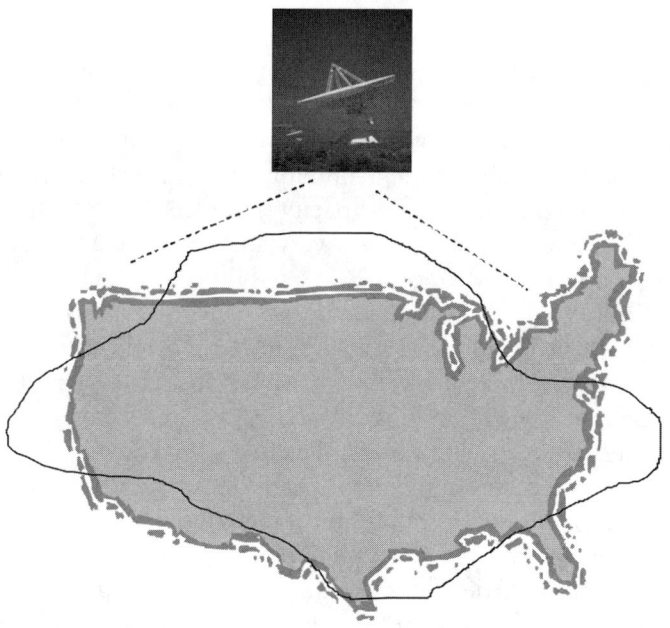

Figure 5.2   Satellite footprint.

responder" locations, sometimes also referred to as executive management team (EMT) locations. From recovery headquarters, other newly established fixed locations could be quickly provisioned and coordinated by satellite communications. These include field operation sites, damaged site recovery teams, and hot sites. *Fixed-to-mobile* configurations, as the name implies, are satellite communications from fixed locations like those described earlier, but this time to mobile units. This might involve insurance adjusters and damages appraisers, catastrophe (CAT) teams, emergency medical units, and a host of other mobile resources. *Mobile-to-mobile* configurations can be used to help nomadic users interoperate with one another (e.g., land-based terrestrial, maritime, and airborne recovery assets). All of these can stay in close communication by satellite due to the vastly reduced size of equipment and antennas in recent years. When one considers that satellite units are routinely affixed to shipping containers and UPS trucks these days, it is little wonder that they have evolved into numerous mobile platforms, each of which has distinct advantages in a disaster area.

Is All Telecommunications Traffic Suitable for Recovery Via Satellite?

The answer varies, but with today's technology, it is usually "yes." When you think about it, satellite systems are designed to interoperate with numerous communications technologies. This is because the satellite providers themselves serve everything from container ships to wi-fi hot spots to network television providers. Because a general-purpose communications medium requires the providers to be ready to traffic in whatever package the customer requires, significant expertise may exist in how to best transport your organization's data after a disaster. At worst, you may be only a *black box* away from a successful recovery. (By a black box, we mean some kind of transcoding or *crossbanding* equipment to take terrestrial-borne data and adapt it to satellite transmission.) More often, recovery planners are amazed to learn that packages and equipment already exist from the satellite providers themselves. Don't be surprised if after describing the need to transport VoIP packets via satellite, for example, the provider says, "We've got it. It's called a Feature Package 10, and here is what it costs." Of course if "Feature Package 10" does not exist yet, it could be a prime opportunity for your provider to nail down its technical specifications and offer it to others. Chances are that if you need it, others do, too. That could be another sales opportunity for the provider and something that you should use in bargaining for a solution.

Mobile Satellite Service

Even a general discussion of satellite service is not complete without an overview of mobile satellite service (MSS). MSS allows the use of portable satellite phones and data terminals. The equipment is often so small that even handheld satellite devices are becoming commonplace. Handheld satellite phones and broadband data terminals in theory at least can provide virtual worldwide coverage. It does not take a whole lot of imagination to envision just how handy this technology can be for command and control after a disaster. There are a few disadvantages. For instance, you generally need to be outside to use these, since they require a clear line of sight to the satellite. If it's a disaster, you will probably need a cigarette anyway, if not something stronger. Seriously speaking, there are already a few vendors tackling this problem by installing antennas on rooftops of enterprise customers and crossbanding equipment that converts a satellite link to POTS service, or two wire loop or ground start trunks that can connect to an inside phone, or PBX. With regard to the data devices, unless the weather is really bad, it is still possible to walk outside, download your email, and then come back in to read it in the comfort of your office. The important thing is to think these issues through in advance, be resourceful, and to mull over where these various technologies fit into your plan.

Mobile Voice and Data Terminals

A long time ago, a good friend of mine in disaster recovery made the following observation about restoring voice as opposed to restoring data communications. He said it didn't make any difference, because voice is really data, and data is really data too. Any questions?

All kidding aside, this is a true statement. A 64-Kbps path can be just as easily used to transmit a 64-Kb data circuit as a human voice. And yes, since your voice is digitized in the process, it is really data. In fact, satellite providers are very clever about this fact. A typical telephone circuit is 64 Kb, but a very understandable human voice can be transmitted in as few as 4–8 Kb. That's an 8–16x advantage. On terrestrial communications with bountiful capacity, who cares? But on satellite, such compression technologies means more capacity can be provisioned to more users for lower cost.

Some users want to use a portion of their satellite capacity for data. Today, mobile broadband *global area networks* can be leased that are capable of moving email, providing Internet access, conducting transaction processing, and a host of other applications that go beyond voice. Data speeds of up to 492 Kbps or more can be provisioned from mobile equipment with a pizza pan–sized dish. Streaming data rates are also possible at up to 256 Kbps. This does not sound like a lot when compared with terrestrial SONET carriers, T1, T3, and the like, but think about what you will be doing with it. You will not be instantly replacing your infrastructure, you will primarily be using this for b4Ci, remember? A 492-Kbps circuit will move one hell of a lot of emails. A 256-Kbps streaming data rate will nail up a fairly respectable video conference or even transmit a critical x-ray image from a disaster area if you are willing to wait a little while. Think about it. First responders would be in a position to speak to their leadership in real time while sending them video feeds of the disaster area. That's b4Ci on steroids, if you ask me! These terminals come in suitcases, backpacks, or in automobile mounted versions. I kind of like the idea of the auto-mounted version. In addition to mobility, there are no worries about batteries as long as you can get gasoline and keep the motor running.

What Is Around the Corner?

I suppose this would not be a Leo Wrobel book without some glimpse of upcoming "Buck Rogers" technology. Accordingly, I will share a couple of exciting morsels that I have encountered "by the water cooler." I would not be surprised to see people using these very soon.

"PRI on the Fly"

This is obviously not a term of art. The idea here is to connect a primary rate interface (PRI, sometimes called a *smart T1*) to a satellite link. "Hey, wait a minute, Wrobel," someone interjects. "You said that streaming rates over satellite were generally limited to 256 Kbps—and PRI is at least 1.536 Mbps." This, of course, is true if you use 64-Kbps channels. But remember from earlier in the article when I talked about voice compression? You don't need all that bandwidth to carry voice. Besides, 256 Kbps divides into 1.536 Mbps very nicely, and equals 6. That means you only need one-sixth of the bandwidth to carry this hypothetical PRI. And 64 Kbps divided by 6 = about 10 Kbps, more than enough for good voice quality with today's technology and no additional delay (other than propagation to the satellite and back, about one-quarter of a second) because it is not based on IP packet or cell technology. In other words, it works on paper, so how does one marry it to a PBX? That part, as it turns out, is pretty easy. A number of companies manufacture black boxes that convert raw bandwidth to the 4-wire circuit and PRI signaling that your PBX expects to see. If you can restore a single PRI and its associated direct inward dial (DID) numbers to a medium-sized hospital isolated by a telecom disaster, you have, for all practical purposes, restored communications to that hospital. The same goes for government offices, businesses, and essential services. That's why I think this particular "PRI on the Fly" idea is really cool. Visit http://www.telecomrecovery.com for more information, as they trademarked the term and have such an offering. You can also find similar services at rentsysrecovery.com

Satellite Resources—Who Is the Global VSAT Forum?

The Global VSAT Forum (GVF) is an association of key companies involved in the business of delivering advanced digital fixed satellite systems and services to consumers, as well as commercial and government enterprises worldwide. The GVF is independent, nonprofit, and nonpartisan. Any companies or organizations with an interest in the very small aperture terminal (VSAT) industry are encouraged to join. More importantly for the contingency planner, however, the GVF provides a terrific repository of satellite service providers worldwide. Since satellite is often the only link left to a twenty-first century telecommunications infrastructure after a widespread disaster, it is a good idea to keep some of this information in your hip pocket just in case.

According to David Hartshorn,[1] secretary general for the London-based organization, a single voice was needed to represent companies involved in the VSAT[2] industry. The promotion of the technology and the services it supports was also seen as an important role for the new organization. For example, Mr. Hartshorn cited GVF's role in educating IT departments that presently use various types of leased line services about the ways in which VSAT could provide tangible benefits at lower cost, while at the same time addressing disaster recovery by providing a totally diverse path. In the way of a cost comparison, consider a large U.S.-based chain of more than 1,000 locations, which recently selected a VSAT solution. Based on a comparison with frame relay services, the old service was calculated at between $400 and $450 per site per month. The VSAT equivalent undercut this by more than 50%. The VSAT solution cost less than $150 per site per month and did not require additional routers/frame relay access service (FRADs) or other external devices. Not only that, but frame relay is not universally available across even the United States, while VSAT services are—even in the remotest areas of Alaska.

It sounds to the authors almost like a sales pitch; however, remember that the GVF is a nonprofit organization and does not push any particular vendor over

---

1.   David Hartshorn is secretary general of the Global VSAT Forum, the London-based nonprofit international association of the VSAT industry. The GVF consists of more than 90 members from every major region of the world and from every sector of the industry, including satellite operators, manufacturers, system integrators, and other service providers. Mr. Hartshorn leads the GVF's efforts to facilitate the provision of VSAT-based communications solutions throughout all nations of the world. In particular, Mr. Hartshorn works closely to support national-, regional-, and global-level policymakers as they formulate state-of-the-art satellite regulatory frameworks. He is also responsible for creating greater awareness of the commercial, economic, political, and technological advantages that VSAT-based communications provide. Mr. Hartshorn also currently serves as a member of the Satellite Action Plan–Regulatory Working Group, the Brussels-based satellite-industry group that provides inputs to the European Commission; he is on the board of directors of the Society of Satellite Professionals International, and he is president of its UK chapter. Mr. Hartshorn has worked in the satellite communications industry for 10 years, serving in sales, business development, publishing, and association offices based in North and Southeast Asia, North America, and Western Europe. He has been published in hundreds of magazines and newsletters, and has spoken and chaired at conferences and seminars in every major region of the world.
2.   VSAT, we will remind you, stands for very small aperture terminal, which is a type of satellite service.

another. And while the numbers are compelling in certain situations, VSAT really turns us on because it is a totally diverse kind of facility that can be deployed very quickly, no matter how widespread the disaster.

While the cost performance can be compelling in developed countries like the United States, by comparison the dominant terrestrial operators in the developing world often shun satellite solutions in general and VSAT in particular. Yet, on close examination, almost all of the major banks in the developing economies of Asia and Latin America now use VSAT service solutions because they are the only way by which they are able to support sophisticated IT platforms reliably and effectively. While the incumbent (and often government-owned) carriers and governments reject the technology, it is VSAT that holds together the very foundation of their economies.

Also consider the fact that if you have a business with multiple locations in Europe, the Americas, Asia, the Middle East, Russia/CIS, or Africa, you have to deal with a variety of carriers and interconnection agreements. You also have to cope with wildly divergent levels of service quality, call two or more contacts to get a problem sorted out, and so on. Many of these problems disappear with a satellite network solution.

Characteristics of First-Rate Satellite Solutions

We would like to thank Steve O'Neal, operations manager of Rentsys Recovery Services, Inc.,[3] for his contributions to the following section. Steve is not a representative of any particular satellite vendor but has researched satellite technology extensively on behalf of his company, which provides various recovery solutions. This section should give you some ideas and raise your familiarity with this useful technology.

Satellite communications during a disaster is sometimes the only communication available in the area, which puts reliability at a premium. A satellite solution should be flexible enough to handle changes in configuration and bandwidth, as a disaster environment is very fluid. The satellite solution is also very flexible, which is absolutely essential in a disaster environment where requirements and situations change hourly.

These satellite offerings meet these requirements. Our hardware consists of various satellite antennas ranging from 1.2m up to 2.4m in diameter. These dishes are either trailer mounted or quick-deploy (QD), which means they can be taken apart and transported within five storage cases (Figure 5.3).

---

3.   Rentsys Recovery Services, Inc., is the premier nationwide continuity provider of true mobile recovery solutions. Our services focus on the recovery of clients' critical business processes using our mobile recovery centers with comprehensive satellite data and voice communications. In addition to our expansive and flexible mobile fleet options, Rentsys also has fixed facilities located throughout the United States and a QuickShip program that includes an extensive inventory of tier-one technology available on site within hours of a disaster. As a leader in the disaster recovery industry, Rentsys Recovery understands the importance of a reliable service provider. All of our services are designed to ensure the seamless recovery of what is most vital to your daily business operations. With locations nationwide and a staff of talented field technicians, we are well prepared to recover your facility and hardware needs in the event of a disaster.

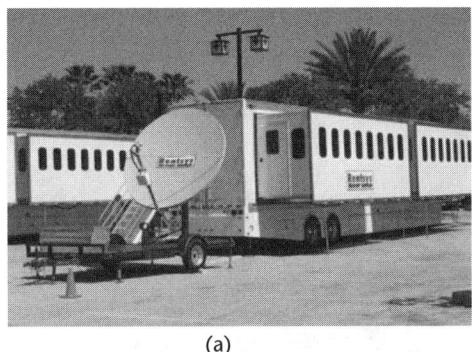 

(a)   (b)

Figure 5.3 Rentsys offers satellite dishes to either (a) trailer mounted or (b) ready for transport depending on the customer's needs.

Regardless of which dish deploys, the electronics remain the same, as pictured in Figure 5.4.

SCPC Versus Shared Satellite Solution

The most common satellite solution in the market today is known as the shared concept (Figure 5.5).

The shared solution consists of a hub-and-spoke approach to providing bandwidth to multiple remote site dishes. The vendor teleport is typically a large facility, with permanently installed large dishes that are 9m or larger in diameter.

The facility houses the hub of the system, and all bandwidth destined for the remote satellite dishes are dispersed through this hub. As vendors bring on more clients, additional cards are added to this hub for expansion.

The remote site dishes are fairly inexpensive systems that are sized according to the company's bandwidth needs. Shared solution systems such as iDirect, Hughes, DirecWay, and Gilat provide solutions to companies via the shared solution.

The SCPC solution is not a shared system and consists of independent point-to-point links.

In this example, the satellite vendor or reseller simply allocates the frequencies needed for the uplink and downlink. The signal does not go through a hub or even traverse the vendor's Earth teleport.

Rentsys Recovery deploys a satellite solution based upon the concept of single channel per carrier (SCPC). This philosophy is much more flexible and redundant compared to the shared solution, which is also used in the satellite offering market.

Although the SCPC architecture is more expensive, the inherent qualities are worth the expense when compared to a shared solution.

Bandwidth

The shared solution does not provide a constant pipe for data interaction between remote sites and the hub. First, it does not provide a full-duplex line of communica-

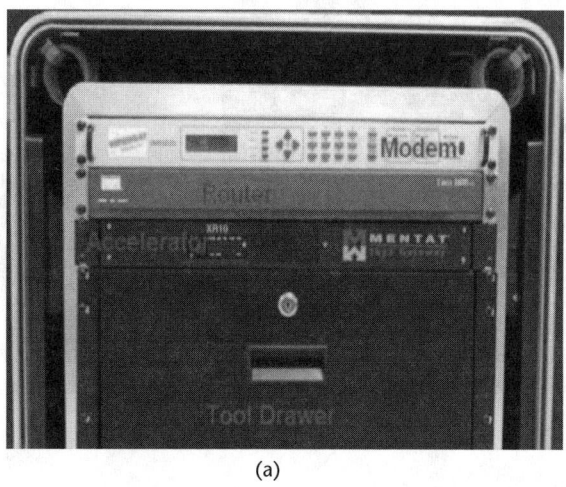

(a)

Shared satellite solution

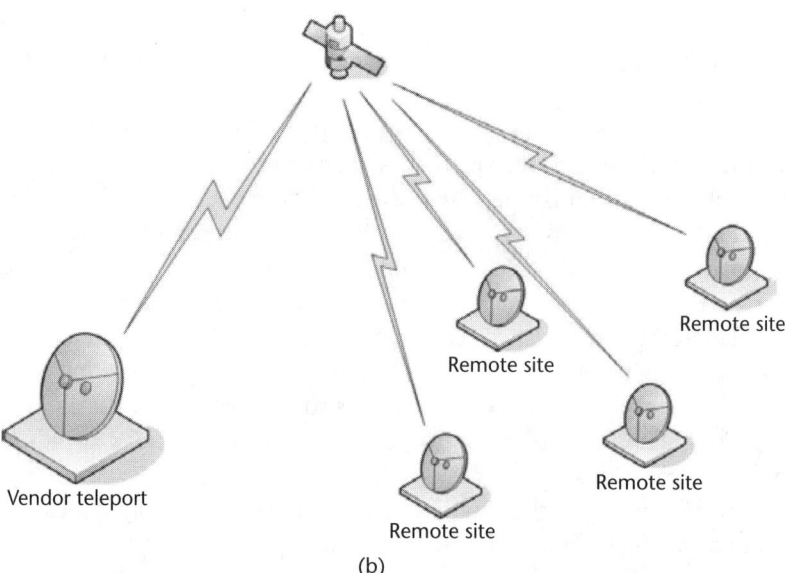

(b)

Figure 5.4 (a) Modem: Comtech 570 satellite modems with IP encoding. (b) Router (optional): Various Cisco routers, depending upon application. Accelerator (optional): Improves satellite bandwidth by up to 30%. Optimizes application performance and improves HTTP response. Unable to use accelerator if company requires VPN connectivity. Tool drawer: Consists of GPS, compass, and spectrum analyzer used to point the dish to the correct satellite in the sky. UPS: Provides battery backup power for satellite connectivity should there be a power loss or fluctuation.

tion between sites, and, second, it allocates a pipe when it detects transmission from the remote site.

Single channel per carrier (SCPC) satellite solution

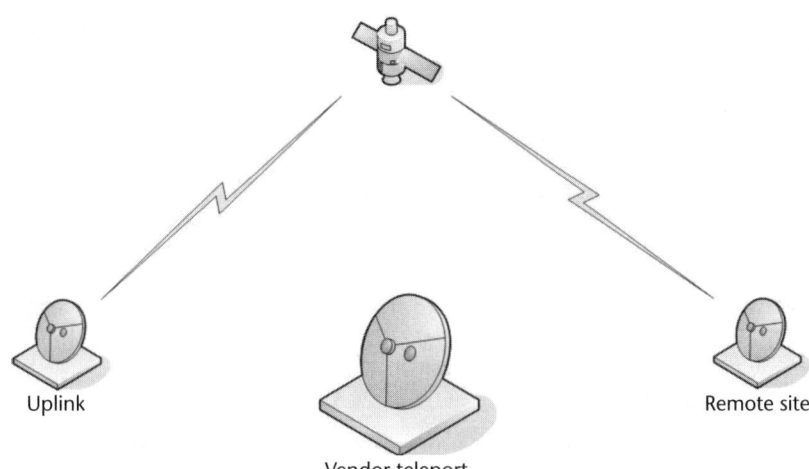

Figure 5.5   Shared concept featuring hub and spoke approach.

This interaction happens quickly but does cause some additional latency that can create problems for some applications. The SCPC circuit is more like the traditional T1 you might receive from your local telephone company (Figure 5.6).

Example T1 = 1.5 Mb

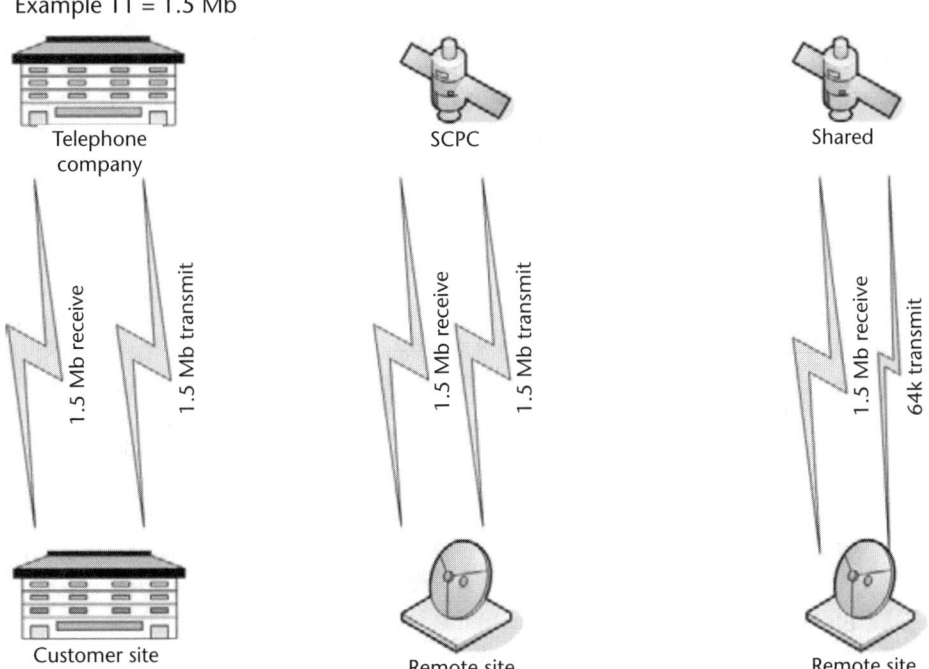

Figure 5.6   Bandwidth size shared versus SCPC.

The shared solution also shares data and voice resources through the hub. As with any bandwidth solution, the number of devices/users utilizing that bandwidth will decrease the amount of bandwidth available to everyone. It is important to identify what size circuit is providing bandwidth for the multiple remote sites. For example:

- 1.5-MB circuit being used by one person = user is using entire 1.5 MB.
- 1.5-MB circuit being used by four people = each user gets 25% or 375K.

Security

Network security is a very important issue in today's world of communications. The SCPC solution provides a more secure connection due to its inherent architecture. Companies routinely use the flexibility of a VPN to provide secure communications through the Internet.

Many companies request a point-to-point connection, combined with VPN, to provide enhanced security. The shared solution would require a double-hop as shown in Figure 5.7, because each device must connect via the hub.

Geo-synchronous satellites, which continuously stay above a specific point of the Earth's surface, are 22,000 miles in orbit. Signals traveling at the speed of light take 250 milliseconds (ms) each way to reach the satellite and go back to Earth, which is known as a hop. This is also called *latency*, and it is the delay you will see on devices using a satellite link.

A point-to-point shared solution would require a double hop to get data from the company data center to a remote site. The 1,000-ms latency will not allow some applications to run efficiently, and many would not perform at all. The resolution in the shared scenario, as shown in Figure 5.8, is to provision a dedicated circuit to the vendor teleport.

Although this would get the job done, it would require an additional monthly expense for a solution that may never be used. The SCPC architecture already has a built-in solution for this scenario (Figure 5.9).

Disaster Recovery

Finally, the shared solution has a distinct flaw with regard to disaster recovery. If the vendor teleport is damaged due to a disaster, all connectivity would have to be transferred to a backup site. This means that companies with provisioned point-to-point circuits would have pay for additional redundant circuits to the backup site as well.

This scenario recently occurred after Hurricane Katrina. The SCPC disaster recovery process is very simple. If the vendor or reseller loses a teleport or their entire infrastructure, the vendor contacts any other provider and receives new frequencies from the provider (Figure 5.10). The settings are changed on the modems and operations continue.

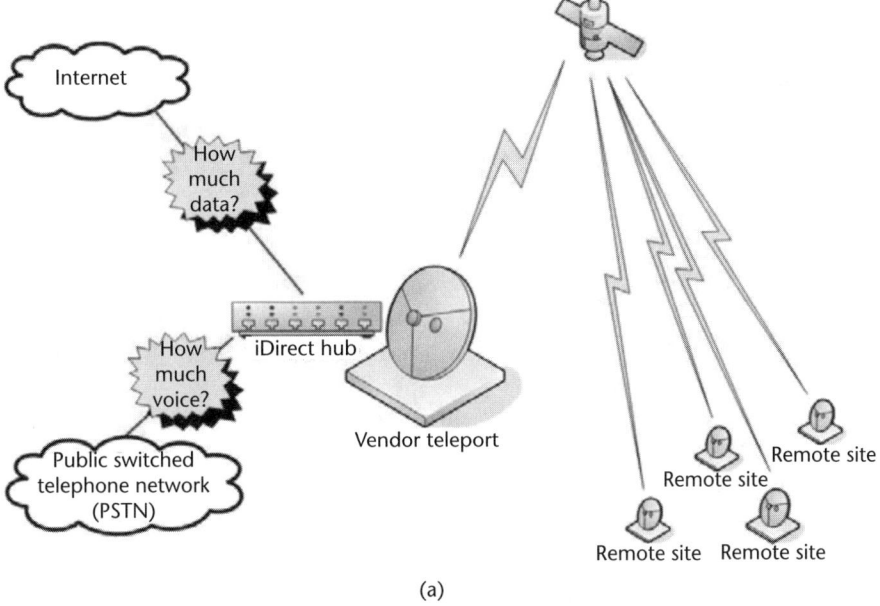

(a)

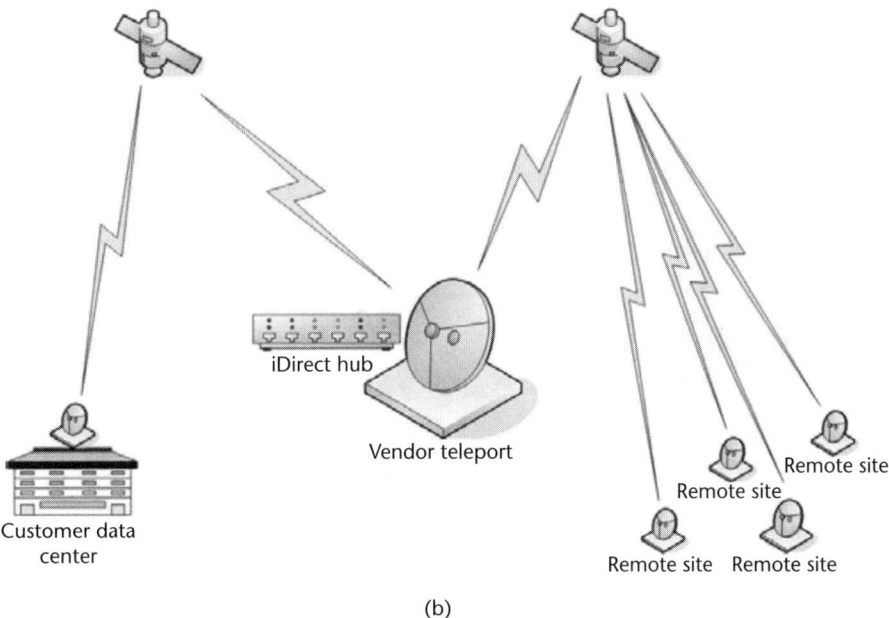

(b)

Figure 5.7 (a) Shared satellite solution and (b) customer point-to-point using shared solution (double-hop).

Summary

These satellite communication solutions were borne out of the need to provide communications to the Rentsys mobile recovery fleet. They have proven themselves

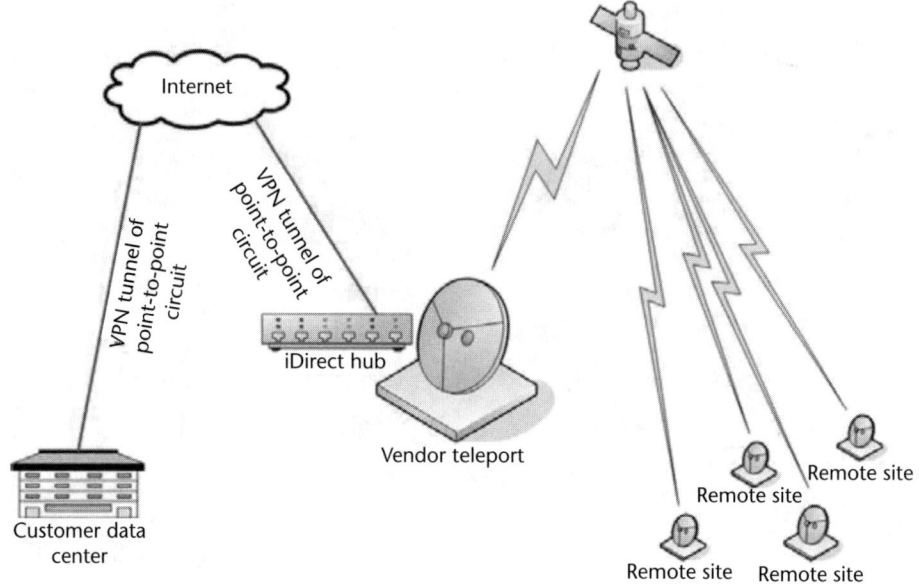

Figure 5.8 Customer point-to-point using shared solution.

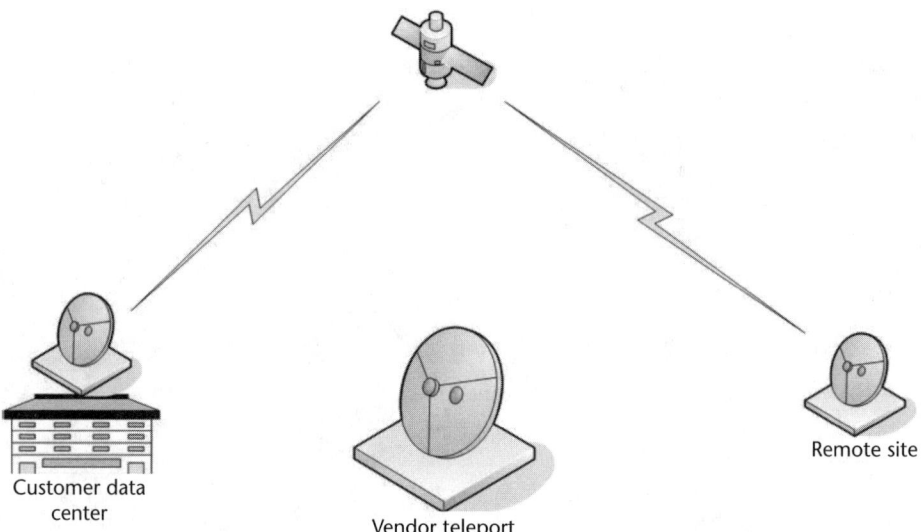

Figure 5.9 Customer point-to-point SCPC solution.

through multiple disaster declarations and multiple testing scenarios by various companies in diverse markets and applications. (For more information see http://www.rentsysrecovery.com.)

The most flexible, redundant, and cost-effective satellite solution on the market today was chosen. The solution is supported through multiple testing scenarios and actual disaster declarations. Technicians have seen many scenarios and can offer insight and solutions for companies in every type of business.

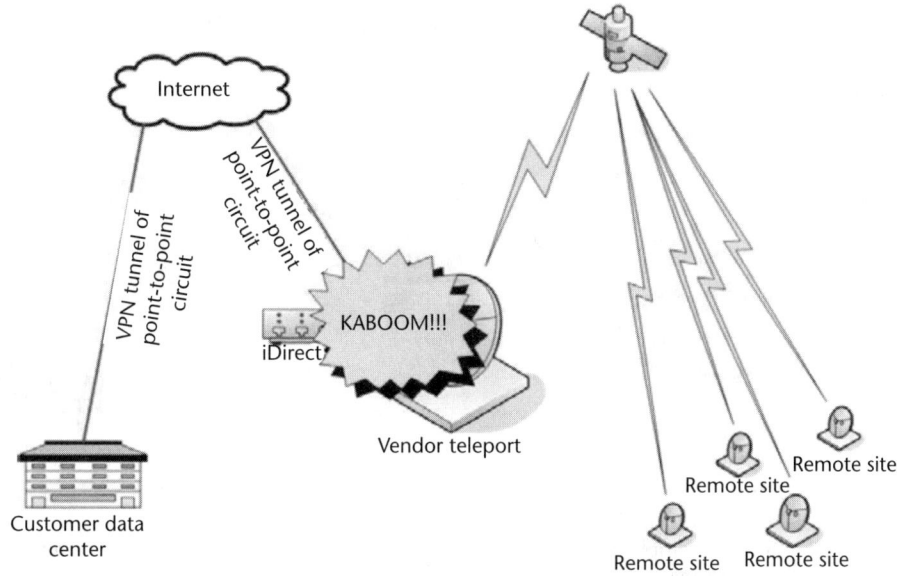

Figure 5.10   Shared satellite solution.

Data and Voice Communications over Satellite

I have stated elsewhere in this book that data is data, and that voice is really data, too. Remember the old fried chicken commercial where the guy says, "parts is parts"? Well, "bits is bits." If you can establish digital connectivity and move bits, you can move data or voice. This has also never been truer then in a post-IP environment where the IP protocol easily adapts to voice or data. So "where's the beef" in your telecom recovery plan? (Sorry, can't resist these food puns; it must be almost lunchtime.) Let's see if we can dish you up a good helping of ingredients next.

According to Steven O'Neal, operations manager for Rentsys, data connectivity via satellite can be accomplished by three various methods: VPN over satellite, point-to-point satellite, and voice over satellite. The end user determines the best choice for their applications, depending on security, accessibility, and internal support.

The first option to consider is for users needing plain old Internet access and simple dial tone for voice. The use of MNTAT SkyX Gateways can be incorporated in such configuration to accelerate general Web access and boost file transfers by some 30%. This helps reduce the effect of the satellite latency on sensitive applications by simply throwing bandwidth at the problem. However, it cannot totally resolve the issue. Satellite latency (the delay caused by data moving at the speed of light as it travels to the satellite 22,000 miles in space and back to Earth) averages 500 ms, regardless of the solution used. The speed of light is still the speed of light. Ask Albert Einstein.

Figure 5.11 is an example of the typical topology connecting a remote mobile recovery center (MRC) to a Rentsys recovery facility ground station. Access to the Internet is provided at the Rentsys recovery ground station.

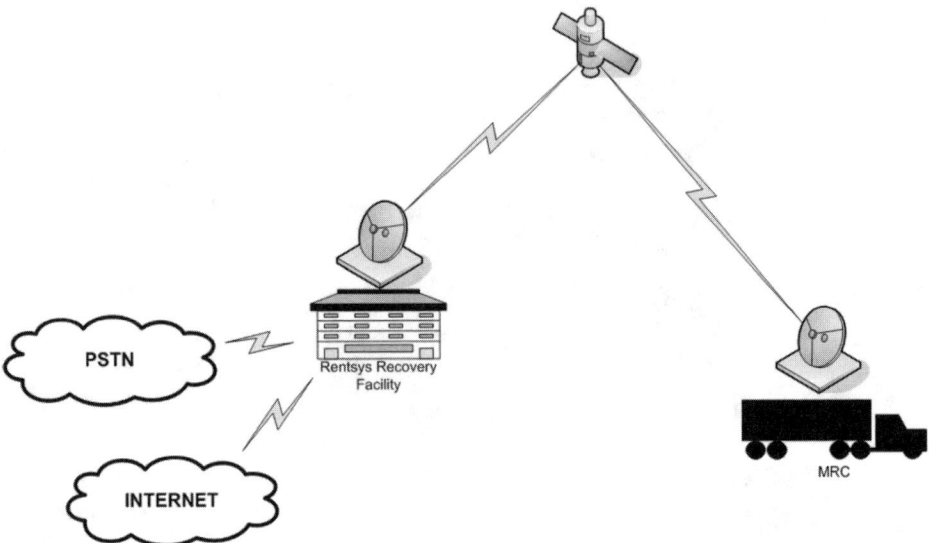

Figure 5.11 Example of the typical topology connecting a remote mobile recovery center (MRC) to a Rentsys recovery facility ground station.

VPN over Satellite

For end users and recovery planners who need to add another level of security to their configuration in addition to that which data via satellite provides, VPN meets this requirement. VPN uses encryption to scramble the user's data signal in order to secure the link from unauthorized access. The encryption, however, does not allow the IP accelerators previously discussed to manipulate the packet header and make actions run faster. Even so, according to Rentsys, customers using the in-house VPN solution today without the IP accelerators have found the application/network response acceptable for a disaster scenario.

VPN solutions require additional preplanning in order to work through IP address issues and determine global and private IP scheme. This proactive setup can occur without using satellite bandwidth time and can be accomplished using Internet circuits and Rentsys satellite equipment set up back to back. Obviously, users have the ability to test with live satellite bandwidth if requested. This option is described in Figure 5.12.

Point-to-Point Satellite

Some users require an even higher level of security than a VPN solution can provide. In that case, a secure, dedicated point-to-point service is deployed at the time of a disaster declaration. This solution requires some preplanning to identify satellite dish logistics at the customer's selected site(s). Issues such as power and data/voice connectivity handoff must be identified. Interestingly enough, the author notes that Rentsys provides this capacity via a "hitching post" arrangement. The hitching post contains power plugs of a predetermined configuration as well as access to telecom and other supporting infrastructure. The hitching post also supports a mobile

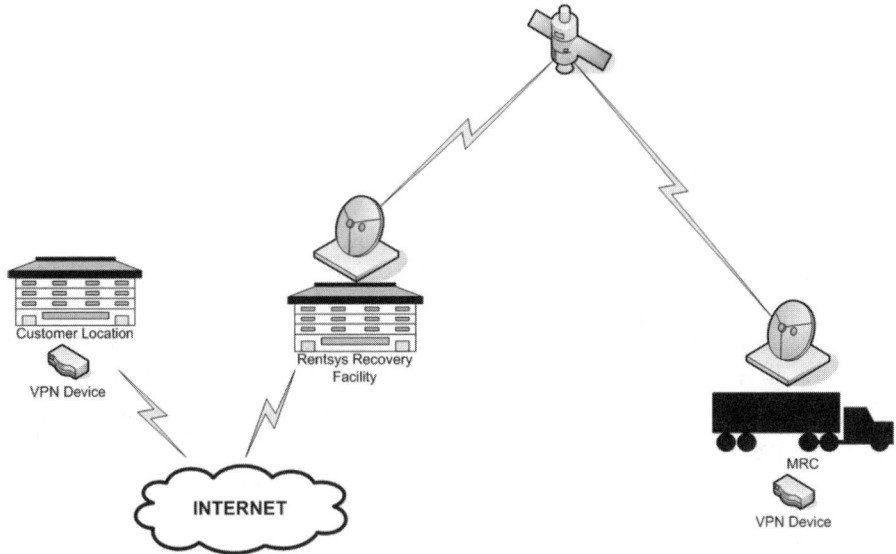

Figure 5.12   Example of a typical VPN over satellite setup.

recovery center, satellite connection, and/or other connections with a predetermined configuration that is documented in advance and included in the recovery plan. Very cool technology.

The drawback to this solution is that the customer needs to have data and voice circuits available to handle the extra needs of the recovery. This means additional monthly costs are incurred to provision circuits that may never be used for a disaster event.

We discuss how to make circuits do "double duty" to reduce telecom costs in Chapter 7. Figure 5.13 shows a point-to-point configuration.

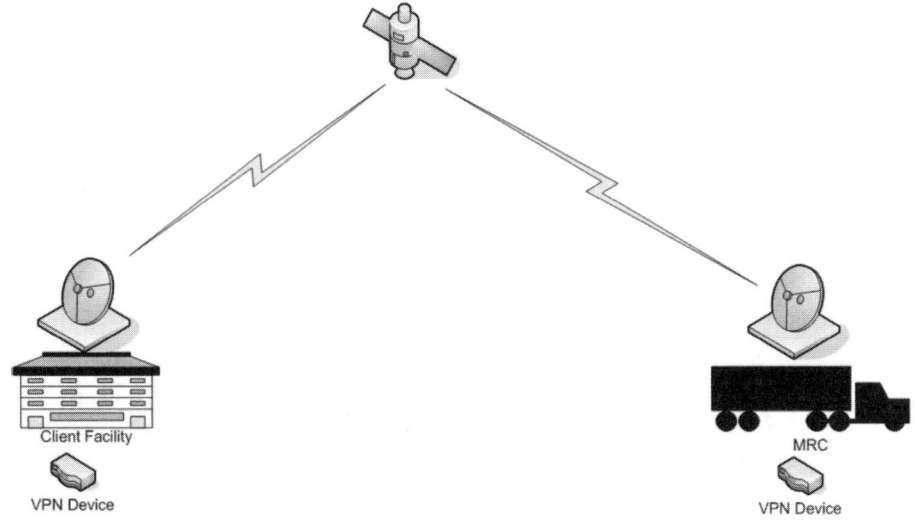

Figure 5.13   Point-to-point configuration.

Restoring a User Workstation

In addition to mobile recovery sites and communications, restoring displaced employees' desktops to a working state is also a concern. This issue can be handled in a couple of ways.

The first option is for an IT employee with certain access rights to connect a PC to the client network and run various installations and scripts. Users can then log in on their local desktops, and the system will build a local profile on that specific machine. Many times after the local profile has been configured, the network will then automatically run department setup scripts that load applications and settings particular to that user's department or job function.

Completing a setup like this in a normal environment can take a while and is obviously a very bandwidth-intensive operation. Under normal conditions this is done on a high-bandwidth (typically 10/100 MB) local LAN or across a WAN segment that has no latency.

Replicating this setup process over satellite, however, can be very slow. The bandwidth capacity of the satellite system, together with latency and all users trying to restore their desktops on site at the same time, can cause the initial setup times to take up to a half of a day, depending on application loads. (This may or may not be acceptable, but consider it might take that long just to get people back to work, depending on the nature of the disaster.) Once this initial setup is complete, bandwidth usage is reduced considerably, and remaining installs finish faster. This additional bandwidth can then be dedicated to actual communications, work, and production.

If customers confirm this is their choice for the desktop install due to security or financial reasons, there are some other ways to help speed up the desktop recovery time. These solutions might include using generic department log-ins during the initial stages of the recovery or using separate images for departments that have already executed the required install scripts. There are many ways to skin this cat. However, the best strategy to accomplish the recovery must be determined in collaboration with the customer's IT staff, as well as core business executives, who mandate what the recovery time must be and what is reasonable to pay for it.

A second solution to the issue of workstation recovery is the use of terminal services or Citrix, an American multinational corporation with a focus on software, services, virtualization, and remote access. The applications and profiles for users are actually held on server(s). Once network connectivity has been established to the corporate network, the desktop users then log in to the server that provides access to the applications. This eliminates the need for local machine profiles, and it allows desktop users to move easily to different systems and not run though lengthy installs in order to work.

This method also uses considerably less bandwidth than the installation process described earlier—as much as 75% less bandwidth is required, and users are able to perform their jobs on a new desktop within minutes instead of hours.

Many users who have employed the first solution are still very pleased with the outcome of this second solution. While they understand the delay and initial time to get users operational, such concerns are outweighed by cost reductions on bandwidth and other factors.

Voice Via Satellite

Once again, "bits is bits," data is data, and voice is data too.

Therefore, voice provided via satellite can be offered as analog or VoIP. It is important for the customer to identify their voice communication needs, both in terms of 4Ci and customer calling, early to help speed up the quote process. Very often, 4Ci is almost exclusively voice communications. It is a fatal mistake not to consider this fact during the planning phase.

Figure 5.14 shows a typical customer connectivity to the PSTN. The PSTN could be considered a cloud just as the Internet is a cloud of multiple vendors and networks that allow communication from multiple entry points.

The telephone line used within a home or small office is called a plain old telephone service (POTS) line. It is capable of carrying only one conversation. If two conversations at one time are required, an additional line is necessary. Businesses may have multiple POTS lines, but at some point they reach a cost point where the next level of telephone circuit is needed. ISDN-PRI's provide 23 or 24 channels of voice capability. One of these channels is used to coordinate the other 23 lines and keep them synchronized. Therefore a PRI can carry 23 simultaneous conversations.

Identifying voice bandwidth needs is based on identifying the number of simultaneous conversations required through the PSTN (Figure 5.15). Simultaneous conversations are also referred to as trunks, concurrent lines, or voice lines. Telephone lines within a facility have no restrictions on the number of simultaneous conversations; however, conversations connecting through the PSTN are limited based upon PRIs provisioned by the company. Customers evaluate their normal call usage and determine the proper amount of lines needed. Rentsys Recovery needs to under-

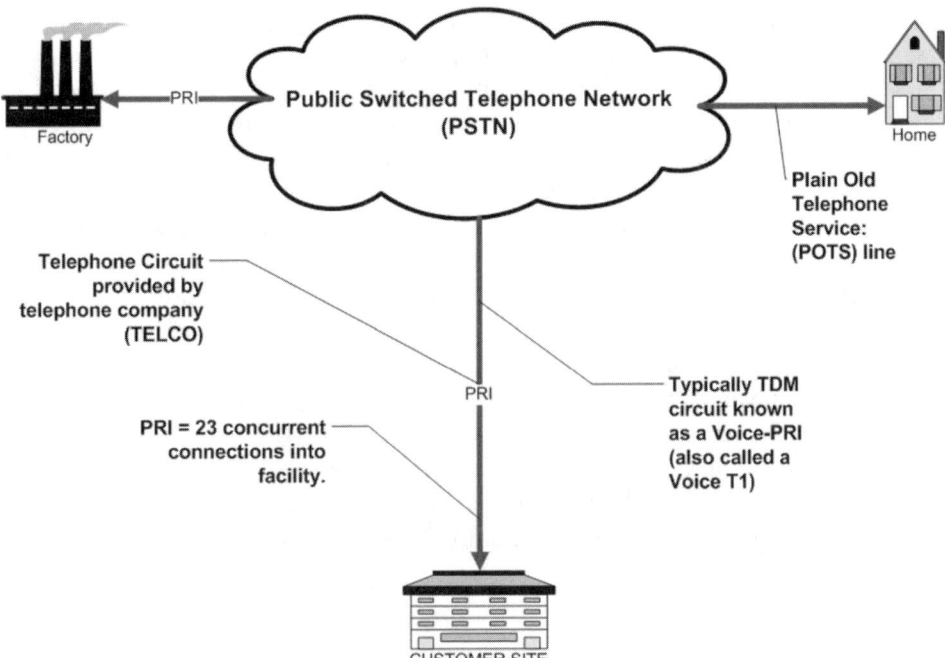

Figure 5.14 Typical customer connectivity to the PSTN.

CONCURRENT CONNECTIONS
Although 100 users work in the building, if only
(1) PRI provisioned, then ONLY 23 people will
be able to communicate outside of the building
at any one time.

SAME THING:
Concurrent Line(s) = Trunk(s) = Voice Line(s)

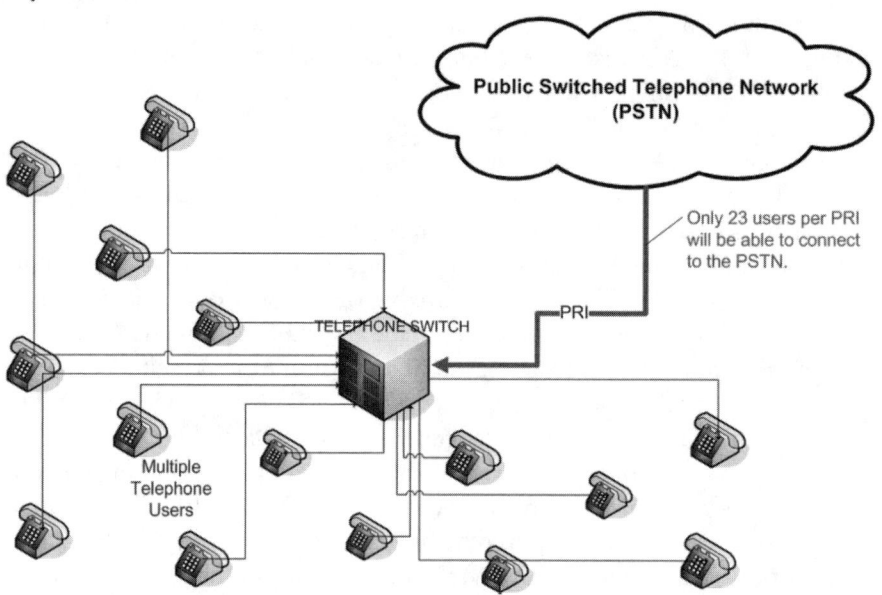

Figure 5.15   The ratio of voice bandwidth needed is determined by the amount of conversation
at any given time.

stand the voice lines required by the customer's type of business to properly size the
satellite bandwidth needed.

Note: Think about combining solutions like these with those described by
TeleCom Recovery, Telecontinuity, and Dell MessageOne in Chapter 2 for a truly
optimal solution. A backup voice communication system in another geography can
offload some of the capacity that might otherwise overtax a satellite solution.

Analog uncompressed telephone conversations require a minimum of 64K per
concurrent line. Multiply 24 channels by 64K and the bandwidth of a standard T1
of 1.536 Mbps is reached. (The remaining 8K is for framing, bringing the total to
the familiar speed of 1.544 Mbps.)[4] This is why PRIs are sometimes referred to as
voice T1s.

Regardless of whether analog or VoIP is used, bits are still bits, and there are
about one and a half million of them available for your use every second. They can
utilize the benefits of VoIP technology to transfer the call across the satellite link or
take advantage of compression capabilities like 8-, 16-, or 32-Kbps TDM voice
technology to provide more efficient use of the link. Analog calls can be translated
with a VoIP gateway known as an Ericsson Webswitch from analog into IP. There

---

4.   Europe uses an E1 standard of 2.048 Mbps or 32 channels. Interestingly, this speed is available in the United
     States via many satellite providers and could be useful to restore international operations. A call center in
     India perhaps?

are various methods that break voice samples into pieces and then packetize them—these methods are called codecs (Figure 5.16).[5]

VoIP gateways and VoIP PBXs take voice from the PSTN and convert it from an analog signal into IP packets that are sent across a computer network. Codecs define the procedure of how this translation is done. Variables are affected, depending on the type of codec used. Some codecs provide very good compression but lose quality of signal; other codecs have good signal quality but do not provide much compression. It is a balance to identify the best codec to use based upon user feedback. The two most popular codecs are G711 and G729ab. Many VoIP manufacturers support these codecs, and typically bandwidth required per concurrent line is 24K to 32K. The voice packet itself averages around 8K, and the IP header information adds an additional 16K. Figure 5.17 shows a picture of a voice IP packet.

It is important to understand that many vendors will declare that the voice is compressed to 8K; however, this is only the payload of the IP packet. It does not include the important to/from information required to properly route the IP packet across the Internet. Regardless of manufacturer, O'Neal believes the best compression usable for any PBX manufacturer is G729ab, which requires an average of 24K per concurrent line.

In an effort to further reduce the amount of satellite bandwidth needed, O'Neal cites the RAD VMUX product as an excellent solution for this problem. The RAD VMUX provides maximum compression without sacrificing quality. Figure 5.18 explains how the RAD's use of encapsulation reduces the need for bandwidth.

Instead of sending every voice packet with a header, the RAD encapsulates or bundles multiple voice packets and then attaches a single header and sends the

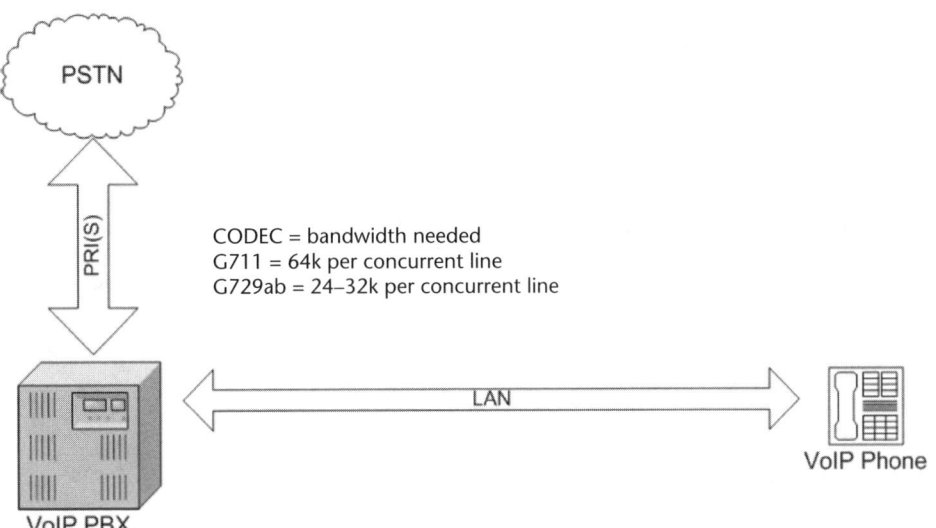

Figure 5.16 Voice sampling compression codecs—variable compression techniques that capture voice and then transmit the voice via data IP packets.

---

5. Codec means coder-decoder. It converts signals to digital the way a modem (modulator-demodulator) converts digital to analog.

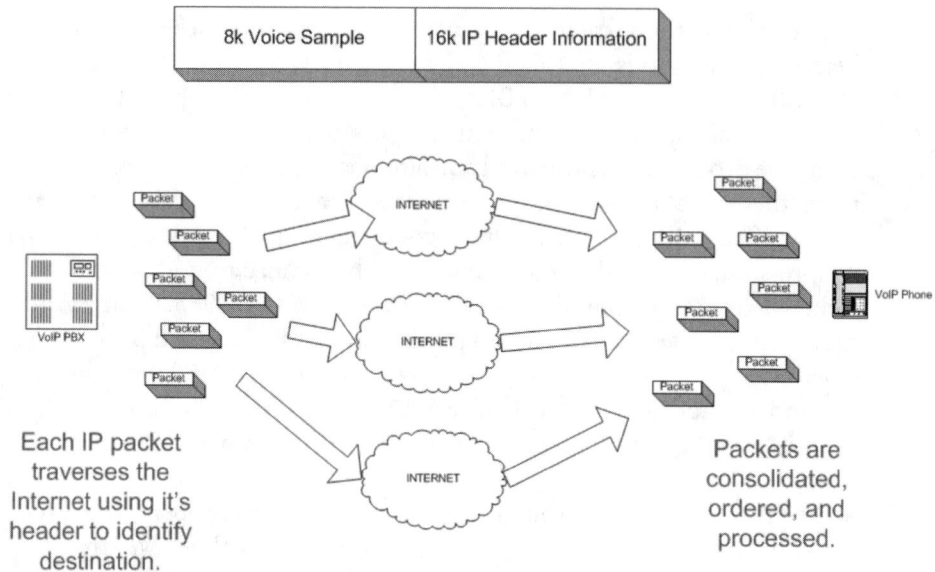

Figure 5.17   Typical voice packet using G729ab compression.

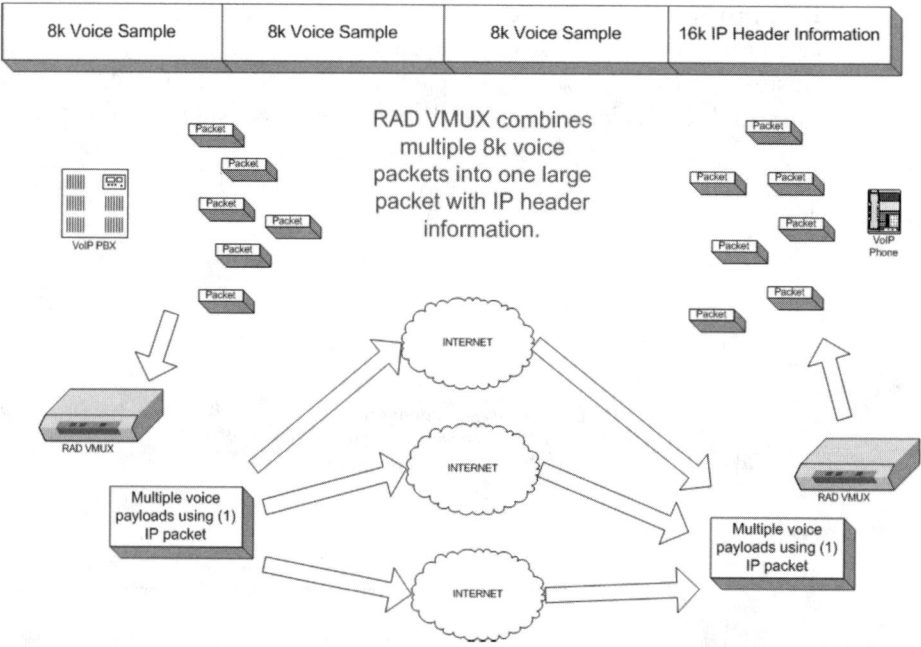

Figure 5.18   RADs reduce the need for bandwidth.

packet. This results in much less bandwidth required at approximately 5.5K per concurrent line or 128K per PRI, as well as in a significant bandwidth savings, which translates into large cost savings as well for satellite bandwidth.

The most common delivery of VoIP services to clients across the Internet or at a disaster site is a managed approach that involves a central VoIP PBX that connects via the Internet to the VoIP phones. Although accessibility is excellent because all that is needed is an IP connection, the best voice compression will be no better than 24K per concurrent line. An example of this in a hybrid arrangement with satellite is shown in Figure 5.19.

Besides compression, another benefit of the RAD VMUX product is that it allows placement of anything (PBX, VoIP, or analog) in the mobile recovery center (MRC) with the customer (Figure 5.20). This approach reduces PBX echo and provides the ability to make changes locally without having to communicate with another remote office. Automated call distribution (ACDs) units can be placed in the MRC with the PBX to allow local administration by people who may also be actually answering calls in the MRC. In essence, users can effectively extend the demarcation point (DeMarc) of the telco company to your disaster site.

Some users already have a VoIP infrastructure and are confident that infrastructure will be available following a disaster. These companies can use "softphones" on their laptops or desktop images that connect via an Internet connection back to the VoIP PBX. A major benefit of using this solution is that the users, extensions, and call routes are intact and behave the same way regardless of whether the users are connecting from the MRC. This reduces administration and the need to learn a new system. The drawback, however, is the amount of bandwidth required, which again averages between 24K to 32K per concurrent line.

The topology diagram depicted in Figure 5.21 shows a typical configuration of how voice and data are provided within a Rentsys Recovery mobile facility.

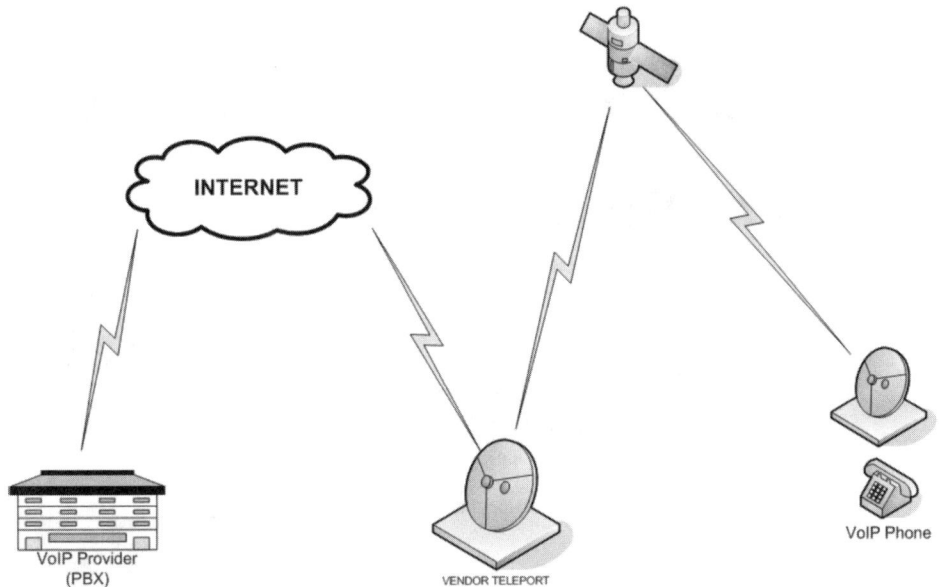

Figure 5.19   Voice provided by Managed Service.

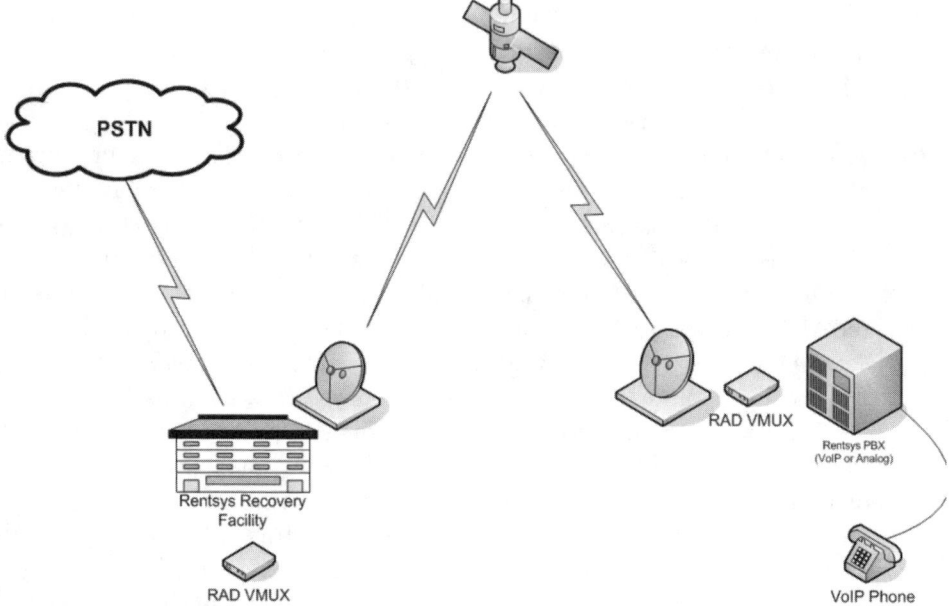

Figure 5.20   Rentsys extends the demarcation point.

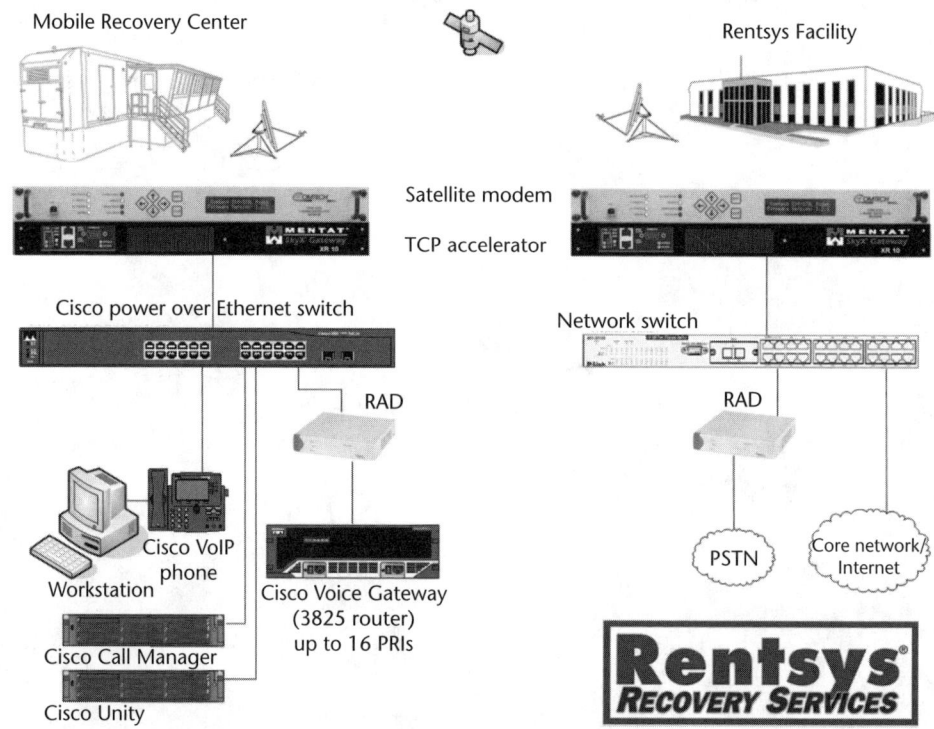

Figure 5.21   Data/voice via satellite.

CHAPTER 6

Justifying and Funding the Planning Effort

Justifying and Funding the Recovery Planning Effort

Probably the most common mistake made by the recovery planner is attempting to plan in a vacuum. Most technologists lack the visibility of the organization as a whole, which means in large part their perspective of what is required is in doubt to responsible management. The approval of executive management is essential for the planning effort. Without it, endorsement of the plan (to say nothing of funding the effort) is impossible. Even so, technologists are from Mars, and executives are from Venus. They speak different languages. In order to cut through this divide and place both parties to the discussion into the same frame of reference, conducting a business impact analysis (BIA) is essential. A BIA, to put it bluntly, is the way you prove to executives and policymakers that you truly know what you are doing and that the money you are asking for is not money that will go down a rat hole. Executives must see the value proposition of recovery planning to the business. Without it you will not get commitment from the bosses, let alone funding.

I have often described this issue using an old saying about drunks and lawyers. The saying goes that lawyers use figures the way a drunk uses a lamppost for support rather than illumination (Figure 6.1). It is kind of like this when one decides to take the plunge and ask management to fund a disaster recovery plan. The figures you use and the presentation you make should support you and illuminate management, but only if you do it correctly. If you put together a compelling BIA and conclusively convince management of the vulnerability that exists in his or her environment, you will get your recovery plan funded. I repeat you will get your plan funded.

Don't get me wrong; nobody is more acutely aware than I am that money is very hard to come by these days, particularly after the meltdown in the U.S. and other financial markets. Even so, Sharon and I are not starving to death in our profession—well, most months anyway. I like to think this is because after 25 years in the disaster recovery business, I know a thing or two about convincing skeptical management of why we must plan for disasters. Therefore, what I would like to do in this chapter to impart a few tricks of the trade that might help you to finally get some funding and commitment for your recovery plan. It's not as tough as you may think it is, provided you do your homework in advance. Let's start by dispelling a few myths.

Figure 6.1   An age-old axiom about lawyers states: "Lawyers use figures the way a drunk uses a lamppost—for support rather than illumination." It is rather the same thing in disaster recovery. Your findings and conclusions in a business impact analysis support the need for your recovery plan and illuminate responsible executive management.

Myth #1: Management Does Not Care About Disaster Recovery

Wrong, wrong, wrong. Disaster recovery, contingency planning, risk mitigation, or whatever you want to call it is a fundamental and fiduciary responsibility of executive management. The topic often comes up at board of directors meetings and transcends any well-run organization right on down the totem pole. I know from experience; I have been a CEO for a corporation before. When it comes to disaster recovery, however, top executives truly feel like Christian Scientists with abdominal pains. To put that more professionally, top management is clear on the mission (e.g., the organization must be able to recover from adverse events). They are just not sure if subordinates are throwing money down a rat hole in pursuing that goal due to lack of understanding of the big picture within the organization. Accountability of top executives for the well being of an organization is well understood these days, as we pointed out in Chapter 1. Some high-profile stories of top executives actually going to jail or being subjected to shareholder and stakeholder suits can be found very quickly with any rudimentary Web search for them. This is because, under U.S. law anyway, corporations are funny animals. Under the law, a U.S. corporation can do anything a person can do—except commit a crime or engage in gross negligence. If and when that were to happen, the people who lost equity in a corporation due to oversight, mismanagement, or gross negligence would not stop at suing the corporation, they could come after the CEO as well. The reason? Because they can.

This thought is never far from management's consciousness, particularly these days after passage of the Sarbanes-Oxley Act of 2002, which provides a whole lot of ways CEOs and CFOs can be held accountable for the way they conduct their business. This actually works in the recovery planners' favor when proposing a plan. In fact, you will be every bit as pressed to find an executive who says a disaster recovery plan is a bad idea as to find a person who says a preneed funeral plan is a bad idea. Most people agree that a preneed funeral plan in principal is a good idea. Having said that, do you have one? How many people do you know that do? Answer: not too many. The same problem exists with recovery plans, which are delayed for many of the same reasons, centering on competing priorities, lack of time, or a sense that it will be too expensive. All of these kinds of concerns can be assuaged through a compelling BIA and crisp presentation that puts the problems in terms the executive can understand.

So what can one expect from such a BIA and resulting presentation? I have mentioned in many lectures, books, and articles I have written that there are three answers one can get from management in response to a funding request:

1. "Yes."
2. "No."
3. "Let's study this some more."

Now, which of the three do you suppose is the most common answer? If you chose "Let's study this some more," you are correct and no doubt have been through such a request before. So what can be done to get a "Yes?" Myself, I like to stack the deck before going into the meeting to request funding. I'll show you some of those tricks in the sections that follow, but first let's continue myth busting with another common misconception.

Myth #2: Management Does Not Understand a Disaster's Impact on the Business

Wrong again! Management has a very good idea of the impact a disaster would have on the business. Think about it. Senior executives get paid based on how the business performs (in theory anyway)! A VP of marketing or sales, or a controller, knows where the money comes from because it is part of their scope of responsibility and probably part of their compensation plan. It is possible to play on this fact a little. If you are successful in showing executives the impact of a disaster on the business in terms they understand (read: dollars), you will have their attention and, more importantly, gain their support. One of the first things you will obviously need is a crisply presented BIA. If you can present such an analysis—in nontechnical terms that management can understand—you will get your plan funded. If there are any ambiguities or doubts, however, management will want to "study this some more" and you could walk out of your meeting with a longer "to do" list than you had when you went in.

Myth #3: Management Will Never Fund a Recovery Plan

It has been my experience that people who strike out on funding year after year have something very wrong with how they are asking for money. Everyone will agree that the last thing management wants is to squander money. However, if you don't have your facts straight that is just how your funding request may be taken. If you can apply a proposed expenditure to a problem that everyone in the room agrees needs to be addressed, you will get funded. Executives are not in those seats because they are bashful, and under the right circumstances they will spend money.

Answering Management's Four Most Important Questions Convincingly Almost Always Means Success

So how do you convince management of the legitimacy of your request? What facts are most relevant and important to management? Fundamentally, management needs to know only four things (see Figure 6.2) in order to make a decision about whether or not to fund your plan:

1. What can happen (fire, flood, hurricane, sabotage, and so on)?
2. What is the probability it will happen (expressed best in a percentage describing the probability of the event in a given year)?
3. What does it cost when it happens (in terms of lost sales, market share, employee productivity, and customer confidence)?
4. What does preventing it cost (best expressed as a high-level overview of the proposed protective system, procedure, or function) (see Figure 6.3)?
5. There is a possible number 5 you can also use: What are the other factors (such as legal liability, government requirements, and so on)? Refer back to

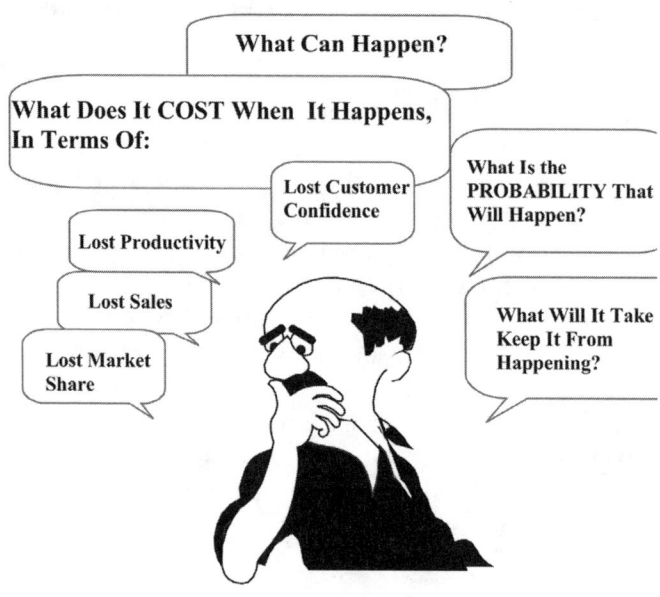

If this is explained in terms understandable to management, you will get support and funding!
The key is to build a case to which they can't say NO.

Figure 6.2 Use business terms that are understandable to management and build a case to which they can't say no.

Figure 6.3 Hot buttons and themes that are effective with senior management include lost sales, lost market share, lost productivity, lost customer confidence, and legal/regulatory exposure.

Chapter 1 of this book for some great material in this regard. Pick your industry and use it! Your boss should already know about many of the issues (legal and otherwise), but it will impress him or her that you know. It will build your credibility, elevate your standing, and make for a much more professional presentation.

What Should Be in a BIA?

First it is important to consider that your company probably does not do only one thing. It has different businesses or, at a minimum, different business dynamics in each business unit. They all have different pain thresholds. Some can last a month after a disaster; others scream in hours. If you don't take this into account, management will discount your BIA. Also note, screaming is not the test of being mission critical. Often the people who scream loudest are the least mission critical, while others whom you may hardly ever hear from can cripple the company if they are affected.

You will need to learn about each business unit, including its "pain threshold," to have a valid analysis. Talk to responsible managers and executives that management holds in confidence. That way, when the time comes for the big funding request meeting, management will have seen these figures. How can management disagree with its own figures? See what I mean by stacking the deck beforehand?

If produced and presented properly, a handful of slides can be the cornerstone of your executive presentation for support and funding. They will also serve as a springboard to fruitful discussion and thoughtful technological planning. As a litmus test, we will ask the question whether your nontechnical wife, husband, father, mother, or grandmother would understand the message your slides carry. If they could, then your presentation is ready for management. Don't laugh! This kind of approach in presenting to executives can be used in a wide variety of companies and for purposes that go beyond disaster recovery. If you can "sell" management on a disaster recovery project, you can sell them on other things you need. They key is learning how to communicate in their terms (Figure 6.4). They are not going to learn yours, and if you wait year after year for them to understand you, all you will get is frustration, not money.

The following is one very simple example. There are literally thousands of things that can happen to an organization, as we pointed out at the beginning of this book. Try to create some slides and spreadsheets like the ones that follow in order to make relative comparisons of risk. This helps communicate the risks of some types of disasters as compared to other disasters.

Stated another way, what good does it do to present a list of 100 disasters to your boss if all of them are deemed critical? Any disaster can be critical—that's a no brainer. The key is to give the executive a means to weigh criticality by type of disaster in order to target what are always scarce funds to where they will do the most good (Figure 6.5).

Consider a failure mode effects analysis (FMEA) (Figure 6.6).

Obviously, a list like this one can get long. Jazz it up a little bit and add color. Make everything below 100 green. Make 100–350 yellow, and those above 350 red. You will truly be surprised how well these work in executive briefings. Everyone begins to speak of "reds" and "yellows" rather than asking stupid questions

Figure 6.4   Hint at how to craft successful presentations: create compelling slides using validated data, then use charts, visualization, and color to present them.

Figure 6.5   Need help calculating an accurate probability of disaster? Try a failure mode effects analysis (FMEA).

Risk priority numbers (RPN)

Component	S ×	O ×	D =	RPN
1. Example	10 ×	10 ×	10 =	1000
2. Air conditioner fails in summer	8	4	4	128
3. Disgruntled employee/sabotage	10	2	8	160
4. Telephone cable cut	7	7	5	245
5. Accidental sprinkler discharge	7	2	7	98
6. Lightning strike	8	4	7	224
7. Etc.	—	—	—	—

Figure 6.6 Basic FMEA worksheet computing RPN.

like "What is a UPS?" (the answer to which is uninterruptible power supply). All they need to know is that a UPS handles power—and that if they don't have one they are vulnerable to the tune of S=10, O=10, and D= 7 or 700. (Power failures happen all the time!) With a good UPS, the numbers drop to S=1, O=10, D=1, or a FMEA score of 10 versus 700. Management will probably know what a UPS is, but realize that you can use the same equipment they don't understand. For more info on this topic, including some great sample slides, see a couple of my Artech House books, *The Definitive Guide to Business Resumption Planning* (1997) and *Disaster Recovery Planning for Telecommunications, Networks, and LANS* (1993). Also, you might consider *Business Resumption Planning, Second Edition* (CRC Press, 2009). All three of these books are more information technology–related but are a super source of ideas for your executive presentation.

The following is an example of a successful executive presentation derived from all three sources. If you are interested in greater detail, consider one or more of these three books.

We have already stated previously in this book that the BIA is probably the most important component of your recovery project. This is the part that defines and quantifies all the reasons you are going through all the trouble of producing a business resumption plan. More importantly, the more factual, understandable, and informative your BIA is, the better your chances are for success. If your BIA clearly communicates the inherent vulnerabilities of the systems you are trying to protect to executive management, you will win the endorsement, support, and funding of your higher ups. The key is how you communicate your findings to executive management.

Let's demonstrate one for you by laying out a hypothetical company. Since we have no idea what kind of company you work for, we will make a few assumptions:

- You work for the parent company of ABC Polymers, a snack food companyspecialty films company that makes bags for potato chips and snack foods as its primary business.
- Your company also has two other types of businesses.

- One is a snack food company (a natural outgrowth of the core business) and a second is more speculative venture, ABC Real Estate Investments, Inc.
- ABC Real Estate Investments, Inc., basically invests money and collects rents for the parent.

Refer to Figure 6.7, which depicts how quickly each of your company business lines' daily revenues decline in the event of a major disruption in comparison to the other two.

Note that the snack food retailer loses almost all revenue after the first day. This is because in a "commodity" market, a caller will order from a competitor almost immediately. If ABC Snack Foods cannot step up, they are going to call Frito Lay. In this case, it is also more difficult for this customer to get customers back after it loses them, since there is little room to discount the product or make other concessions. The margin on sales is next to nothing. This company would be in the most serious trouble after a disaster.

The second company, as we said, is involved in real estate ventures. No worries here, mate. Most people pay their rent monthly. Therefore, except in the event where a disaster occurred on the 31st of the month (thus preventing cash posting), the effect would be nominal. This is not to say these folks would not scream if their system was down, but they would in fact survive.

The third "core manufacturing" company, on the other hand, is more complicated. It has several distinct classes of customer. About one-third of these customers buy plastic meat wrap from your company. This is an easily attainable product from

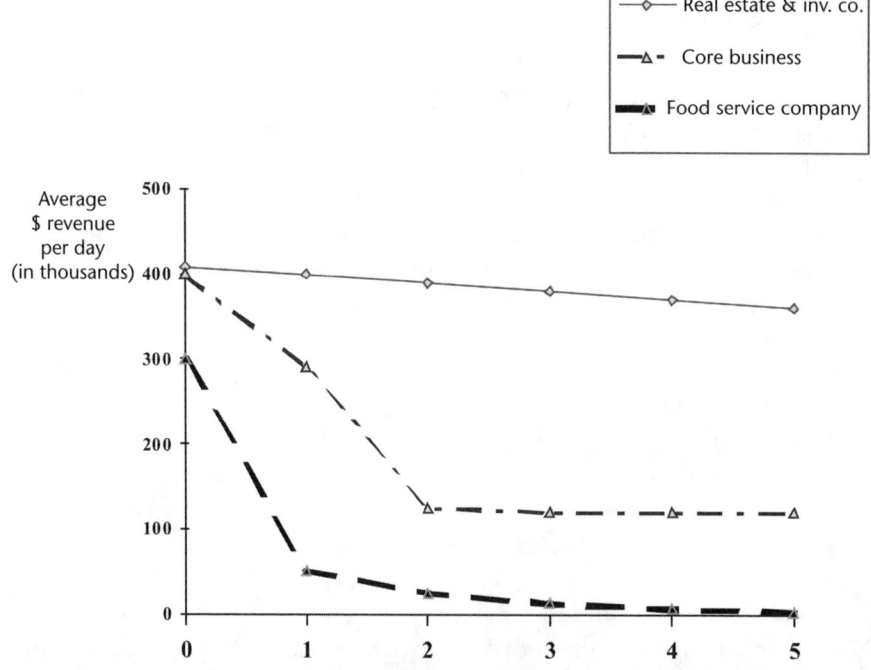

Figure 6.7   In a "commodity" market the snack food retailer loses almost all revenue after the first day.

other suppliers. These "commodity product" customers could order elsewhere on day one.

Another third of the customer base is comprised of "aligned distributors" who could order elsewhere but would probably wait a few days because of an existing business relationship or commission structure. In other words, they might wait longer since your company holds them with a lucrative commission or bonus structure. This particular customer mix produced the center line, showing a significant, but less immediate, revenue loss.

Finally, a third of the base of companies, representing your customers, orders specialized or customized products with nowhere else to go for them. I made a tongue-in-cheek reference to Frito Lay a few paragraphs ago. Maybe this customer is Frito Lay. Maybe they buy a custom bag that is the only kind in the industry to package a specialty product. Perhaps it would take too long for them to transfer complex graphics used for the bag labels to a competitor. Either way, they are stuck with you, and their ultimate success depends on how quickly you get back in business.

Given these differing dynamics in the same company, the scenario of a major disaster affecting ABC Polymers, Inc., core manufacturing would transpire like this:

- Nonexclusive customers bail out day one (one-third of your business).
- Exclusive customers jump ship about day three (another third).
- Monopoly customers take days to leave, but years to get back (the last third).

These "monopoly" customers are the ones who staked their business fortunes on your company and were let down most severely. In the words of one executive in a business not unlike this one, "They are the hardest to lose, but the hardest to get back if you ever do."

In general, what types of companies have the shallowest graphs? These include companies with a high rate of word-of-mouth referrals, or, in other words, companies perceived to be really good deals.

So who has the steepest graph? Typically, people are really gun shy when it comes to money. Banks are high on the list. Brokerages, maybe even more so. If your broker is out of service, and you just have to make that trade today, you will call a competitor no matter how good the relationship with your present broker is. Other companies include operations like catalog sales organizations, home shopping networks, and the like, particularly those geared toward impulse sales. Nobody calls back tomorrow to make an impulse purchase.

What would the graph of an energy company look like? It would look very shallow compared to the others shown here. Imagine a hurricane affecting a major refinery. There would still be only so many places to buy gasoline. There would, however, be other concerns, such as a bunch of new government regulations after the disaster by an angry Congress, which perceives an industry that can't regulate itself. This fact would have to be stated in the presentation, since a shallow graph might lull the decision-making executive into a false sense of security.

To summarize, a very common (and fatal) mistake in the business impact analysis phase and executive presentation is to homogenize all users together and assume that the "pain threshold" and willingness to pay for protective systems is the same for each. It isn't, as this section has clearly illustrated.

It is also necessary to consider different classes of employees in the organization, their pain thresholds, and precisely what their options are when an automated system fails. As a rule, knowledge workers (i.e., attorneys, engineers, managers, and so on) can usually find something else to do if a system is down. They may be on the phone or in meetings, for example. They can even work at home. Production workers on the other hand (customer service reps, telemarketers, inbound sales centers) are completely dependent on the system to do their job and are generally tethered to a desk at the office. Thus, the hit in productivity is more pronounced with the latter.

So how does someone communicate something as complex as a business vulnerability analysis to peers as well as to the broad cross-section of competing interests in the typical large corporation? Let's discuss a few of our favorites, using the fictitious ABC Polymers company once again as our guide. Try this. Illustrate each of the three business divisions with at least three high-level slides, in the following fashion:

A. A "focus" slide showing high-level issues and business overview;
B. A "dynamics" slide designed to show how long the business could operate in a major automated system;
C. A "solution" slide, or cost/benefit slide, designed to represent the line of business need versus perceived benefit (i.e., some systems desired by the food company may be somewhat passé to the real estate and investment company).

Then color-code the solution slide to illustrate the relative cost/benefit by business unit. It's an instant icebreaker, since each business unit feels like you are sensitive to their unique needs, making them all the more eager to support you. Note again that we opt for color, simplicity, and a compelling story.

Tables 6.1, 6.2, and 6.3 show the relative cost based on the associated risks to the components in the environment. They should be coded red, yellow, and green, for example, to illustrate relative risk.

- Where the risk is high and the costs are very low, the decision is not a costly one and requires little thought. These should be indicated in red. Examples include "Keep spare parts on site by all manufacturers for router cards, switch cards, and hub cards," and "Change and monitor passwords on network routers and issue separate passwords to users. Keep a log of all major changes."
- Where the risk is low and the costs are high to very high, it is unlikely that the business line would want to go through with the action. These areas should be indicated in green. These are considerations that will probably never be acted upon unless in combination with another project in order to mitigate the cost. Things like: "Want to eliminate all possibility of cable failure? Rewire the whole building and run two cables to each workstation."
- The bulk of the costs in the moderate range and the moderate risk areas are the ones worthy of further discussion. These should be indicated in yellow.

These are "let's talk about it" alternatives, like: "Replace or duplicate all MAUs with managed MAUs to prevent beaconing from taking down entire rings." and "Install water detectors in cable shafts near restrooms."

The recommendations stay the same for each business unit—only the colors change to reflect the differing business dynamics, pain threshold, and willingness to pay of each individual business unit.

It is possible to save some time on your analysis and not sacrifice quality. For example, most of the money figures could come directly from business line financial controllers, since they are in a position to know and generally give a straightforward answer. Operational capabilities for the most part should come from the VP level, in order to provide both a core business and technical perspective.

If produced properly, these slides will be the cornerstone of your executive presentation for support and funding as well as serving as a springboard to fruitful discussion and thoughtful technological planning. The following sections show some examples.

Focus on ABC Polymers, Inc. Parent Company/Core Business

- Answers all new customer inquiries.
- 95% of all first-time inquiries call in to call center in home office complex.
- Well-known in industry—80-year-old company.
- Perceived by customers as reliable and as good value.
- Number of sales (2007).
  - Primary: 154,011.
  - Cross sales: 32,200.
- Average sale = $895.00.
- 2007 incoming calls (includes fax) = 151,684 (95% of orders).
- 2007 incoming mail = 7,983 (5% of orders).
- Customers demand timely customer service—90/10 rule with 3% abandon rate.
- Backup system exists for short outages—paper.
- Technology: PCs now, implementing server virtualization.
- JIT environment for select customers.

Dynamics of ABC Polymers, Inc. (Figure 6.8)

- Revenue: $128 million per year.
- Business days: 320.
- Daily revenue: $400,000.
- Other issues: lost productivity.
- Lost market share.

Table 6.1 Overview of Exposure in Terms of Cost-Benefit Core Business Support Systems

Costs

Exposure	Very Low/None	Low	Moderate	High
Very high	Want to eliminate all possibility of cable failure? Rewire the whole building and run two cables to each workstation.	Move all servers to a computer room environment.	Arrange for duplicate access facilities from the telephone company.	Duplicate all power supplies and logic cards for all routers to prevent single point of failure.
High	Keep the same wiring, but run a second wire to all users.	Duplicate only main fiber and cable runs.	Replace or duplicate all MAUs with managed MAUs to prevent beaconing from taking down entire rings.	Install new, comprehensive network management system for a "Johnson Space Center" level of command and control.
Moderate	Duplicate wiring to the bay areas only; run single threaded out to workstations.	Install water detectors in cable shafts near restrooms.	Duplicate power cards, logic cards, and software in company's Internet firewall server.	"Virtualize" all servers in a computer room environment.
Low		Install copper-line telephones on site in case of a major fiber-optic failure.	Train users and develop standards.	Change routers to accommodate dual porting for mission-critical applications.
Very low/none			Keep spare parts on site by all manufacturers for router cards, switch cards, and hub cards.	Change and monitor passwords on network routers and issue separate passwords to users. Keep a log of all major changes.

Is there a problem?
Estimate of financial or collateral exposure
Inter views with key management
Check, cross-check, then check again
Final inter view: Controller

Figure 6.8 What the executive will need to know.

- Lost customer confidence.
- Number of idled employees in total outage: 1,000.

Table 6.2 Overview of Exposure in Terms of Cost-Benefit Food Service Business Support Systems

	Costs			
Exposure	*Very Low/None*	*Low*	*Moderate*	*High*
Very high	Want to eliminate all possibility of cable failure? Rewire the whole building and run two cables to each workstation.	Move all servers to a computer room environment.	Arrange for duplicate access facilities from the telephone company.	Duplicate all power supplies and logic cards for all routers to prevent single point of failure.
High	Keep the same wiring, but run a second wire to all users.	Duplicate only main fiber and cable runs.	Replace or duplicate all MAUs with managed MAUs to prevent beaconing from taking down entire rings.	Install new, comprehensive network management system for a "Johnson Space Center" level of command and control.
Moderate	Duplicate wiring to the bay areas only, run single threaded out to workstations.	Install water detectors in cable shafts near restrooms.	Duplicate power cards, logic cards, and software in company's Internet firewall server.	"Virtualize" all servers in a computer room environment.
Low		Install copper line telephones on site in case of a major fiber-optic failure.	Train users and develop standards.	Change routers to accommodate dual porting for mission-critical applications.
Very low/none			Keep spare parts on site by all manufacturers for router cards, switch cards, and hub cards.	Change and monitor passwords on network routers and issue separate passwords to users. Keep a log of all major changes.

The All-Important Executive Briefing

The Executive Will Need to Know Certain Things

Summary

There is way more to learn about selling management on your disaster recovery plan and how to justify and secure funding. Unfortunately, we don't have close to all the room we need in this book. If you would really like to delve down into the topic further, might I suggest the following Artech House book: *The Definitive Guide to Disaster Recovery Planning* (1997) by Leo A. Wrobel.

Just so I am not recommending all Leo Wrobel books, look for any book by Regis "Bud" Bates, a dear old friend and a fellow Artech House author. Bud is a fantastic writer and telecommunications disaster recovery guru. Google him or visit http://www.tcic.com.

Finally, if you would like access to dozens of free articles by Leo and Sharon, look at one of the following URLs:

- http://www.informit.com. Get on their home page and run a search on Leo A. Wrobel or Sharon M. Wrobel.
- http://www.techtarget.com. Same thing here, lots of great articles.

Table 6.3   Overview of Exposure in Terms of Cost-Benefit Real Estate Company Business Support Systems

Costs

Exposure	Very Low/None	Low	Moderate	High
Very high	Want to eliminate all possibility of cable failure? Rewire the whole building and run two cables to each workstation.	Move all servers to a computer room environment.	Arrange for duplicate access facilities from the telephone company.	Duplicate all power supplies and logic cards for all routers to prevent single point of failure.
High	Keep the same wiring, but run a second wire to all users.	Duplicate only main fiber and cable runs.	Replace or duplicate all MAUs with managed MAUs to prevent beaconing from taking down entire rings.	Install new, comprehensive network management system for a "Johnson Space Center" level of command and control.
Moderate	Duplicate wiring to the bay areas only, run single threaded out to workstations.	Install water detectors in cable shafts near restrooms.	Duplicate power cards, logic cards, and software in company's Internet firewall server.	"Virtualize" all servers in a computer room environment.
Low		Install copper line telephones on site in case of a major fiber-optic failure.	Train users and develop standards.	Change routers to accommodate dual porting for mission-critical applications.
Very low/none			Keep spare parts on site by all manufacturers for router cards, switch cards, and hub cards.	Change and monitor passwords on network routers and issue separate passwords to users. Keep a log of all major changes.

- http://www.naspa.com. This is the Networking and Systems Professional Association—a significant contributor to Leo's past books. Leo and Sharon's articles can often be found on this site. Better yet, join NaSPA (a nonprofit organization for IT professionals—Leo sits on their board of directors) for only $40 a year and get access to 10 years' worth of archive articles on disaster recovery. Membership also earns you a subscription to their super new publication, *Virtualize!* magazine. We don't cover virtualization in this book in any significant detail, but if you are an IT manager, virtualization is one of the hottest new technologies in IT recovery. Check it out.

- Finally, you can find many of Leo's articles at—predictably—http://www.4ci.us.

CHAPTER 7

Unique Vulnerabilities in Telecommunications Networks

Planning for Telecommunications

We concentrated a great deal on the 4Ci aspects of disaster recovery in earlier chapters. In this chapter, we concentrate on the other kind of communications—namely, telecommunications. Clearly, the world is more interconnected than ever, and, for many businesses, voice and data communications are the cash registers. Like the 4Ci example earlier in this book, today's business requires a level of control and coordination that in the past only the military needed, because minutes of downtime equate to lost dollars. Businesses move billions of dollars, pounds, yen, pesos, yuan, and euros around every day through every manner of electronic commerce. When money moves this quickly, everything moves quickly. How many minutes, for example, is an "acceptable" outage for an online bank or stockbroker? The answer today is "zero." Yes, you can be assured that if an organization that is highly involved in e-commerce could buy a military level of precision in a disaster, it would.

Many of the activities used solely by the military 30 years ago are embraced today by a broad cross section of commercial organizations. Recall that I (Leo) already told a few "war stories" in earlier chapters about my time in the U.S. Air Force. At that time, when it came to military communications, minutes counted. After all, in 1977, who except the military would mandate 10-minute trouble clearance time on international circuits? That was the benchmark in my former environment, and we had to produce this result with 1977 technology. That meant lots of order wires and hotlines not unlike the "red phone" you see in movies. Or, consider that the military constructed hardened disaster recovery facilities and command centers. Remember the 1963 Stanley Kubrick comedy *Dr. Strangelove*? Remember George C. Scott and "the Big Board"? It's no joke any more in today's companies. I actually got to see "The Big Board" of Hollywood and *Dr. Strangelove* fame and even sat in the general's chair. Today I see these "war rooms" turning up in lots of commercial enterprises, often under names like recovery operations centers (ROCs). The falling price of today's flat panel displays, for situational analysis at a glance, is yesterday's dream but today's reality. It is truly an awesome sight to behold. Let's take the analogy a step further: In decades past, who else but the military meticulously documented elaborate restoration plans—down to the level of exactly what alternate routes were to be established, depending on which major hub was lost? We had to do a lot of this with manual patch and cord panels. Today, however, essentially the same function is performed with digital cross connect systems (DCSs) and self-healing networks or without any human involvement at all by virtue of the IP protocol, also a military technology and discussed in more detail later. It is becoming increasingly clear today that not only the military but also commercial organizations demand a "Johnson Space Center" level of control.

Now, fast forward to 2009 to consider the new threats we face. This time the threats are not targeted to just military installations but also to commercial enterprises that require the same level of telecom and 4Ci redundancy used by the military.

You do have one advantage. The basic fundamentals of a workable contingency plan change relatively little over time. The technology, however, is another story all together. Technology is a double-edged sword. On one hand, it helps organize and strengthen your recovery plan. On the other hand, technology changes so quickly that it is sometimes almost impossible to keep up.

I know that as a technologist you understand how to leverage today's technology as you plan. You have a lot of enabling technologies, from cell phones to Blackberries to laptops, that we certainly didn't have 30 years ago in the "brown shoe days." Even so, the same things that will help you can also drive you crazy because technology changes so quickly and often becomes mission critical in and of itself. Think about how much critical information about your organization could fall into the wrong hands if your CEO dropped his or her Blackberry in an airport. Also consider that things like PBXs and routers are beginning to look and act like yesterday's mainframes. They now demand comparable levels of protection because they are just as important today. In short, you can dust off many of the standards, practices, and procedures of the past and (notwithstanding technology changes) integrate them right into your recovery planning effort.

To illustrate, as a Vietnam-era veteran, I learned a lot of things—some useful and some that are maybe a little obscure. For example, I learned that the fastest way to cool off a 12-pack of beer is to put it into a garbage can and give it a good long blast from a CO_2 fire extinguisher. Seriously speaking, I also learned that the military knows what it is talking about with regard to disaster planning. When you get right down to it, that's all they do. I am therefore sure that I am not the only Vietnam-era veteran who has noted that military contingency planning and practices deserve a second look today. This is particularly true since the commercial sector has "caught up" with the military in many regards and demands the same recovery capabilities due to the threat of terrorism. It's not only nostalgic for me to think of all this again, it's useful as well. For your purposes, remember te following:

- The preponderance of call center operations (moving to India and other countries today) do indeed demand 10-minute clearance time on international circuit failures.
- Telecommunications hubs are no longer targets for the Russians, but they do present irresistible targets for terrorists. Today's technicians could also end up airborne, and they are not even in the military. All the more reason to look at hardening these facilities.
- The public switched telephone network (PSTN) itself has responded to business demands for greater reliability. It now performs many disaster recovery

functions on the fly with such technology as fiber-optic rings. It has evolved from a hierarchical structure (class 1, 2, 3, 4, and 5 telephone central offices) to more of a peer-to-peer configuration—just like mainframes have turned into distributed networks. It's smarter now and includes things like SS7 and LNP discussed elsewhere in this book.

- Of course we have the Internet, which at its essence is also military technology of the 1970s revisited. Today, for example, it is easy to make a Dallas (VoIP) telephone number like mine, (214) 888-1300 work in India almost instantly, or vice versa. All you need is a good access to the Internet, and the change is transparent to the caller. Any place you plug in our VoIP adapter can become an office if it has high-speed Internet, even if it's at home or the Holiday Inn.

Compared to what the military had in the 1970s this is all the stuff of dream, as far as technology is concerned. The documentation, coordination, and CCC issues have, as I have already stated, changed surprisingly little. Therefore, if you know what to borrow from the past and can combine it effectively with today's technology, you can implement a most resilient plan for far less money in a surprisingly short time.

Consider that "first alert" procedures are one area where the fundamentals don't change, and there is a lot you can borrow from past efforts in military or even mainframe recovery plans. We can then use leveraging technology like email, laptops, telephone conference bridges, and wireless phones after the principals are in place. Think first about who needs to be involved, how, and why. Then overlay the appropriate technology.

Having made this point, let's move on to dependence on "key" people. If you are like most people, one or two faces just popped into your mind. The fact is that reliance on key personnel is a very common problem, particularly since year after year, the number of things people do becomes proportionally less while the things automated systems do becomes proportionally more. The trouble is, in order to restore automated systems after a catastrophic failure, you still need people and the means to communicate with them effectively. That's why we spend so much time talking about 4Ci. Even setting aside this issue, however, think about what brings money into your organization. Your customers! The ability to communicate with them is just as important. Let's look at Figure 7.1, a typical example of a global, network-enabled company of the twenty-first century, to see precisely why.

A Typical Sales-Driven Company

Your telecommunications disaster recovery plan will be as diverse as your operating environment. As Figure 7.1 shows, it's no secret that your environment has changed drastically over the last few years. Open systems computing has overtaken mainframes, especially in critical revenue-producing applications such as customer support. Concurrent with these changes, many of the technical decisions associated with these open systems are falling into the hands of users who are often nontechnical. On the plus side, these users are largely free of the "glass house" mentality of traditional mainframe computing environments. That fact often makes them enor-

Seamless Solutions in Today's Interconnected Businesses
In this case, a large insurance company

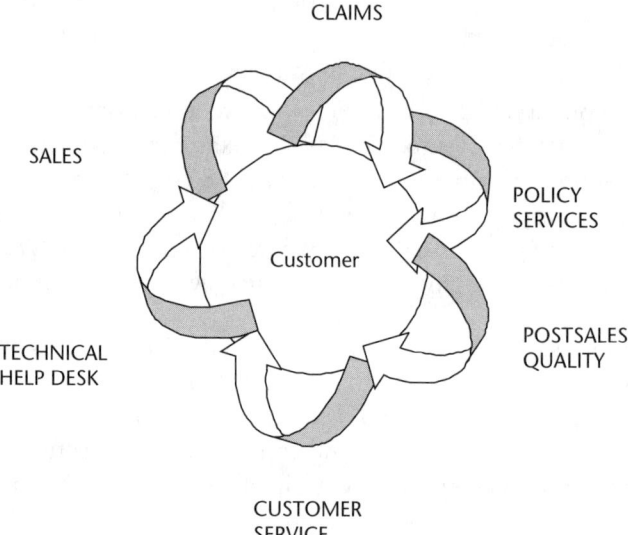

Figure 7.1　How do companies do business today? Answer: With everything interconnected.

mously productive because they are close to the business and know the application better than anyone. On the minus side, since these users often gather their experience in a "seat of the pants" fashion, standards, controls, and disaster recovery issues often lag the deployment of these technologies. If not addressed properly, these trends could place the whole organization at significant risk in a disaster. The first step in a good planning process therefore is to understand what you are protecting. For many companies, here is what the environment to be restored today really looks like. Pay attention; this is what Cisco sold to your boss on the golf course:

Let's start with the cash register—in this case, a large call center. (To fully understand the dynamics in play here, consider the example in Figure 7.1.) The first thing you'll note is that the process, like any good process, begins and ends with the customer. Moving on to Figure 7.2 you see a customer dialing in, probably on an incoming 800 number to a customer service center. As we already stated, voice communication services turn the whole thing into a revenue generator, since the business has no storefront. Let's assume this is a medium- sized bank, which has started to dabble in the money management and retirement market though a separate subsidiary. They have just received a call on their 800 number from an interested prospect responding to a special promotional offer. Look at everything that comes in to play in a properly designed open system and how they each add to the profitability of the organization.

The caller's telephone number—the bank has automated number identification (ANI)—has triggered a file that displays on the banking agent's screen while the phone is ringing. Although the agent generically answers the call, "Thank you for calling ABC Company," the agent already knows the caller. Just about everybody does this these days. That's why a bank's automated prompts will ask you to enter your account number if you are calling from work, but usually only ask you to enter

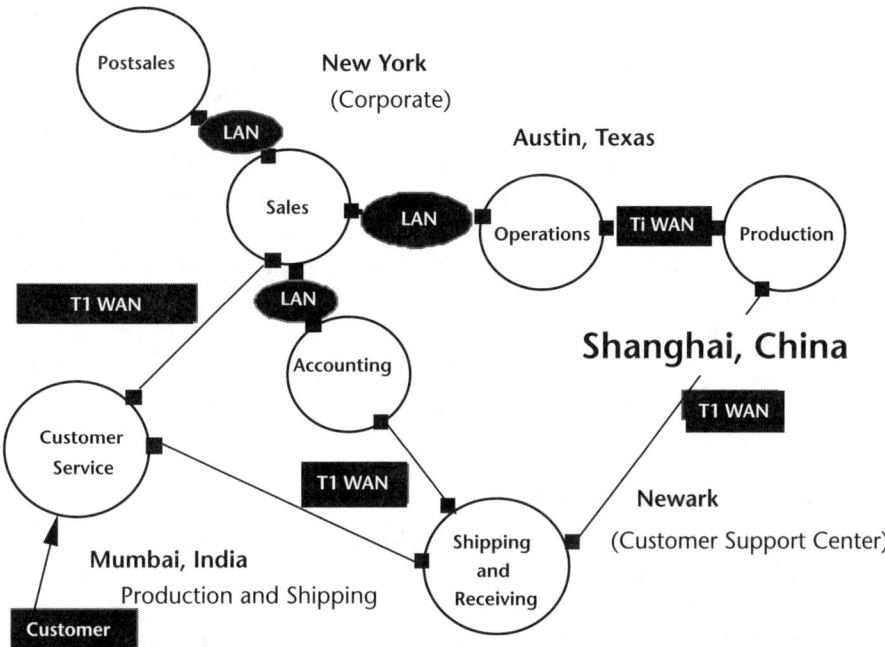

Figure 7.2 The 100,000-ft. technical view of this hypothetical company. (Adapted with permission from *The Definitive Guide to Business Resumption Planning* by Leo A. Wrobel, Jr. © 1997 Artech House Publishers.)

a personal identification number (PIN) if you are calling from home. They screen the data the telco sends them, bounce it off a database, and viola! They already know who you are before they even answer the call.

1. The caller inquires about the terms of a promotional offer or to check an account balance. While on the line with the caller, the agent has also noticed that the caller has a certificate of deposit (CD) coming due. He or she offers to renew the CD during the call. The idea is to be able to sell something else—another profit opportunity—during the same communication. Perhaps the representative adds on someone in personal banking to the call to complete this transaction while the caller is still on the line. That's better customer service.

2. The agent then notices demographic data displayed on the corner of his or her screen. It indicates a birth date of 3/21/43. "Maybe this person has just retired," the screen prompts. "Ask him where he is rolling his IRA." "Ask him who is financing his motor home." You get the idea.

3. The ability to do this is the beautiful part of open systems and one reason why companies deploy them so frequently. By having fast access to information, the agents can respond quickly, efficiently, and accurately to a customer's inquiry, while the customer is on the phone. What would have been "I'll call you later," now becomes "How many would you like to order right now?" Open networks make all this possible.

Look again at Figure 7.2. Which circle would you say is most important in the process, as far as the bank's long-term profitability? Most would answer, "The customer service function," and that answer is probably correct. The "walk-in" market for many banking services is small. When was the last time you walked into a bank? People are depending on the phone and the Internet to access these services. Therefore, the job of today's technologists must include looking at what technical platforms support the customer service center and other critical departments, and make these as fault-tolerant and resilient as possible. This can be difficult, however, as many of these system now reside beyond the direct control of a traditional MIS department. Equally important is the restoration of the voice and data communications services that interconnect these systems and turn them into revenue generators for the company. So where does one start? Well, for starters, begin with the following realization.

Everybody Needs to Communicate—and Telecommunicate—So Telecom Is Paramount!

Technology itself is not enough in today's era of mega-disasters. Consider Oklahoma City, September 11, Hurricane Katrina, and others. In each of these instances, technology was not the deciding factor in the immediate aftermath of the disaster. Communications and coordination were the keys. You need to restore "department to department" communication as well as electronic telecommunication. Sometimes companies don't realize this until it's too late.

For example, how prevalent is the Internet in your organization? What was simply a flirtation a decade ago outside of the military and academia has now often become the mission-critical revenue-impacting system in many organizations. PC home banking, PC-based home shopping networks, and PC-based securities and portfolio trading are but a few examples. Is your Internet platform up to par? Web servers should include redundant common logic boards, hot-swappable parts, inventories of spare parts, redundant T1 access, and other elevated standards because they are mission critical now, too. The most serious communications disasters are the ones that occur outside—beyond your control.

What happens to your organization when the telco has a disaster? Let's look at a few examples.

In today's organizations, telecommunications is king. Whether the application is an inbound call center answering 800 numbers or a big old Web server taking orders from thousands of online users, telecommunications is often the centerpiece of today's "buy it now" business. Stated another way, when the telecommunications network fails, the organization's cash register stops.

In order to understand the dynamics in play here, consider the typical workstation in a service-oriented company. This is the "front line" interface to the customer and the funnel into which cash enters the organization. This workstation can be engaged in any number of activities, from selling concert tickets to trading stock to taking orders for that wonderful Italian sculpture that just aired on the Home Shopping Network.

Taking the last item for a moment as an example, imagine spending tens of thousands of dollars to air that item, only to have the people calling in to purchase the blender greeted by an "all circuits are busy now" recording. Impulse buyers like these will probably not call back again later; indeed, they may not even remember your number 10 seconds after the next item comes on.

Or, consider a stockbroker. Someone that calls in to sell 10,000 shares of stock wants to make that transaction now (especially in these financially volatile times!). If that does not happen, these users can be pretty unforgiving. Someone using "Chuck's" brokerage might just decide that "Joe's" brokerage might be a more reliable provider.

The reason I used these examples is brokerages and home shopping clubs often have elaborate telecommunications recovery plans for precisely these reasons. "That's fine," you may say. "They can afford to have elaborate plans. But what about smaller businesses?" What about nonprofits or a business that is not a WalMart or Microsoft? Surprisingly, having a telecommunications recovery plan may be much more affordable than you think, because a lot of the capital-intensive part of the process has already been accomplished by the telecommunications companies themselves. Did you know that:

1. There is absolutely no reason why your inbound 800 calls should not be able to be answered at an alternate location within 10 minutes of a disaster?
2. There is virtually no reason why your company's main telephone numbers should not be able to be answered at an alternate location within an hour of a disaster?
3. There is virtually no reason why Web-enabled services could not be redirected elsewhere within hours of a disaster?

Disaster Recovery Planning for 800 Calls

Each of these critical services has a readily available solution. In the case of 800 numbers, the name of the AT&T service for "rehoming" 800 services is "Command Routing." Other companies have different names, but the concept is the same. Let's say your organization experiences a fire. In a disaster event like this, even if your building is gone, all of the logic that switches your calls is still intact—in the telco's central office (CO). Recognizing this fact, the carriers have been offering the ability to reroute calls to another working location for more than 20 years. If an organization like yours were to call its 800 provider right now, chances are that the process would only take minutes. We would recommend asking immediately for a supervisor, and making it clear that a disaster has occurred. You should also have an alternate number for a branch office, home office, or disaster recovery center in your hand when you call. (That means figuring out what that number is going to be in advance and putting it in your plan—even if that plan means having the number in your Blackberry!) The supervisor should be able to have your 800 numbers redirected almost immediately, perhaps even while you are on the phone. (Tip: Remember the TeleContinuity and TeleCom Recovery Services described earlier in the book? They pair up nicely with precisely these types of telco services.)

What happens if you try to call your provider to implement these services, and you get the same "all circuits busy" recording that all of your customers are getting? That can easily happen. In a widespread disaster, the network will "block" many calls as users jump off the landline network to wireless phones, for example, thereby saturating the wireless network. Or perhaps the telephone company itself has had a disaster. Hurricane Katrina brings this possibility clearly to mind. This is the reason a disaster recovery plan for telecommunications needs to address multiple contingencies, ranging from a cut cable to your building to the loss of a city due to a hurricane or earthquake. In the following section, we will introduce a few concepts as well as steps you can take right now at little or no cost. The first, remote access to call forwarding (RACF), is also the cheapest and simplest of these services.

Disaster Recovery Planning for "Local" Calls

In much the same fashion as redirecting 800 numbers, "local" telephone numbers can also be redirected to alternate locations. If you are really clever about it, you can arrange it so that you can redirect the numbers yourself. You simply dial the access number given you by the telco. If you want to hear a real one right now, go ahead and dial this number: (972) 224-4900. When you dial it, you get the following voice prompt: "This is your remote access service; please use a touchtone phone. Please enter the number you want to be forwarded."

Since you are not a subscriber, you will not be allowed to go any further—but if you were a subscriber, the service would prompt you further. After entering the number you wanted forwarded, say, (214) 888-1300, the system would ask you for a four-digit PIN. After the PIN is entered, the system would ask you what you want to do. At this point you can enter the same call forwarding code you would enter if you were actually at that phone. Generally this code is *72 or 72#. After entering the *72 or 72#, the system would ask you what number you want your calls forwarded to. This number can be your home phone or your cell phone—in fact, any working number in North America can be used. I have never tried it with an international number, but there is no technical reason why that wouldn't work either. When you hang up, your calls would be forwarded and answered at the alternate location. The cost for this service runs about $1 to $3 per month.

We have recommended and used this service for large and small companies alike. Recall our earlier case study of a major Dallas-based airline that was plagued by winter ice storms. The airlines had figured out the command routing for its 800 service a long time ago and in fact were probably one of the first companies to use that service in the 1980s. The local service, however, remained problematic for years. By installing RACF on its local lines, the airline acquired the capability to forward hundreds of local lines to another call center every time Dallas had a weather event. What's more, as it turned out, the Lucent 5ESS switch that served its location had the capability to switch 99 "paths" at a time to the alternate site. That meant a caller dialing (214) 123-4567 could be behind 98 other callers and not get a busy signal! Moreover, the airline actually improved this service by pointing its local numbers at the 800 numbers, thus allowing the intelligent features in the 800 network to balance the unusual load and distribute it among multiple call centers.

Finally, the data center manager for the airline accomplished all of this from the comfort of his own living room and never had to call the service providers or leave during the ice storm—all for $3.00 a month! You just can't beat something like this.

But what if your organization never answers the phone at all? Think about companies that do all their business online. Ebay comes to mind. I have bought and sold hundreds of items on eBay and have never once spoken to anyone. There are many companies that are set up without the need for a storefront or phones. The Web becomes both. Welcome to yet another concern for telecommunications-dependent organizations: the Internet itself.

Disaster Recovery Planning for "Web-Enabled" Services

Fortunately for us, the Internet itself at its essence is glorified, resurrected military technology from the 1970s. The idea at that time was that if the Soviets got mad enough at us, all of our telecommunications hubs might quickly find themselves in the upper atmosphere. This presented an obvious problem for the war planners of the time. What became necessary was a means of communications that depended not on "hub-and-spoke" technology but lots of different paths. (In a true hub-and-spoke arrangement, where the hub is taken out, nobody else can talk to anybody—see Figure 7.3). What was also necessary was a protocol that was "a little bit smart" in that it did not have to depend on intelligence in the hub to route messages. One descriptive term is "distributed intelligence with nobody in charge." (This is kind of like my office!) This concept became IP.

If our military spending has ever generated such a thing as a peace dividend, IP would have to be a top contender. If your organization uses VoIP, for example, the IP gives your phones a high degree of recoverability. It is possible to unplug an IP phone and plug it in again anywhere that high-speed Internet service exists and begin answering your calls again. Every hotel today is installing high-speed Internet. That's a lot of potential recovery centers, isn't it? Give some thought to some of these ideas!

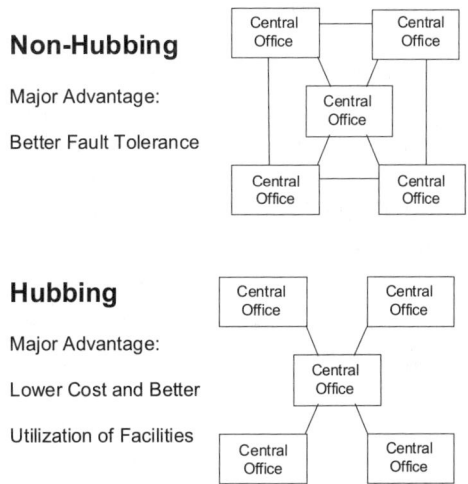

Figure 7.3 Understanding the telco network. (Reprinted with permission from *The Definitive Guide to Business Resumption Planning* by Leo A. Wrobel, Jr. © 1997 Artech House Publishers.)

How well a CO survives a major facility disaster depends largely on how it is engineered into the overall network. Which would you consider worse: a disaster in your own building or one in the telephone company's CO? I submit to you that a total loss of your building is not the worst-case disaster, even though many very good recovery planners make that assumption. The worst case is the total loss of a major telecommunications facility.

If you lose your whole building, say, to a fire, hundreds of vendors and employees will probably pitch in to help you recover. That's because even though your building is gone, hundreds of other unaffected locations and vendors can help you. When the telco loses a building, exactly the opposite is often true. Hundreds of businesses are dead in the water, and only one of them is fixing the problem. Moreover, loss of a major telecommunications center is not as farfetched as you might think. If I understand the vulnerability, chances are a lot of bad guys and terrorists do, too.

There have in fact been dozens of cases of telephone CO disasters over the years. To cite just a few that I am aware of:

- Lightning hit an oak tree 75 yards from a CO in Granby, Massachusetts. It destroyed the central office.
- A broken levee caused a CO to sustain 1.5 feet of water in its switch room—on the second floor.
- An employee for a competitive local exchange carrier (CLEC) in Dallas and his switch vendor inadvertently turned the battery bank of a CO's switch into a 200-amp arc welder while changing connections that were still hot. This forced an evacuation when all of the wires fried and the building filled up with a nontoxic smoke.
- Let's not forget, during Hurricane Katrina, some 34 COs went under water; something like 12 were totally destroyed and dozens more were knocked out to other causes.

It's not just fire and brimstone, either; software causes big problems, too. An improperly installed software upgrade in the AT&T SS7 network blocked 100 million long-distance calls in a single day in 1990 and cost businesses nationwide millions of dollars. It could happen again. In fact, bad guys everywhere would just love to do it. Is a computer virus possible in the SS7 network? It's a question that is worth asking. Some of you might or might not know that the World Trade Center had a CO inside it. What people might not remember is that September 11, 2001, was not the first time the World Trade Center had a disaster. In 1993, someone set off a bomb in the World Trade Center that disrupted communications and knocked at least one large securities broker off the air for a few days. The effect was not as bad as it could have been, however, because that company had a plan. Others in the same World Trade Center building did not. We need not even remind everyone what happened there again eight years or so later.

This list goes on and on, but I think you get the idea.

The causes of CO disasters over the years have run the gamut, including fires, floods, lightning, tornadoes, hurricanes, switch failure, and human error. Consider that one of the most far-reaching CO disasters in U.S. history was on May 8, 1988, in Hinsdale, Illinois. This fire, which started in the main switching room of the CO, the largest switching system in Illinois, affected not only 40,000 local telephone subscribers, but also over half a million other users nationwide (due to disruption of "access" and "tandem" lines discussed later in this article). It took two weeks to completely restore services. All on Mother's Day, the busiest calling day of the year.

Wireless providers are also not impervious to outages. In 2007, one unnamed wireless provider cut off cell-phone service to thousands of customers for 5 hours because a critical piece of wireless connectivity equipment that connects the entire cell sites to the switch had failed.

The same kinds of things happen on the Internet, such as in cases where there are weather disruptions and people stay home to work. Residential areas are not always engineered to take the unusual load, particularly out in the suburbs. That's why the Internet slows down and voice calls carried by the Internet start clipping under those circumstances. In addition, CO disasters can involve a large area and are not limited to dial tone only. Data and voice private lines, circuits, wireless, Internet, special access, paging, long distance, 911 emergency trunks, and other services may also be involved.

Protecting Against Cable Cuts

Closer to home, the cable cut is the most common cause of telecommunications disruption today. In my 30 years in this business, I have yet to find another telecommunications user who has not experienced one. Since planning for something that will happen (rather than what might happen) should be high on the list of any contingency planner, here are a few things you should know about why cable cuts happen and how you can prevent them.

First, consider where everyone lays cable. Virtually all cables, whether fiber optic or copper, utilize public rights of way. Unfortunately, in many cities this same right of way has been in use for many, many years. When someone starts digging, they never really know what they are going to hit on the way down. The issue is even more complicated when one considers that gas, electric, and sewer lines traverse the same rights of way. Gas lines get cut; so do electric lines. Somehow I don't see a lot of sewer lines get cut. Ever wonder why?

Everyone has his or her favorite cable cut story. My favorite happened years ago when I was at AT&T and thousands of people in Dallas lost service for most of a business day. A farmer had dug up the main AT&T fiber optic route while burying a dead cow. Consider the supervisor in New York City who told his technician, "There are two fibers down there in the manhole. Cut the bad one..." New York was isolated for most of the day. Another guy on a backhoe started digging and hit a concrete encased cable about a mile from my office. He assumed the concrete was white rock. He started banging away. First he cut the telephone cable, then broke the water line, then cut an electrical cable. When he heard the telltale hiss of the gas line, he got off the backhoe and ran like hell, right before the huge explosion. It looked like the set of one of the Die Hard movies. We were down several days due to that mishap!

So is there anything you can do, or are these events completely beyond the contingency planner's control? The answer is, there are things you can do. The first step might be a meeting with your local telephone company. There are many telephone companies to choose from so be sure to include both incumbent local exchange carriers (ILECs) and CLECs. They can provide you with helpful information if they take the time to do a little homework on your behalf.

Many times alternate or diverse cable routing is available in your company's service area for little or no cost. Be careful though, because a lot of people use the terms alternate and diverse interchangeably (see Figure 7.4). They are not the same things! Alternate is anything other than what you have now. Diverse implies a totally separate path. Alternate might mean a different "100-count" in the exact same cable. Diverse should mean at a minimum a completely different cable, and if the engineer knows what he or she is doing, a completely different right of way, too. Other technical terms you can use to converse with these folks include count diversity. Count diversity is where different 100 counts in an existing cable are utilized. So what good is that, you might ask? The answer is not much, except for one case. Have you ever had a circuit or number of circuits disconnected in error? Count diversity in some situations could reduce the possibility that disconnection will happen, since the cable appearance will be on two separate physical connection blocks in two different locations on the distribution frame in the CO. Another common term is sheath diversity. Sheath diversity uses separate cables that are still in the same conduct and manhole. Other types of diversity can include things like nonadjacent cable ducts (see Figures 7.5 and 7.6).

Who Is at the Other End of Your Cable?

For most companies, the first point of physical vulnerability other than the cable itself is the local telephone CO or their area. In most cases, all data and voice traffic

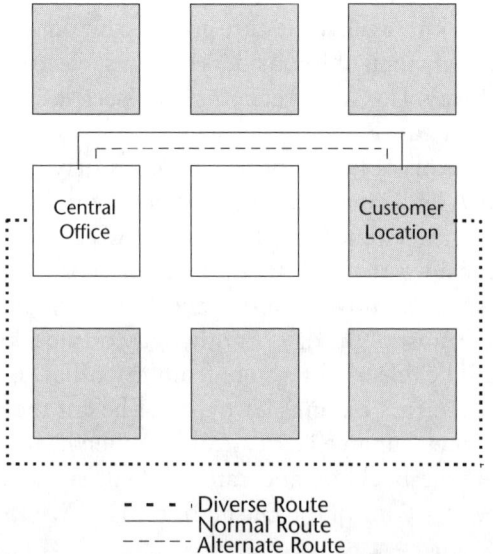

Figure 7.4 Know the difference between alternate and diverse cable routing. (Reprinted with permission from *The Definitive Guide to Business Resumption Planning* by Leo A. Wrobel, Jr. © 1997 Artech House Publishers.)

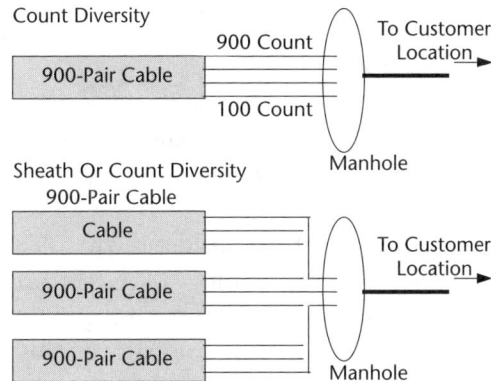

Figure 7.5 Cable diversity from the local exchange carrier (LEC). (Reprinted with permission from *The Definitive Guide to Business Resumption Planning* by Leo A. Wrobel, Jr. © 1997 Artech House Publishers.)

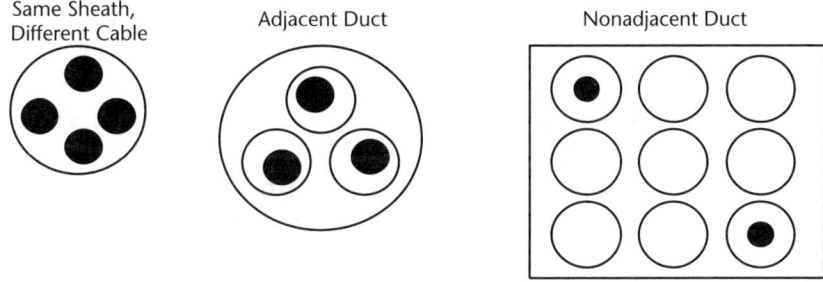

Figure 7.6 Diversity from the LEC. (Reprinted with permission from *The Definitive Guide to Business Resumption Planning* by Leo A. Wrobel, Jr. © 1997 Artech House Publishers.)

must pass through this location as the first leg to wherever it is going. The physical capabilities and configurations of these local offices vary greatly.

Physically, most COs are sound structures constructed of reinforced concrete. Because designs vary, you should drive by the CO or schedule a tour with the carrier. Look for the obvious. Does the structure have a large portion of the surface area covered by windows? Does it appear to be in an area prone to flooding? Is there major construction activity planned or occurring in the immediate area around the CO? Ask some questions of your telephone company's account representative.

1. Is the CO manned 24 hours a day?
2. What type of fire prevention systems, if any, does it employ? How old is the structure?
3. Is the CO a tandem or end office? End offices serve only end customers. Tandems switch traffic from one end office to another. Many COs today perform both functions. This could have a bearing on which systems get recovered first in a disaster, as well as how many people might be affected in a metropolitan area. For example, loss of a tandem all the way across town could still affect you by contributing to network congestion and blocking your calls.

4. What plans exist for restoration in the event of fire and flood? What services would be affected?

5. How difficult or expensive would it be for facilities to be brought in from another CO? Occasionally bringing in alternative facilities is reasonably straightforward, but often it is prohibitively expensive, if not impossible. Sometimes diverse facilities exist from CLECs or the local television cable company. Don't forget to consider them.

6. Try to get an idea of the topology of the entire city's telephone network. Is it a hub-and-spoke or peer-to-peer configuration? Hub and spoke is more economical for the telephone company because facilities are better utilized. It is not as reliable, however, as peer-to-peer, where alternate routes for switching are more likely to exist. Most incumbents these days are evolving to peer-to-peer topologies just like the Internet has evolved, contrasting greatly with the hub-and-spoke networks of the past.

7. Ask about the location of the regional serving offices and access tandems. You may find that many CLECs and long-distance companies are congregated in just a few "carrier hotels" or that they derive all of their connections to the Bell network in the same access tandem downtown. This is not always the competitor's fault. Often the monopoly incumbent makes it difficult for competitors by making switch diversity, route other than normal (ROTN) cable routing, fiber-optic ring technology, or "dark" fiber cost prohibitive or difficult to get. You will obviously get two different stories on this issue, depending on whether you are speaking to the incumbent monopoly or competitor.

Where Does One Start?

The size and density of the serving area for the office varies substantially from serving area to serving area. So where does one start, and how does one prepare? First, you might consider taking a drive to find the site of the local CO for your area. It should be no more than a few miles away, with very few exceptions. In cities, it may be only blocks away. Out in the country, it could be miles away. Suffice it to say that probably 90% of them are within 18,000 cable feet or less of a subscriber, or less than 3.4 miles. Once you have located it, drive the route between it and your company location. Look for construction, digging, or other activity in the right of ways along the street. After a while, if you are so inclined, you will even learn which boxes and pedestals contain fiber-optic equipment, xDSL equipment, T1 repeaters, and other components. It is not like any of this is going to help you do anything about potential disasters, but if you spot a big construction crew working adjacent to your company's fiber terminal, you can at least have some possible warning that something may happen. Another easy way to become familiar with the service area of a particular local CO is to pick up a local telephone directory and look at the first few pages. Sometimes, but not always, there is a diagram that illustrates the local calling area and exchanges served in the immediate area (see Figure 7.7). The scale may not be exact, but it gives a good indication of the relative size of the service area. It might even list the telephone prefixes (NXX codes) for the area. For example, by looking at Figure 7.7 (specifically at the 6 o'clock position), notice that a telephone subscriber with a 223 telephone prefix is served out of the DeSoto CO. By comparing this diagram to a regular city map, it is possible to get an idea how many other companies are served out of your area's local telephone office. This could have a bearing on recovery time in the event of a severe CO outage. You can also get an idea of how far you would have to drive to find a working phone if the CO is destroyed or incapacitated. Beware of a few changes since we first published these tips in earlier books. With the advent of local number portability (LNP), you might actually find "223" numbers (as in this example) in totally different parts of town. LNP allows subscribers to take their numbers with them where they move. Therefore, be sure to look at a prefix that is "native"—assigned primarily—to the CO that you are evaluating!

While the diagram in Figure 7.7 one is a good starting point, there are other better and more detailed references available. One such source in the United States

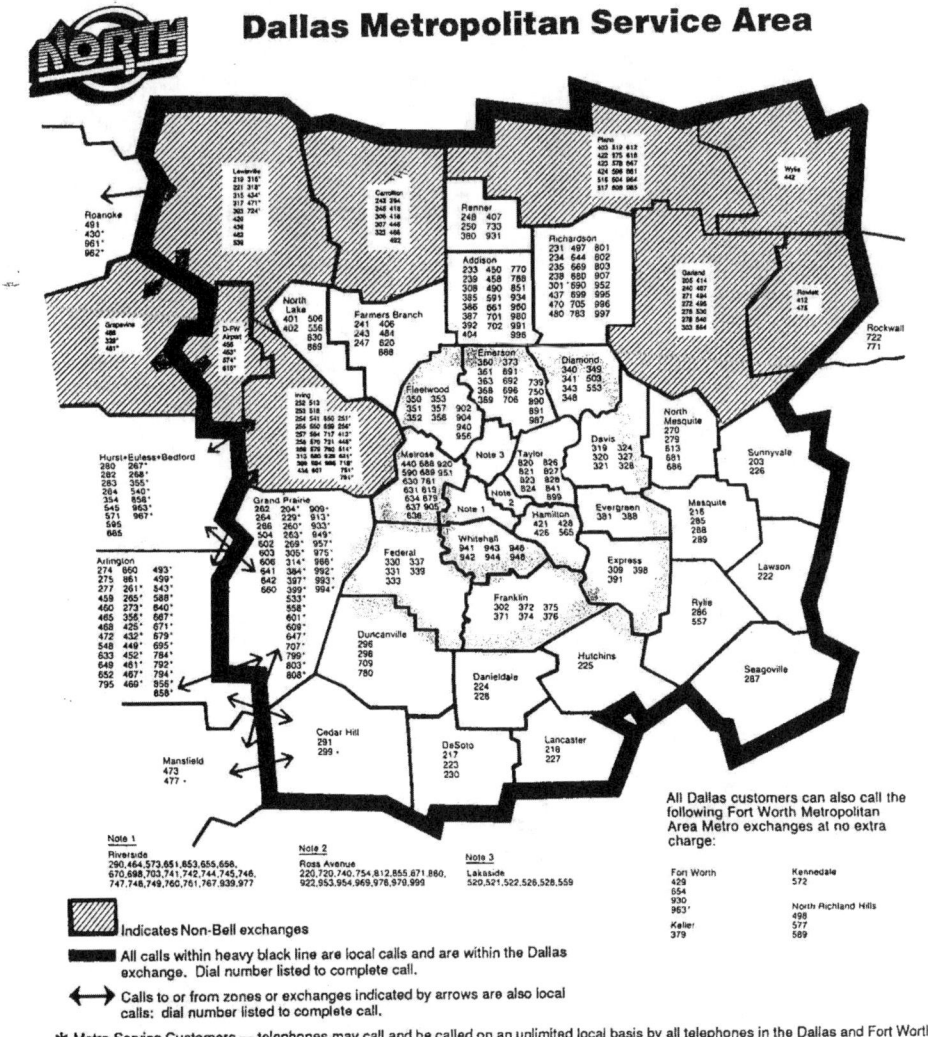

Figure 7.7   Map of the Dallas metropolitan exchange area.

is the public service commission (PSC) for each state. The PSC is repository of a wealth of information on this subject for those willing to some digging—in the sense of research, that is, not backhoes.

Local exchange companies are required to file detailed maps, diagrams, and other information regarding their franchise areas with the individual state's PSCs.

Useful data on file can include the following:

1. Definition of the serving area for every CO in the state, down to the particular street. In Figure 7.8, for example, you can see that the DeSoto CO serves not only the city of DeSoto, but also Glenn Heights and portions of Ovilla as well.

2. Distances to other serving offices can often be approximated through the use of available documents, which could be important for companies considering diversity through "special construction" of facilities at another serving office. Quite naturally, the cost of such a project will vary with the distance and availability of existing cable or fiber.

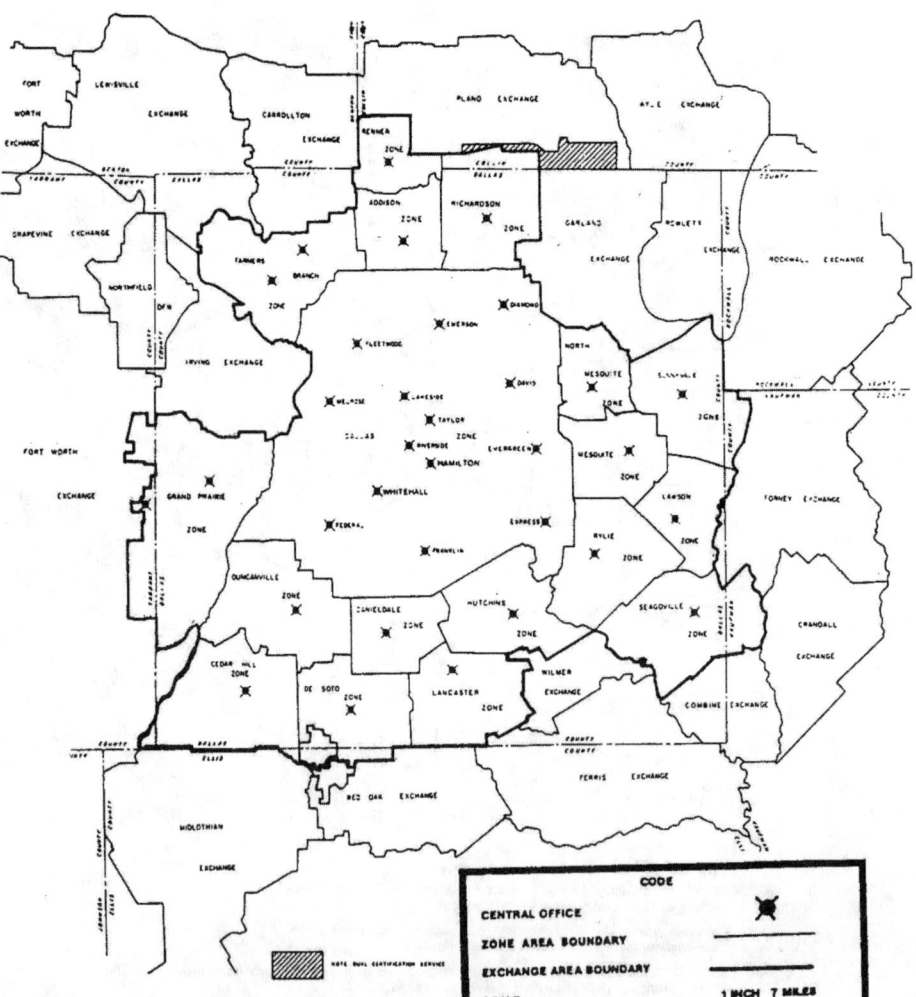

Figure 7.8   Map of the Dallas zones and base rate areas.

3. Distance is not the only consideration. The closest CO, for instance, may not contain the technology needed by the customer. A different local operating company—a difficult prospect in both the construction phase and in future circuit coordination—may operate it.

Information can be secured from the public service or public utility commission for the cost of making copies of it. For the address of your commission, take a look at the website of the National Association of Regulatory Utility Commissioners (http://www.naruc.org) and click the button at lower left that says "State Commissions."

Wireless Alternatives for Diversity and Disaster Recovery

Many wireless technologies exist today to diversify networks against the all too prevalent cable cut or to recover after a disaster. Wireless technologies that are useful to the enterprise user for network diversity and disaster recovery include the following technologies:

- Infrared point-to-point links;
- Microwave radio;
- Satellite communications;
- Point-to-multipoint systems (unlicensed).

Each of these technologies is comprised of inherent strengths and limitations. The application for the technology you choose for backup (voice, bursty data, Internet, and so on) also plays a role in selection of a technology. Let's begin with infrared links.

Infrared Point-to-Point Links

Point-to-point infrared links are not really radio but invisible light, a conceptual cousin to the infrared remote control for your television. Advantages include the fact that they are inexpensive, do not need to be licensed, and come with a variety of interfaces including T1 and Ethernet. Infrared requires line of sight (i.e., one end must be physically visible to the other end of the link). Since these systems operate in a much higher frequency range than other alternatives (light, even infrared light, operates in the tetrahertz range, where radio operates in the megahertz or gigahertz range), there are distance limitations as well. Infrared is also much more affected by fog, rain, snow, birds, and, practically speaking, anything that will interfere with the proprogation of light.

On the other hand, the equipment is compact and does not require licensing. The equipment can easily be mounted in a building, does not require any special power or environment, and the transmitter/receiver can operate through window glass with few problems. If you have an application that requires you to get a T1 across the street or across a small campus, infrared may be your least expensive solution. These are also often used for LAN interconnection in buildings that are in close proximity but separated by public rights of way (like streets) where cabling between buildings is impractical. If you consider the use of an infrared link yourself, be sure you don't exceed a mile or so (less if you are prone to periodic fog or heavy rain) and that you have clear line of sight.

Interestingly enough, I have run into several cases where the line of sight becomes an issue. One user we worked with had clear line of sight to a neighboring building all winter long. In spring, trees grow, they get leaves on them, and, in the case of this user, they block out your infrared link. Another user lost service periodically every time the window washer came by on their high-rise building. This only happened now and then, so it was no big issue. On one such pass, the window washer just happened to leave a large streak across the window. As luck would have it the streak refracted the light just enough to cause problems with the link. They had to get the window washer back out the next day to clean off the streak. Despite the limitations, the cost, diversity of applications, ease of use, and fact that a license is not required still earn infrared links good marks.

Microwave Radio

Microwave has broad applicability, high reliability and availability, relatively good ease of use, and relatively low cost. On the minus side, you do need a license to operate most microwave systems, and the more popular frequencies are congested and difficult to get licensed, especially in the major cities. Microwave started originally as a telephone company technology. The telcos especially liked the ability to span bodies of water, valleys, and other terrain without having to lay cable. Indeed, up until the 1980s AT&T still had microwave links in its network (maybe it still does), some of which spanned from mountain top to mountain top at distances of 30 miles or more. Like infrared, microwave requires line of sight. This is problematic within major cities. If you are lucky enough to get a frequency licensed, all of your work could be undone because someone builds a building between the two points on your microwave link. This occurs more often than you might think.

Microwave enjoyed most of its popularity in the 1980s as a "bypass" alternative to the local telephone companies. At that time, the "last mile" of circuits (known as loops) got expensive. At the same time, long distance got cheap. The logical response of enterprise users was to dump the local telephone company and use microwave to connect directly with long-distance carriers of the time, like MCI and Sprint. While the financial motive was the primary driver, it only took the enterprise user until their next cable cut to realize that microwave also had use as a disaster recovery technology. Microwave provides the ultimate diverse route because one cannot dig up air (Figure 7.9).

As a general rule, the higher the frequency of a microwave link, the shorter the path must be. I have personal experience with installing a 2-GHz (gigahertz or billion hertz per second) link a number of years ago. This link was rock solid on an 8-mile line of sight for years until the 2-GHz band was phased out by the FCC for another wireless technology. Afterwards, it was replaced by a 6-GHz link that boasted comparable reliability. Another popular frequency is 23 GHz, but this has

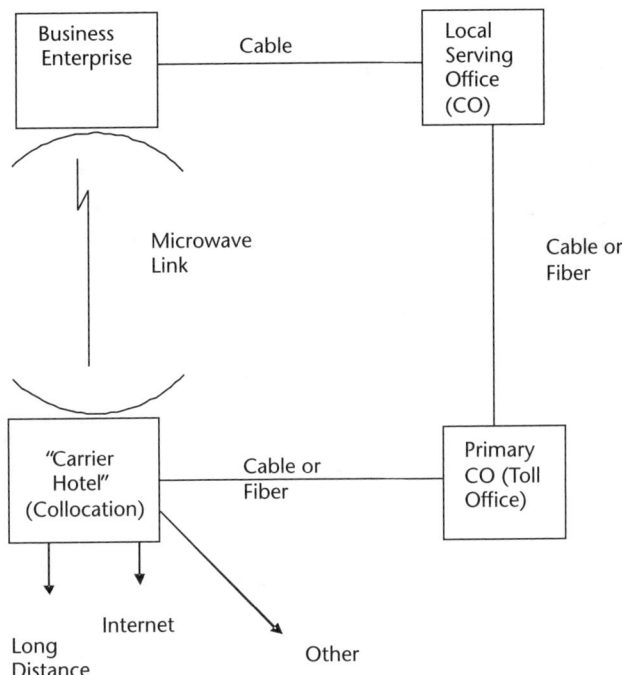

Figure 7.9 Use of microwave facility diversity and carrier collocation in a complete enterprise telecom solution.

more of a tendency to "wash out" in a heavy rain since a higher frequency is more easily absorbed. Even so, it has been more than a few years since I had actual hands on experience with these technologies. I would not be at all surprised if 23-GHz performance rivaled what we achieved years ago with 2-GHz systems simply due to advancements in microprocessor and antenna technology. Check with the manufacturer and stay within the design limits of the equipment and you should be fine.

As microwave has gained favor with enterprise users over the years, the feature richness and reliability has increased. Many systems today employ fault protection switching, which is in effect two radios in one. When one fails, the other cuts in almost instantly with no interruption in service. There is also a greater choice of interfaces today, with Ethernet and T1 interfaces commonplace.

As stated earlier, however, you will be required to secure an FCC license to operate a microwave system. The manufacturer can help you do this, and there are also numerous consultants that can be found in the yellow pages to help with the same. If you are looking for true diversity at reasonable cost and at higher reliability than infrared, microwave may be the ticket.

Microwave is a useful technology to diversify cable and fiber. It is difficult to dig up air!

Satellite Communications

This discussion would not be complete without a brief overview of satellite communications. Like the previous two technologies, satellite communications have taken leaps and bounds over the last few years. Satellite is essentially microwave radio aimed upward—it uses essentially the same frequencies. As such, the same rules hold true regarding tendency to wash out. There are also two times every year when the satellite receiver will be aimed directly at the sun, right around the spring or fall equinox. At that time there will be a brief outage. These outages can be planned for, since the service provider will know precisely when they will occur.

Satellite has gone from elaborate teleports and 16-ft dishes in years past to pizza pan–sized dishes that fit on the side of a building, to units that literally fit in your hand (in the case of GPSs and freight-tracking technologies). Notwithstanding timing delays (it takes a significant fraction of a second for the signal to go from the Earth, 22,300 miles to a geosynchronous satellite and the same distance back), which the service provider can explain to you, satellite is a clean, reliable, and cost-effective disaster recovery solution (see Figure 7.10). For more information on this medium, see Chapter 5, where we have dedicated an entire section to it.

Point-to-Multipoint Systems and Wireless Internet Service Providers

The challenges of setting up an effective primary and disaster recovery system have always involved tradeoffs between cost, complexity, reliability, and time (such as in licensing). Concerns about designing, installing, and maintaining complicated networks and unfamiliar technology also drive the selection process. In many ways, this makes the newest entrant of the technologies discussed in this course,

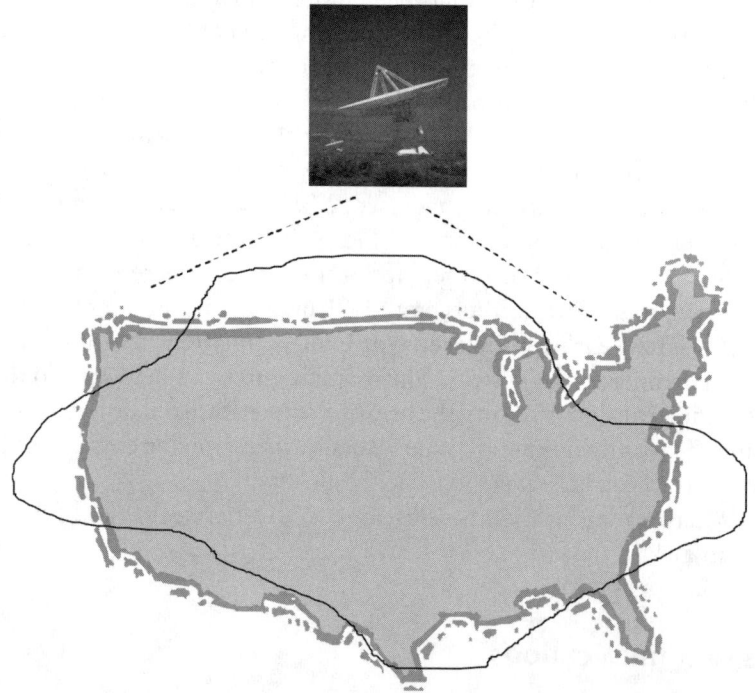

Figure 7.10 Satellite service has the advantage of the ability to deploy anywhere within the coverage area, or "footprint" of the satellite.

point-to-multipoint radio systems (P-MP), the technology of choice. This technology is used to serve wireless Internet service providers (WISPs). P-MP marries microwave radio technology with enterprise communications applications and in many ways makes deploying and delivering telecommunications technologies of all types faster and easier than ever before. P-MP provides the performance, versatility, ease of use, and affordability that enable enterprise environments—including corporate, municipal, healthcare, education, and more—to improve communication, productivity, security, and return on investment (ROI). This technology is also becoming widely used as a disaster recovery technology. Indeed, two users we are familiar with, both of which are county governments, have scrapped their AT&T T1s altogether and now use P-MP as the primary technology, with a few T1s held back as the backup path. Here is how the technology works: P-MP products operate in the 900-MHz, 2.4-, 5.1-, 5.2-, 5.4-, and 5.7-GHz frequency bands. Since these frequencies are lower than many microwave frequencies, wash out and restrictions on range are not as much of an issue. Like the other technologies, a variety of interfaces are available, including T1 and Ethernet. Startup costs are low. We have seen central unit costs as low as $2,500 and "per rooftop" costs in the $500 range. Typically a small antenna (about a foot long and 4 inches wide), or in cases where the range is greater a small dish about the size of a satellite TV antenna, is installed on the roof.

Obviously the need to run cable is obviated. The equipment also does not require a FCC license. The equipment itself is streamlined, with the radio built into the antenna in the same 12-in x 4-in x 2-ft unit on the roof. It's very easy to get up and running. Most P-MP platforms also include the most common interfaces that enable them to easily integrate with standard network management tools and systems. Any system that traverses an airwave should be encrypted. Look for a system that provides security with over-the-air data encryption standard (DES) encryption or advanced encryption standard (AES) encryption capabilities.

P-MP systems serve numerous enterprise locations of virtually any size and can be used for distances up to 15 miles (24 km). Point-to-point links can traverse much greater distances by augmenting the antennas at both ends (a dish similar to a microwave dish is used in these cases) and in fact approximate the "mountain top to mountain top" links described previously. Most P-MP systems require line of sight, although some of the ones that use the lower frequencies (like 900 MHz) do not. The lower frequencies, however, generally limit throughput to T1 speeds (1.544 Mbs) if you are lucky.

To summarize, P-MP systems, in the opinion of this humble author, represent the best tradeoff of cost, performance, ease of use, and variety of interfaces available to the enterprise user seeking disaster recovery and network availability. They are easy to use, flexible, have a wide variety of standard interfaces, and don't require a license. They also provide higher speeds than T1 and are less expensive.

Consider Collocations and Carrier Hotels

What exactly are carrier collocations? Collocations, also known as carrier hotels or collos, are exactly what the name implies. They are generally well-hardened facilities and almost without exception allow carriers to connect with one another under day-to-day or emergency operations. In a disaster, they can be indispensable. A typical collocation can house one or more of the following kinds of technologies:

- Cable and fiber-optic connections to the ILEC like AT&T, Qwest, or Verizon;
- Connections to all major long-distance companies;
- Connections to CLECs;
- Connections to wireless providers;
- Connections to WISPs;
- Connections to VoIP carriers;
- Connections to satellite communication providers.

It is most prudent to position your organization in advance to where it can avail itself of any of them, depending on the circumstances. If connections to the ILEC went down, for example, connections could be quickly purveyed from a CLEC or long-distance provider. If everything went pear shaped, like it did during Katrina, satellite and VoIP might still survive. You would obviously need a place to connect to them and again collocation fits the bill. Satellite is often the only link left to a twenty-first century communications infrastructure after the largest disasters such as hurricanes, earthquakes, or tsunamis, and it's particularly effective when combined with other technologies like the IP.

The IP, you will recall, dates back to a time when U.S. war planners envisioned scenarios where every AT&T primary central office would be in the upper atmosphere due to a nuclear attack. The IP was designed so packets of information could bypass these lost hubs and get through on the facilities that survived in the outlying areas. This technology is equally useful today when major communications hubs are affected by large disasters or even terrorism. Even better, the technology allows voice to be carried, hence the term voice over IP. After all, voice is really data. Data is data, too. The same bit stream is used to carry voice or data, whether it's 64 Kbps, 1.544 Mbps, or another speed. Like the old chicken commercial, "parts is parts." In the case of telecom capacity, "bits is bits," and these bits can be used to carry voice or data. In this regard, carrier collocations are literally the Kentucky Fried Chicken restaurants of cyberspace.

When you sum it all up, it is possible to maintain 4Ci on both a data and voice basis by cobbling together surviving transport communication facilities, then employing IP. There is no better place to do this than in a collo. Interestingly enough, this is precisely how the military planners of the past envisioned the IP would be used, except now you can carry voice over IP. Today, civilian organizations demand a military level of disaster recovery.

Summary and Take-Aways from This Chapter

Here is a quick review of considerations on what will happen after a large telecommunications disaster affects you. (Notice we did not say "if" it affects you.) Pay particular attention to the following services, particularly those that can be accessed easily in a collocation:

1. Paging systems, including two-way paging, should work because they are often satellite based. This includes Blackberries.

2. Satellite services, in general, should not be impacted. You should at least know where to connect to a satellite carrier (e.g., in a collo) if you do not use this technology day to day. If satellite is your primary technology, however (such as if you are a TV network), you must of course plan for a backup in the event of a satellite outage.

3. WISPs should be back on the air relatively quickly, since there is no licensing requirement and the equipment is inexpensive and portable. WISPs often beat the phone companies in establishing Internet connectivity to affected areas. If you have access to the Internet, you have access to IP. If you have access to IP, you can restore 4Ci in terms of voice and data. WISPs can often be found in collocations.

4. Anything traversing a cable (aerial or buried) has strong probability of being affected for reasons explained earlier. Check into route diversity on cable facilities. Many different providers of wireline services can be found in the typical collo.

5. Electric power will be impacted if the experiences of New Orleans and Houston are any indication. Plan for backup power, including –48 volts for telephone and PBX equipment. Also plan for a reliable fuel supply for your generator. Why make a big capital expense for these when they come with the monthly rent in a carrier hotel?

One final tip: If you are an essential service (hospital, government, and so on), look into Government Emergency Telecommunications Service (GETS), Wireless Priority Service (WPS), and Telecommunications Service Priority (TSP). These government-sponsored programs allow for priority when phone lines are saturated and wireless frequencies fill up after a major disaster. They are described in more detail in an insightful white paper at the end of this chapter. One big surprise from Hurricane Katrina was just how few qualifying agencies and essential services actually signed up for these restoration priority programs. Don't wait until you need them—look into them now.

Summary

There is a lot more to learn about protecting telecommunications networks that could go into a chapter like this one, but unfortunately we don't have close to all the room we need. We end this telecommunications section with something rather unusual that was not included in any of our previous books. Have you ever considered the impact of a pandemic on telecommunications? TeleContinuity, Inc., offers a fascinating insight into this latest vulnerability to telecom and offers some great tips as to what you should be doing about it.

Maintaining Telecommunications During a Pandemic: An Interesting Perspective, by TeleContinuity, Inc.

Introduction

Business continuity plan (BCP), continuity of operations plan (COOP), and telecommunications planners and executives face a different challenge when planning and preparing for an influenza pandemic. The H5N1 "bird flu" virus is of immediate concern because it has already shown itself in parts of Asia and Europe. There is fear that it may bring sickness and death on an unimaginable scale, devastate cities and towns, and bring business to a halt. While the H5N1 virus has killed only a few people, there is great fear that it will mutate into a form that is easily transmissible between humans resulting in a global pandemic. The 1918 Spanish flu epidemic swept around the globe, resulting in 40 million deaths in just three years. The government estimates the potential for this flu pandemic to be up to 207,000 deaths, 776,000 hospitalizations and outpatients, and a staggering $166 billion in economic losses.

This section discusses the practical problems of maintaining operations when people can no longer meet in their usual locations and may have to work from multiple locations or homes. This paper examines the current telecommunications infrastructure, its weaknesses, and the keys to a successful telecommunications plan that has the flexibility and survivability to maintain a company over many months of dislocation.

The Pandemic Will Come in Waves

According to the U.S. Homeland Security Council, the pandemic will come in waves, each one lasting months, and will cause massive absences and dislocations. The Financial Services Sector Coordination Council (FSSCC) states, "Staff absences may increase to 40% as companies deal with illness, family demands, or fear of contagion. The current 'all-hazards' approach to BCP may not be sufficient for the organization to function during an outbreak of 'avian flu.' Even backup remote sites hundreds of miles distant may be just as affected by the outbreak they are intended to back up. Current plans are for cities to lock up buildings where people gather and close streets to vehicular traffic. Millions of people will flow from the central cities to outlying areas, straining the resources of these outlying areas."

Pandemic Threat—What Makes It Different

A pandemic will have a direct impact on a company's most valuable resource: people. It will affect staff, executives, emergency service workers, regulators, customers, suppliers, service providers, and so on. Contingency planners have developed a strong expertise on protecting buildings and dealing with regional and technical outages. In planning for the unique "people" outage that will occur in a pandemic, planners will have to deal with the following:

- Unprecedented rate of absenteeism due to illness, caring for others, and psychological fears—up to 40%;

- Unpredictability of who will be unavailable—will it be executives, critical staff, or workers necessary to maintain production and operations?
- Uncertainty and soft planning assumptions—how long will employees be unavailable? The normal turnaround period for a healthy person is two weeks, but the widespread effects of the pandemic will force critical people to stay home to nurse sick family members.

A pandemic makes it impossible to move business operations out of harm's way and to obtain support from a company's remote branches and locations. Vital government and community services like schools, transportation, banking services, and emergency response services will also be impacted, rendering support from these services limited or unavailable.

The U.S. government projects that the upcoming pandemic will probably appear in three waves, each lasting 8–12 weeks. For business, continuity, and contingency planners and company executives, this is an extremely long time in which to manage a crisis—especially a crisis that reaches far beyond the local area and whose impact will contain many unknowns.

Strategies to Meet This Threat

The pandemic will have a direct impact on the workforce. Every company will need to incorporate a special measure of care toward employees with the aim of providing a reasonably safe workplace by taking measures to reduce the spread of infection. Three ways to accomplish this are by

- Organizing a split workforce that includes provisions for home working;
- Establishing social distancing practices (separating the employees and groups);
- Developing a viable solution for maintaining telephone voice continuity.

Pandemic Preparedness: The Telecommunication Challenges

No matter what the mode of telephone communications a corporation uses—landlines, cell phones or VoIP—a pandemic will create a crisis in telecom connectivity. While the carrier's physical facilities may be undamaged, the dislocation of thousands or even millions of workers from their usual work location to remote sites or even their homes will put tremendous strain on the carriers' telecommunications infrastructure—both telephone and Internet.

Three problem areas will make telecommunications extremely difficult. The company's difficulties, the carrier's imbedded infrastructure problems, and the carrier's potential personnel problems may well all adversely contribute to the situation.

Pandemic problems are unique in that they may not impact the physical telecom circuits, PBXs, switches, or central offices. The problems will occur because of carrier personnel absences and the shift of data (IP) and voice traffic from downtown areas where the carriers have installed sufficient capacity to outer and residential areas where the carrier infrastructure is not robust enough to carry business traffic because it has been sized to carry only light residential traffic.

Under normal conditions, call traffic begins when the calling party lifts the telephone and dials the number of the person being called. The call is connected to the caller's local CO where the carrier's SS7 network latches together telephone lines to pass the call to the called party's local CO. That CO identifies the specific circuit that will send the call directly to the called party's desk telephone. If it is a cell phone call, the call is sent to the called party's local cell tower from which it is trunked to the person's cell phone. It is important to remember that cell towers, like COs, are clustered according to the usual traffic density and are thickest in downtown and other business areas and sparse in residential areas.

VoIP calls are calls placed to or from a computer that travel over the Internet to another computer, landline phone, or cell phone. VoIP calls utilize a soft phone to translate voice into packets that are sent over the Internet to the called party's telephone or computer soft phone. VoIP calls compete for space over the Internet circuits with the data that is also being transferred over the Internet—a source of potential voice quality problems during the pandemic. Also, voice over the Internet, unlike data, is extremely sensitive to delay, jitter, and packet loss during transmission. Computers can wait for data to arrive without displaying any problems, but packet loss, delay, or jitter (common Internet problems) can make the VoIP message unintelligible to the listener as sounds and letters fall out of their proper sequence.

Normal Types of Telecom Outages

Companies can suffer telecom outages whenever their PBX fails, people are displaced, the cable connecting the company to the local CO is cut, or the company's local CO is shut down. This span between the company's telephones and the CO is called the last mile of telephone service. This last mile is the most fragile part of the telephone system because there are many single points of failure between the CO and the user's telephone instrument—single points that can cut off all telephone communications.

During natural disasters or terror attacks, telecom outages are caused by the destruction of key points in the last mile such as the COs. The FCC reported that nine COs were destroyed during Katrina, 141 suffered damage and outages, and 214 million incoming calls were blocked. Cell towers were also severely affected, as well as both above-ground and underground transmission cables. In addition, companies and their personnel may be displaced to planned or unplanned remote locations because of damage to their building or the building's critical service structures such as HVAC or telecom switches, PBXs, and connections. The result is that staff and executives are no longer at their normal business telephone numbers—and if the cellular system suffers tower, cable, or computer damage (as is usual during severe weather or other natural disasters), cell phone communication will be disrupted, cutting off all incoming and outgoing telephone traffic. Cell towers can handle only tens of simultaneous calls, and during any large-scale disaster cell towers become easily overloaded with telephone traffic, blocking calls and giving callers fast busy signals.

Telecom Problems Caused by a Pandemic

As companies voluntarily move to remote facilities or are forced to close their offices by the health authorities and shift toward tele-working, most will immediately activate their call forwarding to move calls to their desk phones, to their cell phones, or other phone devices. This will result in a tremendous strain on the carrier's forwarding services and infrastructure. Forwarding takes place in the company's local CO switch, which means incoming calls have to go all the way to that switch before being rerouted to a new CO servicing the employee's new location. This will mean doubling the number of lines used in connecting each call—possibly creating a severe shortage of available call handling lines—resulting in many blocked calls.

Other causes of blocked calls or fast busies will be the overwhelming congestion in carrier infrastructure as facilities designed to handle light suburban and residential call traffic are now inundated by very heavy and consistent business traffic. Carriers do not have the density of COs, switches, and cables capable of handling the sudden increase in traffic. Adding to this problem is the parallel leap in Internet traffic that will accompany the move of business to remote locations and home operations. Infrastructure will be overloaded, causing a massive backup of both voice and data traffic—and that is when circuits are available. For people shifting to VoIP, other problems may derail their efforts at voice communications. Real-Time Transport Protocol (RTP) and Session Initiation Protocol (SIP) may be blocked by firewalls or network address translations (NATs), while packet loss, jitter, and delay from congestion may cause conversations to become unintelligible.

Of course, as discussed earlier, peoples' first moves will be to go to call forwarding for their desk phones and first shift their calls to their cell phones. As the workers move away from the city or other business areas, the increase in cell traffic will overwhelm the cell system, which has not been provisioned to handle the rapid increase in call traffic.

The pandemic of 1918–1920 attacked the United States in three devastating waves that lasted three years. The Homeland Security Council is predicting the next pandemic will also arrive in multiple waves, with each wave lasting for many months. Imagine this new series of pandemic attacks lasting months or years and the disastrous strain it will put on all the systems that support modern business. With workers suffering from the flues' effects and central operating locations locked up, the carrier's ability to install new infrastructure and adjust their facilities imbalance will be severely constrained. Companies will have to look to more flexible alternatives to be certain their vital telecommunications are assured.

Telecommunications Solutions

Companies BCP telecom plans will need to meet a number of core requirements to assure the necessary flexibility and survivability:

1. Location independence: Calls must be able to be delivered to any location—sometimes to multiple locations in one day.
2. Network independence: The call traffic must be quickly switchable between all networks (PSTN, cell, Internet) to avoid congestion and outages.

3. Device independence: Calls can be received on any voice instrument, landline phone, cell phone, laptop, PC, IP phone, or even a PDA.

4. Flexibility: Call-handling technology must be able to instantly on demand shift between delivering a large volume of calls to a single location, deliver the calls to a group of locations, or send the calls to thousands of location if people find themselves working from home.

5. Ability to provide conferencing services: Because executives and staff will not be able to physically get together, easy conferencing will be critical to sustaining command and control of operations.

6. Self-healing network: The transport network must be capable of analyzing its own traffic and the traffic patterns of all other networks to reroute traffic around congested locations and around damaged infrastructure and equipment.

The TeleContinuity network was designed with the purpose of assuring corporate telecommunications under all these conditions. TeleContinuity delivers alternate network paths, technologies, and the flexibility to put its clients in control of their communications during any disaster, evacuation, or pandemic.

Formal Programs for Telecom Priority in a Disaster

Other disaster communications programs available to end users and their telecom utility during a pandemic include the following:

- GETS: The GETS program provides critical individuals in vital industries and government agencies a code to assure priority for their outbound calls. While important for managing an intense crisis, GETS provides only outbound calls, which are of limited value when the difficulty is in having inbound calls reach the called party at their new number or location.

- TSP and WPS: The TSP program sets a national priority for having a government agency or critical company's telecommunications infrastructure reconnected following a disaster that has caused an outage or damage in the carrier's facilities and/or network. The WPS provides the same service for wireless communications. These will not be of much use in a pandemic, as the facilities are physically unaffected but the people are no longer able to receive their calls at their previous locations and numbers.

- In addition to these, many carriers have available trucks that contain small switching centers capable of partially replacing damaged or destroyed central office switches or company PBXs. These have been used extensively in natural disasters such as hurricanes, where physical facilities have been damaged and mobile resources have to be trucked in for temporary relief until local facilities can be rebuilt. In a pandemic situation, where facilities are intact but overloaded, the emergency mobile units will be of doubtful utility. And, finally, without doubt, there will be a tremendous surge in calls transmitted over the Internet during the pandemic. In fact as covered earlier, the increase of data traffic to areas away from the usual business centers will add to the congestion problem—resulting in a serious degradation in the intelligibility of Internet voice traffic because of increased packet loss, jitter, and latency (delay).

- Alert and notification services: Alert and notifications services will be widely employed during any pandemic. Corporate management will utilize the services to send important messages and instructions to employees and other executives—especially when they cannot physically get together because of the danger to health. However, notification is a one-way communication service and is not capable of supporting the continuing functioning of an organization—especially during a pandemic where dislocation may last for months. The same is true for message centers, which will also be pressed into service, especially for companies that did not prepare an alternate backup telephone communications service to handle telecom connectivity during an extended pandemic.

Conclusions

The pandemic will not resemble any other disaster companies have planned for in the past. Besides the continuing health issues severely limiting personnel options and mandating a dispersion of the work force, the length of the pandemic will sorely try the resources of every corporation. Telecommunications will be a critical factor in a company's ability to survive and maintain operations when normal facilities are closed, transportation and traffic are restricted, and personnel is continually moving to different locations and voice devices.

A plan for coping with the congestion of the PSTN and the Internet, as calls and data move from fully provisioned central offices and transmission cables to the lightly provisioned outer areas of the suburbs and exurbs, is something the professional planner will have to have in place long before the pandemic strikes. TeleContinuity offers planners the flexibility and survivability necessary to meet the continuing operational demands of any government agency or company—whether meeting the demands of a pandemic or any other disaster.

For more information, contact TeleContinuity, Inc., www.telecontinuity.com, (240) 453-6308

Acknowledgment: Parts of this white paper were published in the Journal of Business Continuity and Emergency Planning, Henry Stuart Publications, LLP. This paper was written in conjunction with Joanne DeLuca, former global head of crisis management of Barclays Capital.

(c) 2007 TeleContinuity, Inc.

CHAPTER 8

Stability Services Inc. (SSI) Disaster Recovery Plan

What does a disaster recovery plan look like? For the planner who is just beginning, this is usually one of the first questions.

The following is an example of at least a dozen plans we have written, all homogenized together to hide the identity of each company. Since this book has a 4Ci focus, the example deals with the backup of a large call center under a mutual aid scenario. This plan is in no way a complete plan but will give the reader an idea of what one looks like and some of the things that should be in one.

(Stablity Services, Inc., is a fictitious company. Any reference to any company expressed or implied or resemblance to any actual company is unintended.)
December 31, 2008

Contents

Section IV—Damage Assessment

Section V—General Procedure References

Section VI—Testing of the Plan

Section VII—Appendices

Acronyms and Definitions

Section I—Executive Summary, Goals, and Objectives

Stability Services (SSI) has long recognized the need to protect all mission-critical operations. In this spirit, a disaster recovery plan has been devised to address the possible loss of an SSI call center to fire, flood, sabotage, and other causes.

This document outlines procedures to be implemented not only in cases of catastrophes such as hurricanes and tornadoes, but also during major network disruptions involving information or telecommunications systems, particularly those affecting the call centers. In this manner, SSI intends to maintain its high standard of service to customers, clients, and stakeholders, under any circumstances.

The specific goals and objectives of this plan include the following points.

Protection of Human Life

The first and foremost concern in this plan is the protection of human life, including the well being of SSI, their families, and all SSI stakeholders.

Minimize Risk of Loss to SSI and Its Stakeholders

SSI recognizes its obligation to protect the assets of its customers, clients, investors, and employees from undue risk of catastrophic loss. It has demonstrated its commitment to this objective by developing recovery plans for critical business functions and by implementing procedures and policies consistent with this goal.

Maximize the Ability to Respond to Any Unfortunate Circumstance

The ability to recover information and telecommunications systems necessary to the support of SSI customers is critical to the well being of SSI and its clients. This is particularly important in the area of communications, where dependence on outside vendors for service is high and where competition for scarce resources in the event of a widespread disaster would be severe. Therefore, another goal of this plan is to codify and document recovery activities to a level where response to telecom and network disasters is possible, with minimal outside assistance.

Preserve Customer Confidence and Goodwill

SSI also recognizes that secure long-term business relationships with customers demand the highest possible standard in customer care. Loss of this capability for even a brief period can tarnish or destroy our hard-won image as a reliable trading partner. In this spirit we invite comments from both internal stakeholders and external customers, since our mutual success is entwined with the success of this plan. Please accept our sincere thanks for your continued trust and confidence.

Basic Assumptions and Policy Statement of SSI

1. Stability Services, Inc., (SSI) shall maintain key components of this disaster recovery plan. SSI commits to regularly updating and testing this plan as

well as tracking equipment and personnel changes that have a bearing on recovery.

2. Inventories of critical personnel, equipment, facilities, and procedures are documented in this plan, as well as operating and security standards for keeping them up to date and current.

3. Policies have been adopted to ensure that all data germane to this plan is regularly stored off site at SSI's principal storage vendor, Safe-Store, Inc.

4. This document defines strategies for recovery of a catastrophic failure in either SSI call center. This plan addresses responsible management, initial notification procedures, damage assessment, and restoration phases for those call centers.

5. Specific SSI positions are defined as Recovery Team Leaders and others as Responsible Individuals. Areas of responsibility are documented and alternate contact persons designated in case of absence or injury of a member.

6. No telephone numbers are contained in this plan per SSI corporate security standard #ABC123SSI. Please refer to this standard as to where to get number information and how to safeguard the devices that contain that data.

7. In the opinion of SSI management (and as a generally accepted axiom), the cost of protecting automated systems should not exceed the cost of equipment replacement, unless the equipment (a) has an unacceptable replacement time, or (b) supports a critical function that demands near-immediate restoral (e.g., a "mission-critical" function). In cases where mission critical is not easily determined, it is the policy of SSI that the system be considered mission critical and protected accordingly. The two call centers fall into the category of mission critical.

8. This call center recovery plan is included in SSI's overall corporate recovery plan, and appropriate references to that plan are contained herein.

Recovery Objective

This plan contains the procedure to be implemented in the event of loss of the principal SSI call center at 901 Main Street, Dallas, Texas. The back up facility for Dallas is the SSI call center at 1305 Willow Avenue, Nashville, Tennessee. The specific time frames for recovery of the Dallas call center in the event of such a disaster are:

- Impaired mode: near-instantaneous transfer of calls;
- Business as usual: within 24 hours of the disaster.

Similar procedures will be implemented to restore the Nashville call center, should circumstances warrant, with Dallas as the backup center.

Section II—The Emergency Response Plan

Depending on the nature of the disaster, either the Dallas or Nashville call center may operate in an extraordinary mode (such as via wireless phones or other technology) for up to 24 hours. During this time, queue and answer times may be longer than usual, but all SSI customers will be answered. It is the stated objective of SSI that NO CALLS are missed and that customers are always in contact with SSI personnel. It is further the objective of SSI to have the call center operating at 100% of normal capacity within 24 hours.

2.01   Activation of the Call Center Recovery Plan

Except under extraordinary conditions, the decision to activate a disaster recovery plan will be made by the most senior SSI executive, who is a vice president or higher, alerted at the time of the disaster. This person will assemble the executive management team (EMT) at a designated emergency operations center (EOC) and conduct the preliminary disaster assessment meeting. This executive also assumes the role of recovery chairman, unless and until relieved by a more senior executive.

> The recovery chairman must receive a situation report from the on-site response team (see Appendix 1) within 60 minutes of arrival at the EOC. Based on this information, the recovery chairman will decide whether or not to activate this disaster recovery plan.

2.02   Activation of Disaster Recovery Teams

Upon declaration of a disaster, the recovery chairman and other members of the EMT will activate the disaster recovery teams or an applicable subset of those teams, as denoted in Appendix 1 of this recovery plan. Contact will be made to each team leader or his or her alternate to implement one of the preplanned disaster recovery plans contained in this document, place his or her resources on alert, or take other appropriate action. Such contact will be recorded by the EMT as to date, time, and success in reaching each team leader. A critical events log, in a format to be designated by the recovery chairman (paper, laptop, PC, and so on) will be opened by the EMT at the time of first notification.

2.03   Establish a Dallas-Based Meeting Place and Command Post

In order to maintain command and control in the recovery process, an EOC will be used in the case of fire, storm, and so on, where the damage is localized to the SSI Dallas call center. A meeting place has been predesignated from which to coordinate EOC recovery activities in Dallas. However, in the case of a major disaster such as a hurricane, major tornado, and so on, where the entire local area is impacted, the operations will be moved to Nashville.

Since this is the staging point for restoration of the call center at 901 Main St., Dallas, Texas, the EOC has been predesignated two miles from the Dallas call center. This is close enough for reasonable convenience but far enough away to probably not be affected by the same disaster. The address and contact numbers are as follow:

EOC Location: Hyatt Regency Reunion Center
106 S. Church Street
Dallas, TX 75202

Contacts:
Sleeping Rooms: Sarah Morgan, Sales Manager (214) 555-0182
Meeting Rooms: Angela DeRossett (214) 555-0183
After Hours: Alyssa Randall (214) 555-0184

2.04   Nashville, Tennessee, Backup Call Center Responsibilities

- The director of telecommunications for the call center in Nashville, Tennessee, will be expected to coordinate activities regarding activation of the Nashville call center, whether or not he or she is in active voice communications with the Dallas command post. The assumption is that no voice communications will be available between these persons.
- The director of telecommunications for the call center in Nashville, Tennessee, will make major command decisions with regard to activation of the backup call center and assume the role of recovery chairman in Nashville unless or until properly relieved by a more senior executive.
- The recovery chairman and recovery management team in Dallas will activate emergency call lists of Dallas operations personnel. Personnel shall be instructed to report to the EOC, damaged site, or Nashville call center. Additional instructions may be directed by phone, e-mail, or instant messenger depending on circumstances.
- In the event of a grave or widespread disaster in the Dallas/Ft. Worth area that isolates communications services, the director of telecommunications for the call center in Nashville, Tennessee, will assume the role of recovery chairman and establish the EOC in Nashville.

Notification of Recovery Teams

All SSI management employees have laptop computers, handheld PDAs, intelligent phones, or in some cases all of these. It shall be the responsibility therefore of every SSI management employee to keep accurate and up-to-date contact information in one or more of those devices in order to perform their duties under this recovery plan pursuant to SSI Corporate Operating Standard #ABC123SSI. This rule is designed for two reasons: (1) to be sure accurate and up-to-date call out data is available when it is needed, and (2) to be sure that in the meantime, data is kept SECURE and CONFIDENTIAL. The best way to accomplish this is through use of existing SSI equipment that already employs the means to keep information confidential, such as encryption on laptop computers, for example. Therefore, please read and heed the following warning:

> Please note that for security and privacy reasons, with only certain exceptions, callout numbers are not included in this plan.

Read the following instructions carefully and contact the EMT as a last resort only if there are questions.

Other Important Assumptions in This Plan Include

- The recovery chairman and EOC shall be responsible for notifying the recovery team leaders designated in Appendix 1 through use of the callout list defined in Appendix 9 and illustrated in Figure 8.1 regarding activation of their respective plans.
- Upon notification from the EOC, the team leaders will activate their respective team members using their own emergency callout lists. Team leaders are responsible for keeping this list up to date and confidential, in their company PDA, wireless device, laptop, or preferably all three. If all media are unavailable, consult Appendix 13 of this document for instructions.
- A recovery management team leader appointed by the EMT will instruct employees to report to the damaged site, or to the EOC, or a third location with a copy of their recovery plan as well as laptop and PDA. The recovery management team leader will also relay any special instructions.

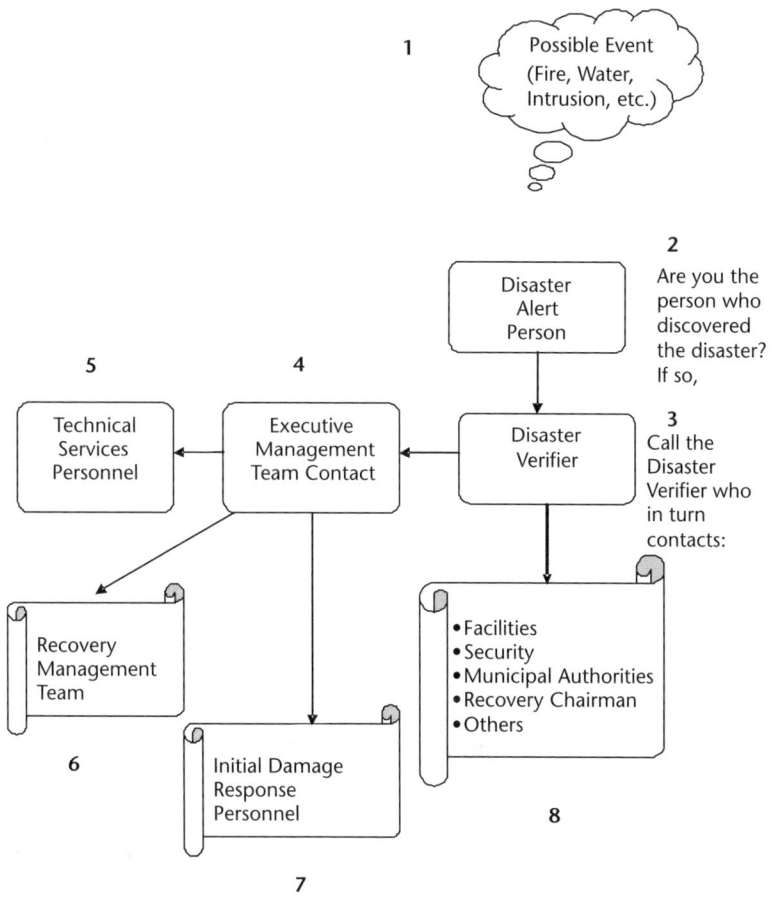

Figure 8.1   Logical callout sequence after an event has been identified.

- The recovery management team leader will be responsible for producing a damage assessment for the EMT within 60 minutes of arrival on site. The damaged site team leader will make a recommendation to the recovery chairman at that time about whether or not the situation warrants abandonment of the call center.
- Based on this recommendation, the recovery chairman or his or her alternate makes the decision on whether or not to activate this plan.

2.05   Designation of Personnel to Man the Nashville Call Center

The backup call center for Dallas is the SSI call center at 1305 Willow Avenue, Nashville, Tennessee. The center is equipped to take on the workload of the Dallas call center, but some initial adjustments will be required. Here is what to expect should you have to use this center to back up Dallas:

The emergency call center in Nashville is normally configured as follows:

1. Thirty-six attendant positions with Intel-based PCs, two monitors each, a VoIP phone, and supervisor position. The center is capable of supporting Dallas call center applications on a transparent basis for the Dallas center.
2. The Nashville center is normally manned by 24 people but has space and equipment for up to 36.
3. Like Dallas, SSI Nashville personnel carry cell phones.
4. Like Dallas, most SSI Nashville center personnel carry or have access to laptops.
5. Like Dallas, the Nashville call center uses Safe-Store, Inc., for off-site data storage.
6. Like Dallas, Nashville has adopted Corporate Operating Standard #ABC123SSI regarding callout numbers and security of devices.
7. The Nashville call center keeps no fewer than two copies of this recovery plan document.

See Related Sections:
"Where to Get Cash" See Appendix 10
"How to Book Travel" See Appendix 10
"How to Purchase Equipment" See Appendix 10

Section III—Continuity of Operations (COOP) Plan

Loss of Dallas Call Center—Detailed Overview

Be aware, in all probability, loss of the Dallas call center may not be immediately noticed by SSI clients, customers, field personnel, and OEMs. This is because automatic "fail over" arrangements have been made to quickly divert calls with Telecom Recovery, Inc., an SSI vendor (see Appendix 3 of this document). It is possible that a degradation of service will be noticed, however, shortly after the disaster. This is why it is expected that the center will operate in an "impaired" mode for up to 24 hours following a catastrophic event.

3.01   Order of Priority and Procedure for Recovery of Critical Numbers

The following procedures are to be followed for restoration of mission-critical 800 numbers that were previously provided by the Dallas (or Nashville) call center(s).

3.02   Auto Fall Back of Telephone Service for Dallas Call Center

As a minimum, the following numbers shall be forwarded to the Nashville call center on a near-instantaneous basis. Within 24 hours, any service queues should be back to a normal wait time. In the unlikely event both centers are affected, the numbers should be directed to telecom recovery, where they can be redirected to wireless, VoIP, or satellite phones, depending on availability.

Dallas Call Center Numbers	Nashville Call Center Numbers
(800) 123-4567	(866) 123-5000
(800) 123-2578	(866) 123-5001
(800) 123-2579	(866) 123-5002
(800) 123-2580	(866) 123-5003
(800) 123-2581	(866) 123-5004
Dallas Backup Numbers for Nashville	Nashville Backup Numbers for Dallas
(800) 123-4567	(866) 123-5000
(800) 123-2578	(866) 123-5001
(800) 123-2579	(866) 123-5002
(800) 123-2580	(866) 123-5003
(800) 123-2581	(866) 123-5004

A recovery coordinator shall be designated by the EOC to contact the MCI business and/or other carrier testboards in Dallas and assure that these numbers have been forwarded to the Nashville call center, using the procedures contained in Appendix 3 of this document. The recovery coordinator shall then verify each number by making an actual call to each number. Even if numbers have been forwarded to a common hunt group at Nashville (or other location), each number must be voice verified in this fashion. The numbers for MCI business and other telecom providers are described but not generally listed in Appendix 2, regularly updated, and stored off site.

3.03   Procedure to Re-Establish Work Order Processing

Geek Legion (GL) is the system SSI uses to dispatch field service personnel to clients. GL is already active at both the Dallas and Nashville sites. In order to accommodate the additional capacity required by Dallas call center personnel, activate or verify GL within 24 hours of the disaster. For additional information, refer to Appendix 6. For further information on application recovery, see Appendix 6, as well. Responsible persons are listed in Appendix 1.

3.04   Procedure to Establish Manual Dispatch of Field Ops Personnel

Depending on the type of disaster, it may be necessary to use other extraordinary means besides GL to communicate with field service personnel, including but not limited to e-mail or manual dispatch. If instructed by the recovery chairman, responsible SSI personnel will enable SMS to wireless devices in order that text messages to wireless devices (carried by all field service personnel) can be employed as a backup. If this becomes necessary, please refer to Appendix 3. Responsible persons are listed in Appendix 1.

3.05   Notify Vendors, Request an On-Site Representative

The recovery coordinator or other member of the EOC must contact all major equipment vendors involved and request an on-site representative immediately to aid in damage assessment and facilitate fast restoral of the equipment.

   A. Equipment inventories are located in Appendix 5 of this document.
   B. Vendor callout lists are located in Appendices 2 and 13.

3.06   Coordination of Vendors at Affected Site

In cases that involve on-site equipment, the individuals responsible for on site damage assessment must contact the vendors involved and request that an on-site representative immediately be dispatched to the affected site.

   A. Equipment inventories are located in Appendix 5 of this document.
   B. Vendor callout lists are located in Appendices 2 and 13.
   C. Refer to Appendix 12 before calling vendors to check for possible roll-in replacement guarantees and other contractual guarantees for SSI equipment.

Section IV—Damage Assessment

4.01   Protection of Human Life

   A. The first and foremost concern in the implementation of this plan is the protection of human life and the physical well being of SSI employees. Protection of human life will be of primary concern when formulating and updating the recovery plan.
   B. SSI will abide by policies set forth in the corporate disaster recovery plan with regard to personal safety in all recovery operations
   C. Hard hats, or other protective equipment, will be utilized by SSI employees during all damage surveys. Protective equipment will be specified by either SSI, federal, state, or local authorities.

4.02   Notification of Police, Fire, Medical

   A. In a life-threatening emergency, don't hesitate: dial 9, then 911 immediately.

B. If time allows, the preferred method of alerting local authorities to an emergency is through SSI security personnel, since the condition may have already been reported or to assist them in evacuating the rest of the building.

C. Any SSI employee however, may dial 911 or the appropriate emergency number if the situation dictates. Be prepared to give:

- Your name;
- Address;
- The nature of the problem (i.e., fire, bomb threat, and so on).

4.03   Determination of the Cause

While the goal of SSI is to completely recover as soon as possible, it is sometimes also necessary to determine the cause if possible to aid later with such things as insurance claims, criminal investigations, and so on. SSI employees are expected to cooperate with both SSI personnel and local authorities involved in determining the cause of the disaster.

4.04   Notification of Management

In the event of significant damage to any SSI location in the Dallas/Ft. Worth (D/FW) area, the vice president operations for SSI shall be notified immediately for possible activation of the company disaster recovery plan.

4.05   Informing Vendors

Vendors who provide the equipment damaged during the disaster shall be notified immediately and an on-site representative dispatched by them as soon as possible. Lists of all equipment vendors are included in Appendices 5, 6, and 13 of this document.

Particular emphasis should be placed on equipment critical to the operation of the call center, including but not limited to:

- Inbound T1 and T3 circuits;
- The PBX and automated call distribution (ACD) unit for the call center;
- Data communications bridges, routers, and gateways supporting the call center.

4.06   Coordination with Local Authorities

All coordination with local authorities (fire marshals, security forces, and so on) will be performed through the EOC, who in turn will coordinate with SSI's security vendor. The contacts for this function are found in Appendices 2 and 13.

Remember, local first responders use a system called the National Incident Management System (NIMS) and may use different terminology than SSI uses in this plan. For example, rather than a recovery chairman their leader is called an incident commander (IC).

Remember, government agencies as well as hospitals may use a NIMS framework in their own internal recovery plan (see Figure 8.2).

Author's Note: A brief tutorial on NIMS is also included in Chapter 9.

All coordination with local authorities will be accomplished through the on-site (damaged site) representative in constant communication with the EOC.

4.07   Notification of Customers and the Media

Do not make any comments to the media! Refer all media inquiries ONLY to the persons designated in Appendix 12 of this document.

ICS positions have different and distinct titles

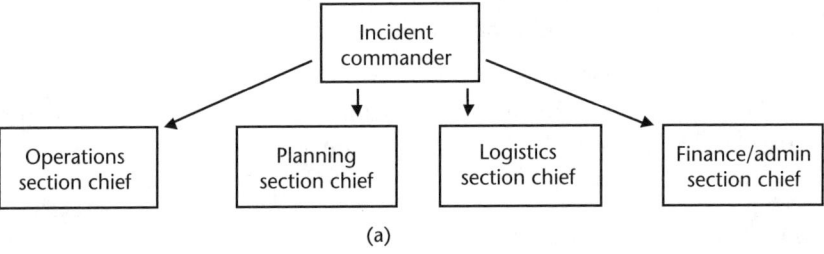

Incident command system (ICS) types of commands

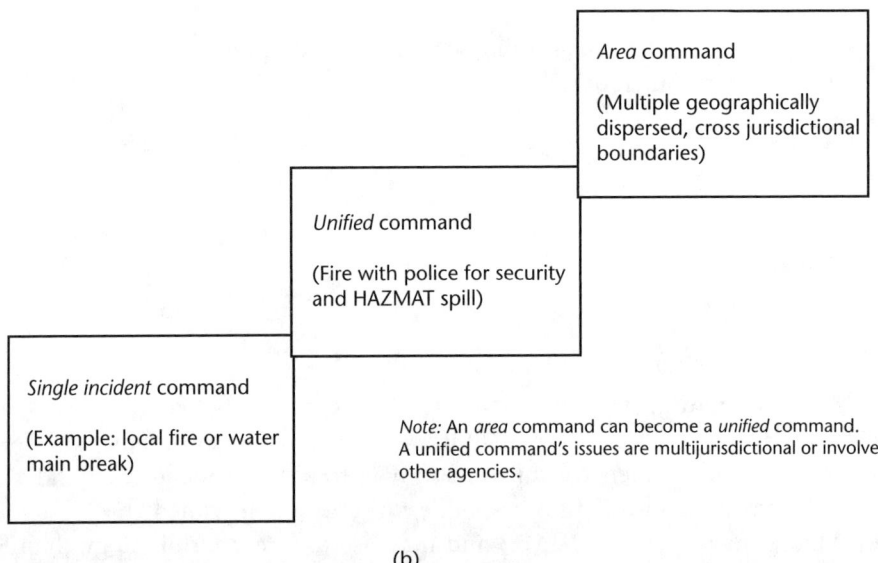

Figure 8.2   (a, b) Using NIMs standard terminology makes position titles easier to remember.

4.08   Tracking Work for Audit Purposes

A daily log detailing equipment purchased, persons assigned, hours worked, important decisions made, and other relevant information shall be kept for the duration of the outage at the network command center. This could be useful later for scheduling compensatory time, for insurance, and for other purposes. It is recommended that you include a small 4 × 6-inch pocket notebook in the back of your plan for this purpose, as it may be difficult to find a scratch pad for this purpose after a disaster.

4.09   Modified Signing Authorities for Equipment Purchases

Modified signing authorities will be in effect for the length of the disaster. Questions by vendors as to the authority of employees empowered to make major decisions should be referred to the team leader of the finance team designated in Appendix 10.

4.10   Where Do SSI Employees Get Cash?

Extraordinary procedures for dispensing cash and accounting for those disbursements will be in effect for the length of the disaster. Requests for travel cash (such as for transporting Dallas employees to Nashville) should be referred to the team leader of the finance team designated in Appendix 10.

4.11   Where Do Employees Make Travel Arrangements?

Travel arrangements shall be coordinated through the SSI corporate travel agent. For further information, see Appendix 10. If this is not possible, contact the EOC.

4.12   Maintaining Physical Security at the Damaged Site

Maintenance of physical security at the damaged site will be coordinated through SSI's security vendor. The damaged site response team will assume responsibility for security at both the damaged and the relocated sites for the duration of the outage, including the hiring of temporary officers if required. All requests for service will be made through the emergency operations center (EOC).

4.13   Responsibilities of the Damaged Site Response Team Personnel

  A. Direct periodic reports to the EOC and any other personnel defined in this plan;
  B. Direct supervision of damage assessment activities related to the building, call center, network, or other on-site facilities;
  C. Determine the status of any personnel who may have been on duty in the area during the disaster and report to the EOC;
  D. Aid with maintaining security in the affected area.

Author's Note: Recovery services are available that offer roll-in replacement equipment and mobile trailers for personnel. Examples can be found in Chapter 9.

4.14   Notification of Clean Up Companies

BMS Catastrophe (BMS CAT) is the SSI vendor of choice for cleanup and restoration efforts. BMS CAT handles recovery of equipment, circuit boards, magnetic media, paper manuals and documents, and other items from damage by fire, smoke, water, sabotage, and so on. BMS Cat will probably have been notified by the EOC as part of the plan declaration, but the number and contact are provided here for your information:

> Primary Contact:
> BMS Catastrophe, Inc.
> Ft. Worth, TX
> (800) 433-2940

Author's Note: See useful information on BMS Catastrophe in Chapter 9.

4.15   Preliminary Liability Assessments and Preparation of Statements

All SSI employees will cooperate as required with internal investigative SSI personnel, insurance companies, and local authorities in any investigations. All information uncovered in the course of these investigations will be kept absolutely confidential. All SSI employees will cooperate with SSI legal in preparation of statements for the media, customers, investors, and other affected parties. Direct any questions to the following or see Appendix 12 for responsible SSI department.

4.16   Reconstruction of the Original Site

- Await directions from the EOC regarding activities to be taken at the original site.
- Make a preliminary assessment of equipment and resources that will be required, and notify vendors of possible equipment acquisitions, but do not purchase new equipment unless instructed by the EOC.
- Cooperate with BMS Catastrophe and other recovery vendors when requested.
- Once cleared, begin reconstruction according to the prearranged and then-defined plan for doing so.
- Once cleared, begin reinstalling new equipment according to the prearranged and then-defined plan for doing so.

4.17   Restoration of Original Software Systems

- Await directions from the EOC regarding activities to be taken at the original site.
- Make a preliminary assessment of equipment and resources that will be required, and notify vendors of possible equipment acquisitions, but do not purchase new equipment unless instructed by the EOC.

- Cooperate with BMS Catastrophe and other recovery vendors when requested.
- Once cleared, begin reinstalling new software according to the prearranged and then-defined plan for doing so.

4.18   Restoration of Power and UPS

- Await directions from the EOC regarding activities to be taken at the original site.
- Make a preliminary assessment of equipment and resources that will be required, and notify vendors of possible equipment acquisitions, but do not purchase new equipment unless instructed by the EOC.
- Cooperate with BMS Catastrophe and other recovery vendors when requested.
- Once cleared, begin reinstalling new equipment according to the prearranged and then-defined plan for doing so.

4.19   Replacement of Fire Detection and Suppression System

- Await directions from the EOC regarding activities to be taken at the original site.
- Make a preliminary assessment of equipment and resources that will be required, and notify vendors of possible equipment acquisitions, but do not purchase new equipment unless instructed by the EOC.
- Cooperate with BMS Catastrophe and other recovery vendors when requested.
- Once cleared, begin reinstalling new equipment according to the prearranged and then-defined plan for doing so.

4.20   Maintaining Security in the Area

- All security for the affected facility will be handled by SSI-designated security personnel. Direct all requests for additional security to the EOC or the responsible on-site SSI person designated.
- Assist security personnel in identifying authorized persons when required, and cooperate with any extraordinary security measures (distinctive clothing, badge stickers, and so on) that must be implemented because of the situation. For further information consult the damaged site response team leader.

4.21   Rewiring the Damaged Facility

- Await directions from the EOC.

- Make a preliminary assessment of equipment and resources that will be required, and notify vendors of possible equipment acquisitions, but do not purchase new equipment unless instructed by the EMT.
- Assist electrical contractors in determining the proper outlets and power ratings, corresponding with the manufacturer's specifications of the equipment being replaced. Use equipment vendors as well.
- Once cleared, begin reinstalling new equipment according to the prearranged and then-defined plan for doing so. Aid and cooperate with the installation contractor.

4.22   Restoring the Original Configuration

- Await directions from the EOC.
- Make a preliminary assessment of equipment and resources that will be required, and notify vendors of possible equipment acquisitions, but do not purchase new equipment unless instructed by the EMT.
- All newly installed primary circuits will be thoroughly acceptance tested before switching over.
- Existing SSI policies regarding major network reconfigurations will apply (e.g., no major switchovers during peak business hours).
- Emergency circuits should be kept in place for a reasonable time after switching back to primary traffic, in case a rapid return to the recovery center is dictated due to unforeseen circumstances.

4.23   Testing New Hardware and Software

- All equipment components will remain in the emergency configuration until specifically instructed by the EMT to return to the primary configuration.
- All newly installed equipment will be thoroughly acceptance tested before switching over, both by the vendor and by responsible SSI personnel.

4.24   Training Call Center Personnel on New Equipment

Managers will be expected to train their own personnel on any new equipment installed as a result of the disaster. SSI employees are expected to cooperate in this effort by explaining any functional or operational differences in the new equipment.

4.25   Scheduling Migration Back to Original Site

- Await directions from the emergency management team regarding the scheduling of the return to the original site.
- Provide the EMT with concise reports in your specific area of responsibility as to the status of the recovery of the original site to aid management in scheduling resumption of business at the restored site.

- Make a preliminary assessment of equipment and resources that will be required, and notify vendors of possible equipment acquisitions, but do not purchase new equipment unless instructed by the EMT.
- Develop a final check list to be conducted to assure the equipment was installed in accordance to manufacturers' specifications and is functioning normally. Vendors will be expected to sign off. Pay particular attention to anything installed "on patch" or in a temporary configuration.
- Remember that SSI policies regarding major network reconfigurations will apply unless otherwise instructed (i.e., no major switchovers during peak business hours).
- Proceed only when the recovery chairman says it is authorized to deactivate the backup call center, and only after ascertaining business is capable of a complete return to normal.

4.26   Preparation of Final Review and Activity Report

Final reports will be submitted in the format requested by the EMT describing recovery operations. This format will not be determined until after the disaster. It would be a prudent practice, however, to keep track of the following items in a daily log for your area of responsibility, for possible inclusion in the report after a disaster:

- Estimates of hours worked per employee, for scheduling compensatory time;
- Any major decisions made by you that affect recovery operations;
- Items regarding outstanding vendor response, performance, or lack thereof;
- Items that could have been stated more clearly in the recovery planning document or misunderstandings that arose by use of the document for use in later versions;
- Anything that might support the company in determining litigation or liability;
- Your honest critique of the disaster recovery plan, personnel, facilities, or response;
- The names of anyone who merits special recognition by executive management.

Section V—General Procedure References

5.01   Fire

In a life-threatening emergency, or when in doubt, dial 9, then 911 immediately.

If time permits, refer to SSI standard emergency response procedures included in this document in Appendix 13.

5.02   Water

In a life-threatening emergency, or when in doubt, dial 9, then 911 immediately.

If time permits:

1. Attempt to locate and shut off the source of the water.
2. If this is not possible, and the equipment has not powered itself down, shut power off to the affected equipment and attempt to shield it as well as possible with an equipment cover, sheet plastic, or any available material.
3. Notify the equipment vendor and request an on-site representative before restarting the equipment.
4. If time permits, refer to SSI standard emergency response procedures included in this document in Appendix 13.

5.03 Sabotage/Terrorism/Vandalism

In a life-threatening emergency, or when in doubt, dial 9, then 911 immediately.

If time permits, refer to SSI standard emergency response procedures included in this document in Appendix 13.

5.04 Bomb Threat

In a life-threatening emergency, or when in doubt, dial 9, then 911 immediately.

If time permits, refer to SSI standard emergency response procedures included in this document in Appendix 13.

5.05 Power Loss

Please refer to SSI standard emergency response procedures included in this document in Appendix 13.

5.06 Security Breach

Please refer to SSI standard emergency response procedures included in this document in Appendix 13.

5.07 Severe Weather/Company Information/Disaster Hot Line

This number is intended to provide up-to-the minute information about major systems failures within SSI, as well as disaster recovery and severe weather information. It is for internal use only and is not to be released to outside sources! Please program it into your wireless phone now.

(866) 123-4567

Author's Note: 4Ci is not only about voice communications! Please see Chapter 9 and the section on ESI and WebEOC for more details on situational analysis and information sharing that can help keep all employees and stakeholders on the same page in a disaster.

Section VI—Testing of the Plan

6.01   Policy Statement

SSI shall maintain key components of this network disaster recovery plan in the overall contingency plan for the SSI companies. The plan will be updated regularly and equipment changes that have a bearing on recovery operations will be tracked. Drills and practice tests of the plan will be scheduled by the vice president and general manager, SSI, to verify the integrity of the plan.

Section VII—Appendices

The biggest problem with recovery plans is that they quickly go out of date. This plan makes provisions to ensure that the most up-to-date information will be in the hands of responders to an SSI disaster. It does so by making the assumption that the most up-to-date information can be obtained from the SSI off-site storage vendor, Safe-Store, Inc.

Many of the following sections make provisions to ensure that critical disaster recovery information is stored off site along with the normal data backup. The filenames for this information are included in the relevant appendix.

It is assumed that a copy of this inventory would be procured and disseminated to the applicable responders immediately after a disaster, since it is expected that any contained in this plan will be out of date. Basically, as soon as an administrative employee can procure a copy of the backup, and have it printed (at a FedEx, Quick Copy, or wherever), the most up-to-date information will be available. In the meantime, the information actually contained in this plan will suffice, when augmented with the information contained in the laptops and handheld communications devices of the SSI responders.

Appendix 1

Overview of Appendices and Response Teams

1. Overview of Appendices and Response Teams (Examples Follow)
2. Emergency Network Activation Procedures, First Two Hours
3. Emergency Network Activation Procedures, First 24 Hours (Includes Carrier Callout and Escalation List Info)
4. Call Center Redundancy Test Procedure
5. Hardware List and Serial Number Inventory
6. Software Lists, Passwords, and License Numbers
7. Recovery of SSI Asset Database
8. Equipment Room Floor Grid Diagrams
9. Employee Inventory and After-Hours Callout Info
10. Cash Disbursement and Transportation Procedures
11. Procedures for Physical Security, Damaged Site

Executive Management Team (EMT)

Team Leader:

Alternate:

Notification Procedure:

Any SSI employee, contractor, or outside security person can report a disaster, normally through their responsible manager or director. The responsible manager or director in turn acts in the role of "disaster verifier" and, if the situation warrants, alerts the EMT through numbers contained in his or her PDA or laptop.

Duties and Responsibilities:

This team will provide executive decision-making authority during a disaster and is responsible for making an actual disaster declaration. Members of the executive management team (EMT) should be familiar with the contents of this plan at a global level as well as notification procedures for the recovery management team (RMT) and damaged site response team (DRT) noted in this document. Team members will include but not be limited to:

- Vice president and general manager, SSI;
- Chief financial officer, SSI;
- Media and public affairs liaison;
- Human resources executive;
- Corporate manager, safety, and security;
- Others, including admin support, at the request of the recovery chairman.

The EMT will convene a conference call using the procedures established in this plan immediately upon arrival at the emergency operations center (EOC) described in Section 2.03 of this plan. If a disaster is declared, the EMT is responsible for directing the activities of subordinate teams, including the recovery management team (RMT) and damages site response teams (DRT) for Dallas or Nashville, depending on the nature of the disaster.

Within one hour of a disaster declaration, the EMT should receive a report from the DRT regarding details of the disaster, including specific recommendations on response and preliminary estimate of time for recovery.

Team Members:

PDA or laptop in possession of team leader and alternate, including all current phone numbers.

Recovery Management Team (RMT)

Team Leader:

Alternate:

Notification Procedure:

The RMT is notified by a member of the executive management team (EMT) or recovery chairman in the emergency operations center (EOC).

Duties and Responsibilities:

Depending on instructions from the EOC, the RMT team leader will instruct specific team members to report to the EOC or the alternate call center site. All RMT team members must be ready to travel on short notice depending on circumstances and instructions. Members of the RMT must be intimately familiar with the contents of this plan in order to provide support to the EMT and coordinate the actual recovery. If a disaster is declared, the EMT is responsible for directing the activities of the actual responders as well as coordinating with the damages site response teams (DRT) for Dallas or Nashville, depending on the nature of the disaster. Team members will include but not be limited to

- Chief executive over the call center (Dallas or Nashville);
- All available direct reports for the executive except call center manpower (who will be traveling or working remotely).

The RMT will designate at least one team member to report to the EOC. This is the team member who will coordinate with the carriers to reroute 800 and other telecom services to the affected call center. The RMT will also designate at least one employee as a liaison to the damaged site response team (DRT). This liaison will assist the DRT and provide periodic reports to the EMT and RMT by whatever technology is available. Finally, if a disaster is declared, the RMT is responsible for assisting call center manpower (along with the resources defined in Appendix 10) in transportation arrangements to Dallas or Nashville, depending on the nature of the disaster.

Team Members:

PDA or laptop in possession of team leader and alternate, including all current phone numbers.

Damaged Site Response Team (Dallas, Texas)

Team Leader:

Alternate:

Notification Procedure:

The DRT for Dallas is notified by a member of the executive management team (EMT) or recovery chairman in the emergency operations center (EOC).

Duties and Responsibilities:

Depending on instructions from the EOC, the DRT team leader will instruct specific team members to report to the damaged site. All DRT team members must be ready to travel on short notice, depending on circumstances and instructions. Members of the DRT must be intimately familiar with the contents of this plan in order to provide support to the EMT and coordinate the actual recovery. The DRT is responsible for providing the first report to the EMT within an hour of arrival at the damaged site. Their report must be accurate, since it will be the basis for whether a disaster is declared by the EMT. Team members will include but not be limited to

- Corporate manager of safety and security;
- Program manager (LSS).

The DRT will designate at least one team member to report to the EOC to brief the EOC and deliver the report. This person will remain at the EOC to assist in coordinating carriers to reroute 800 and other telecom services to the affected call center. This person is also responsible for contacting high-priority clients within an hour of the EMT briefing.

Team Members:

PDA or laptop in possession of team leader and alternate, including all current phone numbers.

Damaged Site Response Team (Nashville, Tennessee)

Team Leader:

Alternate:

Notification Procedure:

The DRT for Nashville is notified by a member of the executive management team (EMT) or recovery chairman in the emergency operations center (EOC).

Duties and Responsibilities:

Depending on instructions from the EOC, the DRT team leader will instruct specific team members to report to the damaged site. All DRT team members must be ready to travel on short notice, depending on circumstances and instructions. Members of the DRT must be intimately familiar with the contents of this plan in order to provide support to the EMT and coordinate the actual recovery. The DRT is responsible for providing the first report to the EMT within an hour of arrival at the damaged site. Their report must be accurate, since it will be the basis for whether a disaster is declared by the EMT. Team members will include but not be limited to

- Director of IT services, Nashville;
- Information security and network infrastructure director/manager.

The DRT Nashville will determine the need to establish an EOC in Nashville. If so, it will designate at least one team member to report to the EOC to brief the Nashville EOC and deliver the damaged site report. This person will remain at the EOC to assist in coordinating carriers to reroute 800 and other telecom services, if required, to the affected call center. This person is also responsible for contacting high-priority clients within an hour of the EMT briefing and for briefing counterparts in Dallas via telephone. Obviously, in the case of a disaster in Nashville, Dallas will be operating normally, so the Nashville EOC can depend on them for resources and support.

Team Members:

PDA or laptop in possession of team leader and alternate, including all current phone numbers.

Off-Site Storage Retrieval Team (Dallas, Texas, and Nashville, Tennessee)

Team Leader:

Alternate:

Notification Procedure:

The OSRT for Dallas or Nashville is notified by a member of the executive management team (EMT) or recovery chairman in the emergency operations center (EOC).

Duties and Responsibilities:

Depending on instructions from the EOC, the OSRT team leader will instruct specific team members to report to the respective Safe-Store, Inc., sites for retrieval of the latest copies of this recovery plan (on magnetic media) and other media that is stored off site. All OSRT team members must be ready to travel on short notice, depending on circumstances and instructions. Members of the OSRT must be intimately familiar with the contents of this plan in order to provide support to the EMT and other teams as well as coordinate the actual recovery.

Finance and Procurement Team (Dallas, Texas, and Nashville, Tennessee)

Team Leader:

Alternate:

Notification Procedure:

The FPT for Dallas or Nashville is notified by a member of the executive management team (EMT) or recovery chairman in the emergency operations center (EOC).

Duties and Responsibilities:

Depending on instructions from the EOC, the FPT team leader will make arrangements to support other teams in procurements, travel funds, and other necessary financial arrangements. In addition, FPT will assist with any travel arrangements and interface with the corporate travel department in paying for travel and lodging. The FPT will take precautions in advance to have emergency purchase orders "in the pipe" and ready to process in the event equipment must be purchased immediately. In addition the FPT team will maintain procedures for fast signing authority under emergency circumstances, where the normal bid process for equipment purchases is not possible. Finally, the FPT will assist the EOC in financial matters and in interfacing with banks and the financial community.

Acronyms and Glossary

Business continuity	The plan used by the SSI organization to respond to or recover from a disaster event.
Command post	See emergency operations center.
Damage assessment	The process of assessing damage—following a disaster—of facilities, personnel, business operations and systems, vital records, and the like, and then determining what can be salvaged, restored or replaced.
Disaster declaration	A declaration by the EMT that a sever disruption to one or more mission-critical business processes has occurred.
DRT	Damaged site response team.
Emergency operations center	A predetermined facility/location (separate from the main facility) at which the EMT shall convene—at a time of their choosing—for the purposes of continuing event response operations.
EMT	Executive management team.
EOC	Emergency operations center; a predetermined location (separate from the main facility) at which the RMT shall convene—at a time of their choosing—for the purposes of continuing event response operations.
First responders	Emergency personnel such as police, fire, and ambulance teams.
FPT	Finance and procurement team.
IT	Information technology.
Mission-critical business processes	Those identified business processes deemed vital to the operation, reputation, or profitability of the organization and that have a recovery time objective of 0–24 hours.
OSRT	Off-site storage retrieval team.
Recovery time	The time elapsed between the onset of an event and its successful resolution (when business returns to its normal operation).

Recovery time objective	The failover or recovery time capability required of business processes defined by the latest business impact analysis.
RMT	Recovery management team.
Significant event	An event deemed too critical to be handled by the event escalation process, but does not justify full escalation to a disaster declaration, according to the judgment of the RMT leader.
Situation assessment	The process of assessing a potential disaster immediately following an event to facilities, personnel, business operations, and systems to record the event and determine if the event needs to be escalated.

CHAPTER 9
Other References

Overview of NIMS

Figures 9.1 through 9.14 provide an overview of NIMS.

Roll-In Replacement Services (Rentsys Recovery)

Figures 9.15 through 9.23 show roll-in replacement services for Rentsys Recovery.

Cleanup Vendors and Tips from the BMS Catastrophe

Figures 9.23 through 9.33 show cleanup vendors and tips from the BMS catastrophe.

Aids to Situational Awareness and Disaster Communication

Even though we are limited in print space to what we can reasonably discuss in this book, we authors believe it is important to make the point that 4Ci is not only about voice communications. Indeed, we have referred in detail throughout this book to the fact that the last "C" in 4Ci, Computers, also includes things in your hand (like Blackberries), on your belt (like cell phones), or in your temporary workspace (like laptops). Here is one more example of how computers can enhance 4Ci.

Augusta, Georgia–based ESi is a global leader in crisis information management technology. ESi pioneered the market with WebEOC, the world's first Web-enabled emergency management communications system.

WebEOC connects crisis response teams and decision makers at national, state, and local agencies, healthcare providers, airlines, and corporations worldwide, providing access to real-time information for a common operating picture during an event or daily operations. This Web-hosted system was broadly employed, for example, during the recent Southern California wildfires and for other, more diverse purposes such as the 2008 Democratic National Convention.

Web-hosted information sharing systems like this one have significant potential as an enhancement to 4Ci. ESi appears to be the de facto standard for this type of resource coordination and information sharing among federal, state, and local

What is the National Incident Management System?

NIMS is a standard.
NIMS is a comprehensive, national approach to incident management that is applicable at all jurisdictional levels and across functional disciplines.

NIMS is intended to be applicable across a full spectrum of potential incidents and hazards, regardless of size or complexity.

NIMS is intended to improve coordination and cooperation between public and private entities in a variety of domestic incident management activities.

Figure 9.1   What is the National Incident Management System?

What is the NIMS, continued:

NIMS provides a framework for interoperability and compatibility by balancing flexibility and standardization.

NIMS provides a flexible framework that facilitates government and private entities at all levels working together to manage domestic incidents.

This flexibility applies to all phases of incident management, regardless of cause, size, location, or complexity.

NIMS provides a set of standardized organizational structures as well as requirements for processes, procedures, and systems designed to improve interoperability.

Figure 9.2   Further information on what the National Incident Management System is.

NIMS Compliance

HSPD-5 requires federal departments and agencies to make the adoption of NIMS by state and local organizations a condition for federal preparedness assistance (grants, contracts, and other activities) by FY2005.

Jurisdictions can comply in the short term by adopting the incident command system. Other aspects of NIMS require additional development and refinement to enable compliance at a future date.

Figure 9.3   NIMS compliance.

NIMS Components

NIMS is comprised of several components that work together as a system to provide a national framework for preparing for, preventing, responding to, and recovering from domestic incidents. These components include:

Command and management
Preparedness
Resource management
Communications and information management
Supporting technologies

Figure 9.4   NIMS components.

NIMS Command and Management Structure

NIMS standard incident management structures are based on three key organizational systems:

The Incident Command System (ICS), which defines the operating characteristics, management components, and structure of incident management organizations throughout the life cycle of an incident.

Multiagency Coordination Systems, which define the operating characteristics, management components, and organizational structures of supporting entities.

Public Information Systems, which include the processes, procedures, and systems for communicating timely and accurate information to the public during emergency situations.

Figure 9.5   NIMS command and management structure.

The Incident Command System (ICS)

ICS is a proven, on-scene, all-hazard incident management concept. ICS has become the standard for on-scene management. ICS is interdisciplinary and organizationally flexible to meet the needs of incidents of any size or level of complexity.

ICS has been used for a wide range of incidents, from planned events to hazardous materials spills to acts of terrorism.

ICS has several features that make it well suited to managing incidents:

Common terminology
Organizational resources
Manageable span of control
Organizational facilities
Use of position titles
Reliance on an Incident Action Plan
Integrated communications
Accountability

Figure 9.6   The incident command system.

emergency responders, based on the sheer number of users. The ESi Web site is http://www.ESi911.com.

Other Interesting Stuff

Figures 9.34 through 9.37 show other interesting stuff.

ICS Positions Have Distinct Titles

Only the incident commander is called commander—and there is only
one incident commander (IC) per incident.

Only the heads of sections are called chiefs.

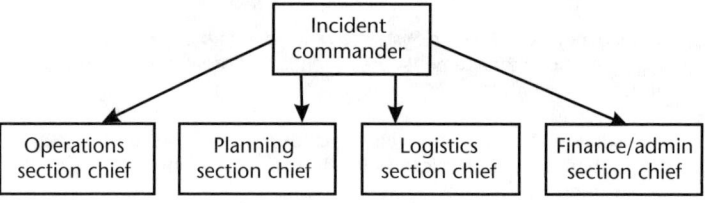

Learning and using standard terminology helps reduce confusion
between the day-to-day position occupied by an individual and his or her
position at the incident.

Figure 9.7   The distinct titles of ICS positions.

Incident Action Plans (IAPs)

Incident Action Plans (IAPs) communicate the overall incident objectives
in the context of both operational and support activities.

IAPs are developed for operational periods, usually 12 hours long.

IAPs depend on management by objectives to accomplish response tactics.

These objectives are communicated throughout the organization and are
used to develop and issue assignments, plans, procedures, and protocols,
as well as direct efforts to attain the objectives in support of defined
strategic objectives.

Results are documented and fed back into planning for the next
operational period.

Figure 9.8   Incident Action Plans (IAPs).

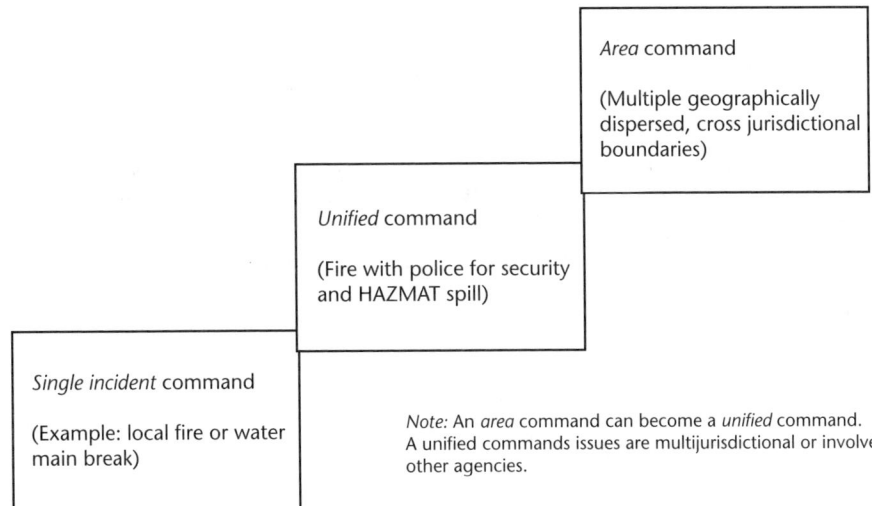

Figure 9.9   Single incident, unified, and area command.

Area Command Structure

An Area Command is organized similarly to an ICS structure but, because operations are conducted on scene, there is no Operations Section in an Area Command.

Other sections and functions are represented in an Area Command structure.

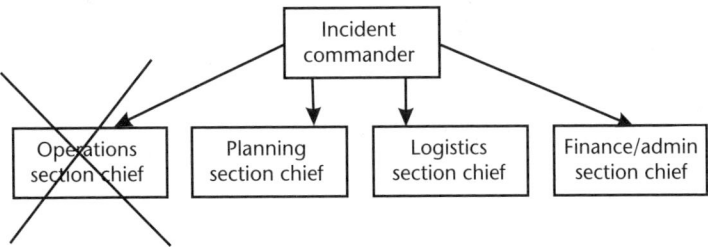

Figure 9.10   Area Command Structure.

Multiagency Coordination Systems (MCS)

Multiagency Coordination Systems (MCS) include Emergency Operations Centers (EOCs) and, in certain multijurisdictional or complex incidents, Multiagency Coordination Entities, also called MCEs.

EOCs are the locations from which the coordination of information and resources to support incident activities takes place. EOCs are typically established by the emergency management agency at the local and state levels.

MCEs typically consist of principals from organizations with direct incident management responsibilities or with significant incident management support or resource responsibilities. These entities may be used to facilitate incident management and policy coordination. EOC organization and staffing are flexible, but should include:

Coordination
Communications
Resource dispatching and tracking
Information collection, analysis, and dissemination

EOCs may also support multiagency coordination and joint information activities. EOCs should be staffed by personnel representing multiple jurisdictions and functional disciplines. The size, staffing, and equipment at an EOC will depend on the size of the jurisdiction, the resources available, and the anticipated incident needs.

Figure 9.11   Multiagency Coordination Systems (MCS).

Duties and Responsibilities of an EOC or MCE

Regardless of their form or structure, MCEs or EOCs are responsible for:

Ensuring that each involved agency is providing situation and resource status information.
Establishing priorities between incidents and/or area commands in concert with the Incident Command or Unified Command.
Acquiring and allocating resources required by incident management personnel.
Coordinating and identifying future resource requirements.
Coordinating and resolving policy issues.
Providing strategic coordination.

Following incidents, multiagency coordination entities are typically responsible for ensuring that revisions are acted upon. Revisions may be made to:

Plans
Procedures
Communications
Staffing
Other capabilities necessary for improved incident management

Revisions to recovery plans and procedures based on lessons learned from the incident should be reviewed after the fact with the emergency planning team in the jurisdiction and with mutual aid partners.

Figure 9.12   Duties and responsibilities of an EOC or MCE.

Who Talks to the Media?

Under ICS, the *public information officer*, or PIO, is a key member of the command staff.

The PIO advises the Incident Command on all public information matters related to the management of the incident, including media and public inquiries, emergency public information and warnings, rumor monitoring and control, media monitoring, and other functions required to coordinate, clear with proper authorities, and disseminate accurate and timely information related to the incident.

The PIO establishes and operates within the parameters established for the Joint Information System, or JIS.

The JIS provides an organized, integrated, and coordinated mechanism for providing information to the public during an emergency. The JIS includes plans, protocols, and structures used to provide information to the public. It encompasses all public information related to the incident.

Key elements of a JIS include interagency coordination and integration, developing and delivering coordinated messages, and support for decision makers. The PIO, using the JIS, ensures that decision makers and the public are fully informed throughout a domestic incident response.

Figure 9.13   Who talks to the media?

Types of Plans

Jurisdictions must develop several types of plans, including:

Emergency Operations Plans (EOPs), which describe how the jurisdiction will respond to emergencies.

Procedures, which may include overviews, standard operating procedures, field operations guides, job aids, or other critical information needed for a response.

Preparedness plans, which describe how training needs will be identified and met, how resources will be obtained through mutual aid agreements, and the facilities and equipment required for the hazards faced by the jurisdiction.

Corrective Action or Mitigation Plans, which include activities required to implement procedures based on lessons learned from actual incidents or training and exercises.

Recovery Plans, which describe the actions to be taken to facilitate long-term recovery.

Figure 9.14   Types of plans.

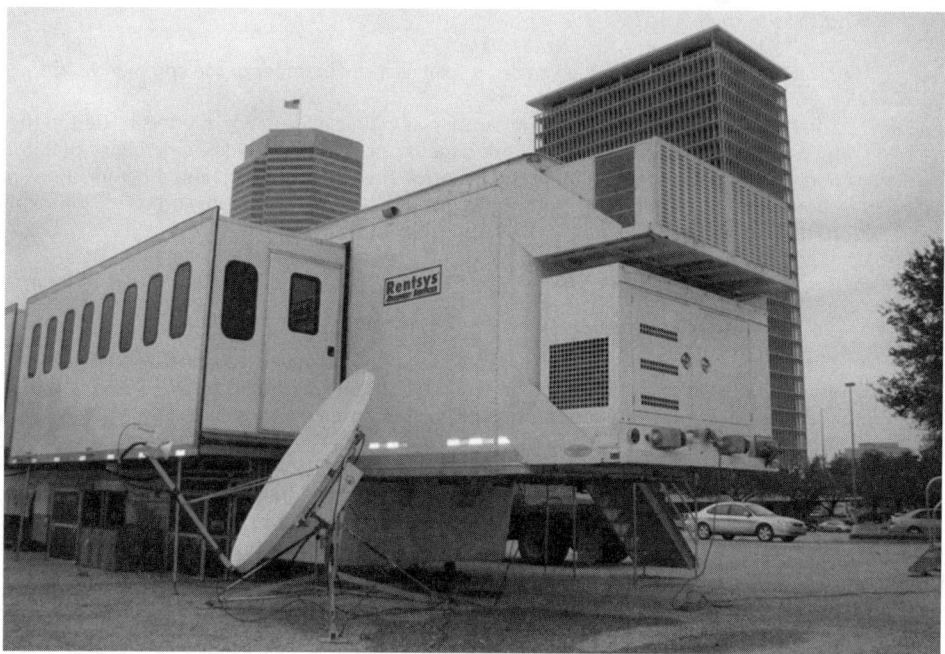

Figure 9.15 A Rentsys Recovery Services, Inc. Mobile Recovery Center (MRC) deploys to Houston, Texas in the aftermath of Hurricane Ike. The MRCs are completely self-sufficient with generator power and satellite. Most MRCs were onsite and operational within six hours of declaration. In some places the MRCs were the only structures with power, air-conditioning and communication for several days

Figure 9.16 Personnel work within a Rentsys Recovery MRC. This MRC was configured as a call center for an insurance company, allowing the company to respond almost immediately to its customers that were also severely impacted by the hurricane. MRCS can also be configured as command centers, mobile banks, work stations and data centers.

Figure 9.17 An MRC travels south on I-45 to deploy to Galveston, Texas, after Hurricane Ike. Rentsys Recovery is headquartered in College Station, Texas, but has hubs in Nevada, Florida, Wisconsin, Ohio, and Massachusetts, providing the ability to support anywhere within the continental United States within 48 hours of a declaration.

Figure 9.18 Days after Hurricane Ike, there was still significant water and wind damage. This heavily traveled underpass in Houston remained closed due to significant flooding.

(a)   (b)

Figure 9.19 Galveston bore the brunt of Hurricane Ike's force. (a) All that remains of a structure along Galveston's beachfront. (b) A pickup truck was dragged out onto a pier.

Figure 9.20 Other mobile recovery solutions for 4Ci include portable microwave systems and telephone cell sites called COWS (Cellular on Wheels). This unit was on display outside Dallas City Hall during an annual business resumption planning forum.

Figure 9.21 Sharon gets a tour of a mobile command center outside Dallas City Hall. Such mobile resources are invaluable after a major disaster, but remember, mobility might be a problem so take that into account in your plan.

Figure 9.22 Another view of the mobile command center. Set on a full size bus chasis, such a center has room to get almost as elaborate as necessary.

Figure 9.23 Another view of the same mobile command center.

Figure 9.24 Mold can be a serious problem in water-related disasters.

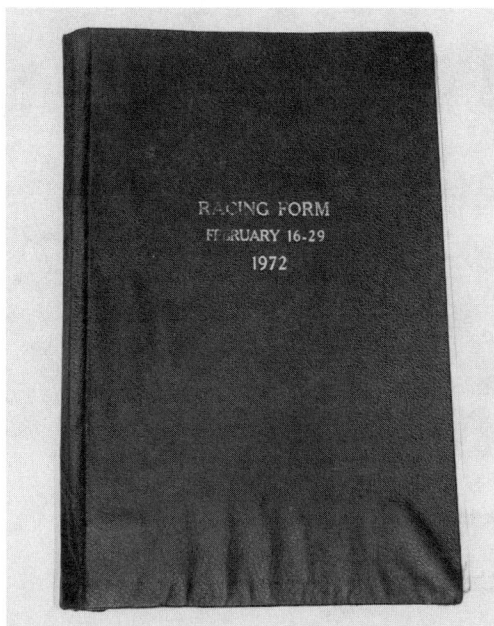

Figure 9.25 Fortunately, such a disaster is recoverable if the affected party acts quickly. The recommended course of action? Place water damaged items (paper and magnetic media) into a commercial freezer until they can be professionally recovered!

Figure 9.26   Before pictures of water-damaged paper.

Figure 9.27 After pictures of freeze-dried paper.

Figure 9.28 The inside of a BMS CAT freeze dry chamber.

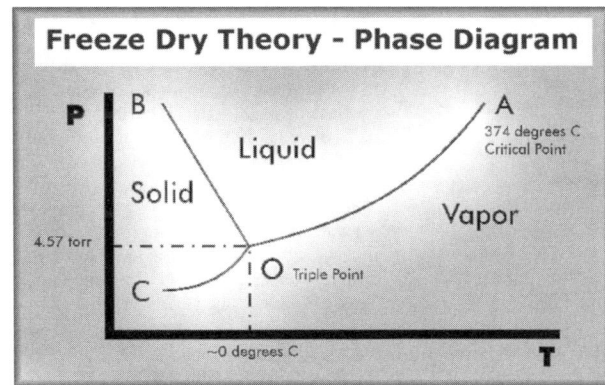

Figure 9.29   Freeze dry theory phase diagram.

Figure 9.30   Smoke damage—sometimes subtle but always costly.

WHAT TO DO IN THE FIRST 24 HOURS!
BMS CAT Special Technologies Division
Documents & Vital Records
Preventing Damage to Paper, Books and Microfilm

One of the most daunting tasks faced by record managers is recovering wet documents. Fire suppression, floods, rain, sprinkler pipe breaks and other disasters can leave paper records, microfilm and microfiche soaked with water. While you might think otherwise, 100 percent recovery is possible if you respond quickly. The basic strategy is to keep photographic media from drying and blocking, and to freeze paper documents to prevent further damage (which is the only way to save gloss finished paper). Here are the steps you should follow for recovery:

1. Seal the film. Photographic media (microfilm, microfiche and x-ray film) should be your first priority. Prepare a list so the contents can be tracked. Then, box and seal to prevent drying, refrigerating (at 35¼ - 40¼ F) if possible.

2. Freeze the paper. Puckering, swelling, ink smearing and blocking occurs as long as paper is wet. Inventory these documents, pack in boxes with plastic liners, palletize, and freeze. Once frozen, the damage ceases and the loss is in stasis until restoration can be accomplished.

For critical documents and special collections, blast freezing is best but seldom available in the time frame required. The freezing process is usually accomplished with refrigerated trucks that will transport the documents to the closest freeze dry facility for restoration (see step five below).

If documents are covered with silt (as a result of flooding), they are rinsed and cleaned on site before freezing. Remember that once frozen, documents become blocks of ice; if the unit you wish to consider is less than a box, separate the documents into modules with plastic or wax paper before freezing.

3. Separate vellum and leather bound documents. Vellum and leather are derived from animal skin and should be carefully separated from the rest of the documents. Drying should be done slowly and in a controlled fashion. Unlike other materials, they should not be heated during the freeze dry process.

4. Reprocess the microforms. As implied above, the emulsion layer on film will stick to contiguous substrate if it is allowed to dry, resulting in tearing and loss of data if you subsequently attempt to separate the film. Restoration involves machine reprocessing of wet microfilm and manual processing of fiche and other photographic film. Film may also be frozen for indefinite storage without further damage. For restoration, it must then be thawed and wet processed.

5. Freeze dry frozen paper documents. The next trick is to dry the paper without exposing the documents to the liquid phase. This can be accomplished by forcing sublimation (solid-state to vapor-state drying) in a freeze dry chamber with sufficient vacuum.

Call BMS CAT and state "This Is An Emergency!"
Contact: Lindsay Childs, BMS CAT 1-800-433-2940

Figure 9.31   Cleanup tips for paper, books, and microfilm.

WHAT TO DO IN THE FIRST 24 HOURS!
BMS CAT Special Technologies Division
Electronics

THE FIRST 24 HOURS of exposure to the effects of Fire or Water could determine if electronic equipment can be saved. This plan is intended only as a guideline for deciding if professional restoration services maybe required. Testing will help to determine the extent of exposure from the following potential sources of damage:

- **Fire**
- **Soot**
- **Water**
- **Heat**
- **Smoke**
- **Chemicals**

Examples of electronic equipment that may have a limited life if not treated within 24 hours:

- **Computers**
- **Telephone Switching**
- **Test Equipment**
- **Fax Electronic**
- **Controls**
- **Medical Equipment**
- **Copiers**
- **Processing Equipment**
- **Sound Equipment**

THE FIRST ACTION IS TO DE-ENERGIZE

The first action is to immediately de-energize and disconnect all equipment including any battery backups. Not only is there a danger to personnel working in the area and a danger of fire from electrical shorts, but electrochemical action can plate contaminants onto printed circuit boards and associated connectors and backplates. The reverse action may permanently remove metals.

CONTROL THE HUMIDITY

The first objective of restoration is to **remove the contaminants**. If all of the equipment cannot be cleaned simultaneously, it is important that immediate steps be taken to arrest the corrosion process. ***The most important step is to control the humidity!!*** Corrosion occurs very slowly if the relative humidity is below 50%. Testing will help determine which dehumidification process is best suited; Refrigerant or Desiccant, both have advantages if properly utilized.

- Maintain the electronics in an area where the relative humidity is below 50%.
- Move to another area if necessary to maintain the humidity and temperature balance.
- If you cannot move the electronics, seal each piece from the outside elements. Be careful not to trap moisture inside the chassis. Desiccants may be required.
- If water or liquids from the fire suppression systems are visible, perform the steps outlined under the heading "Water Damage" first.

TESTING TO ASSESS THE DAMAGE

Two basic measurements should be made to assess the corrosion potential in a loss involving electrical / electronic equipment. The first is a surface concentration of **halogenides** to determine the chloride corrosion potential. The second test is pH, a measure of **acidity** of the contaminant. These tests are run on hard horizontal surfaces not disturbed by cleaning efforts. The importance of these tests and their interpretation is critical for establishing a baseline and a cleaning protocol.

(a)

Figure 9.32 (a, b) Cleanup tips for electronics.

CORROSION CONTROL

In cases of severe contaminant concentration, a special non-petroleum preservative may be sprayed on the equipment to exclude moisture and air. This very thin film is designed to be removed easily in the restoration process later.

ESTABLISH A CLEANING AND QUALITY PROTOCOL

Once the corrosion process is stabilized, the appropriate cleaning and quality protocols will be designed and applied by BMS Special Technologies. A written Scope of Work will detail specific concerns to assure quality compliance to industry and Mil Spec. Standards.

FIRE DAMAGE - EFFECTS AND PROCEDURES

Equipment which has suffered thermal damage as evidenced by cosmetic aberrations in plastic components may not be restored. Besides heat, a fire generates combustion byproducts. These byproducts are locked into the soot which condenses on all cool surfaces. Smoke exposure during the fire for a relatively short time does little damage, but the particulate deposited may contain active corrosive components. These components in the presence of humidity and oxygen will corrode metal surfaces. Irreversible damage can occur in the time period of a few days.

WATER DAMAGE - EFFECTS AND PROCEDURES

WARNING: DO NOT ENERGIZE ANY WET EQUIPMENT - REMOVE POWER

It is a popular misconception that electronic equipment exposed to water is permanently damaged. Water which has sprayed, splashed, or dripped onto electronic equipment can be easily removed. Even equipment which has been totally submerged can be restored. The most important issue in the amount of damage is whether the equipment was powered at the time of exposure. As in the case of fire created corrosives, ***immediate countermeasures are imperative!!***

- Open cabinet doors, remove side panels and covers, and pull out chassis drawers to allow water to run out of the equipment.
- Remove standing water with wet vacs. Use low pressure air (50 psi) to blow trapped water out of the equipment. Absorbent cotton pads (diapers?) can be used to blot up water. Use appropriate caution around header pins and backplane wire wrap connectors to avoid bending.
- Vacuum and mop up water under any raised computer room floor.
- Equipment which contains open relays and transformers will require a special bake out before application of power.
- Water displacement aerosol lubricant sprays may be used to protect critical components.

Call BMS CAT and state "This Is An Emergency!"
Contact: Lindsay Childs, BMS CAT 1-800-433-2940

(b)

Figure 9.32   (continued)

WHAT TO DO IN THE FIRST 24 HOURS!
BMS CAT Special Technologies Division
Magnetic, Optical, Information Media

The First 24 Hours of exposure to the effects of Fire or Water will usually determine if emergency steps need to be taken to clean, restore and preserve magnetic data storage media. This plan is intended only as a guideline for deciding if professional restoration services maybe required due to exposure from the following potential sources of damage:

- Fire
- Soot
- Water
- Heat
- Smoke
- Chemicals

The media under consideration includes but is not limited to:

- Magnetic Media
- Optical Media
- Magneto Optical Media

DO NOT USE MAGNETIC MEDIA THAT HAS BEEN EXPOSED TO CONTAMINANTS
The most important asset, which must be preserved after a disaster, is the critical data on magnetic media. A professional should examine media that has been exposed to contaminants before any attempt is made to use them. A **95% to 100% restoration success is possible predicated on 72 to 96 hour response time.**

Removable hard disk platters exposed to smoke often have particulate matter on the surface that must be professionally cleaned prior to activation. **Hard disk data** can be partially saved - even after a head crash. This process is very labor intensive and requires special equipment in a clean room. Contaminated media is replaced with clean media. Restoration of data is a process involving the emergency cleaning of the media so that data may be copied onto other media. The original media will be discarded (or archived).

Disk read/write heads are subject to severe damage if an attempt is made to operate with media which is not clean. A "head crash" caused by particulate on the surface of the hard disk media will not only damage the drive, but may cause loss of data.

Floppy disks with hard particulate matter on the surface could cause damage to the oxide layer and may destroy data as the floppy spins. Water can dissolve the adhesive between the substrate and the magnetic oxide coating resulting in loss of data. If the media is wet, keep it wet until professional restoration.

Magnetic Tapes must be dry and clean before any attempt is made to copy the data.
- **Initial mitigation** includes removing standing water. Open cartridge access door and shake out water. Use clean low pressure air to force water off. Remove rings on reel tape for the same purpose.
- **Keep the media wet** until restoration. Use zip-lock bags, plastic trash-can liners, etc.
- Do NOT attempt to dry with heat!!!!

Optical Media: A CD disk contains from 650 mb to over a gigabyte depending upon format. This data layer is read through the optically clear substrate material by a low power laser diode. Anything such as a scratch or particulate can interfere and cause a data error. The protective lacquer coating on CD-ROM

(a)

Figure 9.33 (a, b) Cleanup tips for magnetic, optical, and information media.

disks can be very thin. The lacquer coating side of a CD is the side with the printed label, artwork, and/or track information. If this layer is scratched, the data layer is directly exposed to the environment and degrades at an accelerated rate. The surface must be protected until restoration can be used to remove contaminants. If optical media is wet, keep wet until it can be cleaned.

<u>**Magneto Optical**</u> (CD RW) is processed the same as other optical media.

<u>**Microfilm, Microfiche, X-Ray Film:**</u> The most important thing to know about microfilm is that once the film is wet do not let it dry!! The film must be processed while still wet or the gelatin coating will stick to the next layer and the document information will be torn from the film. Here again, speed is of the essence:

•   For short time storage, five gallon buckets can be used to store film with enough (preferably distilled) water to cover the film. Zip lock bags or plastic cellophane wrap can also be used to package the film and prevent drying. (keep film cold 35 to 45 degrees)

•   Use gloves when handling wet materials and wash hands thoroughly to prevent infection from flood bio-contaminants

•   For longer storage than a few days, a conservator must add special gelatin hardening chemicals to water.

•   For long term storage, the film may be frozen without damage. Film can never be freeze-dried - it must be thawed and wet processed.

Call BMS CAT and state "This Is An Emergency!"
Contact: Lindsay Childs, BMS CAT 1-800-433-2940

(b)

Figure 9.33   (continued)

Figure 9.34   Disaster Recovery in China. Leo A. Wrobel addresses the Chinese Academy of Sciences about "4Ci" in November 2008.

Figure 9.35   Disaster Recovery in China. Leo A. Wrobel pictured with Dr. Chen An, (third from the right in the picture) Chinese Academy of Sciences, and distinguished professors from Korea, China and Japan.

Figure 9.36   Leo and Sharon Wrobel tour the Olympic Park in Beijing.

Figure 9.37   Chinese Cellular on Wheels (COW) at the Olympic Park in Beijing.

CHAPTER 10

Directory of Disaster Recovery Information Sources

One of the more popular sections in our previous books has been sources of free, public domain, or low-cost information available to the contingency planner. This sentiment has been further reinforced by the prospect (at least at the time of this publishing) of a sluggish economy and attendant tight budgets. Even in tough economic times, however, sources of good disaster recovery information exist for disaster recovery planners forced to take a "guerilla warfare" approach due to lack of funding for the initiative. Indeed, disaster recovery planning often becomes even more important in lean economic times, when a major disaster could be the last financial straw for many organizations.

With these thoughts in mind, we present a comprehensive listing of places, organizations, and Web sites that you can visit to sort out your planning effort (Table 10.1).

Edwards Information publishes the *Edwards Disaster Recovery Guide*, which is in use by thousands of organizations of every size and shape. With more than 7,000 listings in 450 different categories, it is probably the most comprehensive resource of its type on the market today. We illustrate three different categories from the guide in this chapter:

- Disaster recovery associations;
- Disaster recovery consultants;
- Disaster recovery publications and Web sites.

These three categories were specifically selected by the authors in order to aid contingency planners in finding low-cost, public domain, or reasonably priced resources, particularly for those on a tight budget due to the economy. We thank everyone at Edwards Information and specifically CEO Bob Bulick for their thoughtful and useful contribution, and encourage you to visit them at http://www.edwardsinformation.com.

Table 10.1   Disaster Recovery Associations

Association of Washington Business P.O. Box 658 Olympia WA 98507-0658 USA 360-943-1600 www.awb.org	County Executives of America 1100 H St. NW Suite 910 Washington, D.C. 20005 USA 202-628-3585 www.countyexecutives.org	International Association of Business Communicators One Hallidie Plaza Suite 600 San Francisco, CA 94102 USA 415-544-4700 www.iabc.com
Security Industry Association 635 Slaters Lane Suite 110 Alexandria, VA 22314 USA 703-647-8484 www.siaonline.org	Society for Human Resource Management 1800 Duke St. Alexandria, VA 22314-3499 USA 703-548-3440 www.shrm.org	
American Biorecovery Association P.O. Box 828 Ipswich, MA 01938 USA 987-356-4606 www.americanbiorecovery.com	Association of Specialists in Cleaning and Restoration 8229 Cloverleaf Drive Suite 460 Millersville, MD 21108-1838 USA 410-729-9900 www.ASCR.org	Association of Specialists in Cleaning and Restoration International 9810 Patuxent Woods Dr. Suite K Columbia, MD 21046 USA 443-878-1000 www.ascr.org
The Carpet & FabriCare Institute P.O. Box 2160 Mission Viejo, CA 92690 USA www.carpet9.org	Carpet and Rug Institute 730 College Dr. Dalton, GA 30721 USA 706-278-3176 www.carpet-rug.com	Carpet Cleaners Institute of the Northwest PMB 40, 2421 S. Union Ave. Suite L-1 Tacoma, WA 98405 USA 503-612-9692 www.ccinw.org
Cleaning Management Institute c/o Nat'l. Trade Publications 13 Century Hill Dr. Latham, NY 12110-2197 USA 518-783-1281 www.cminstitute.net	Institute of Inspection Cleaning and Restoration Certification 2715 E. Mill Plain Blvd. Vancouver, WA 98661 USA 360-693-5675 www.iicrc.org	International Fabricare Institute 14700 Switzer Lane Laurel, MD 20707 USA 301-622-1900 www.ifi.org
International Society of Cleaning Technicians SCRT Headquarters 200 Vantage Way Franklin, TN 37067 USA 615-591-9610 www.isct.org	Laundry and Dry Cleaning International Union 307 Fourth Ave. Suite 405 Pittsburgh, PA 15222 USA 412-471-4829	MasterPros Emergency Services Consortium 270 Sheldon Avenue Suite 405 Toronto, ON M8W 4M1 Canada www.masterpros.com
Mid-Atlantic Cleaners and Launderers Association 7430 Little Chatterton Ln. King George, VA 22485 USA 540-775-2525 www.macla.net	National Cleaners Association 252 W. 29th St. Second Floor New York, NY 10001-5201 USA 212-967-3002 www.nca-i.com	New England Institute of Restoration and Cleaning 253 Low St. P.O. Box 313 Newburyport, MA 01950 USA 978-779-0950 www.neirc.org
North East Fabricare Association 580 Main St. Reading, MA 01867-3107 USA 781-942-7630 www.nefabricare.com	Restoration Alliance Suite 1 1305 Fraser Street Bellingham, WA 98229 USA 877-693-0111 www.restorationalliances.com	Rocky Mountain Fabricare Association 11166 Huron St. Suite 27 Denver, CO 80234 USA 303-433-4446

Society of Cleaning and
Restoration Technicians
200 Vantage Way
Franklin, TN 37067 USA
615-591-9610
www.scrthq.org

Southwest Drycleaners
Association
1800 N.E. Loop 410
Suite 308
San Antonio, TX 78217-5251 USA
210-826-4684
www.sda-dryclean.com/

Western States Drycleaners and
Launderers
P.O. Box 31838
Phoenix, AZ 85046-1838 USA
602-253-9186
www.wsdla.org

AFCOM
742 East Chapman Avenue
Orange, CA 92866 USA
714-997-7966
www.afcom.com

American Society for Industrial
Security (ASIS)
1625 Prince Street
Alexandria, VA 22314-2818 USA
703-519-6200
www.asisonline.org

American Society for Information
Science and Technology
1320 Fenwick Lane
Suite 510
Silver Spring, MD 20910 USA
301-495-0900
www.asis.org

Association of Service and
Computer Dealers International
131 N.W. First Ave.
Delray Beach, FL 33483 USA
561-266-9016
www.ascdi.com

Computer Security Institute
600 Harrison Street
San Francisco, CA 94107-1387 USA
866-271-8529
www.gocsi.com

Independent Computer Consultants
Association
11131 S. Towne Square
Suite F
St. Louis, MO 63123-7817 USA
314-892-1675
www.icca.org

Information Systems Audit and
Control Association
3701 Algonquin Road
Suite 1010
Rolling Meadows, IL 60008 USA
847-253-1545
www.isaca.org

Information Technology
Association of America
1401 Wilson Boulevard
Suite 1100
Arlington, VA 22209-2318 USA
703-522-5055
www.itaa.org

ISSA Information Systems Security
Association
7044 South 13th Street
Oak Creek, WI 53154 USA
414-908-4949
www.issa.org

IT Productivity Center
6900 Westcliff
Suite 801
Las Vegas, NV 89145 USA
435-940-9300
www.itproductivity.org

Software and Information Industry
Association
1090 Vermont Ave., N.W., 6th Fl.
Washington, D.C. 20005 USA
202-289-7442
www.siia.net

The SANS Institute
8120 Woodmont Avenue
Suite 205
Bethesda, MD 20814 USA
301-654-7267
www.sans.org

Wall Street Technology
Association
241 Maple Avenue
Red Bank, NJ 07701 USA
732-530-8808
www.wsta.org

Academy of Certified Archivists
90 State St.
Suite 1009
Albany, NY 12207 USA
518-463-8644
www.certifiedarchivists.org

ARMA International
13725 West 109th Street
Suite 101
Lenexa, KS 66215 USA
913-341-3808
www.arma.org

Association for Information &
Image Management (AIIM)
1100 Wayne Avenue
Suite 1100
Silver Spring, MD 20910-5616 USA
301-587-8202
www.aiim.org

Data Management Association
International
P.O. Box 5786
Bellevue, WA 98006-5786 USA
425-562-2636
www.dama.org

Database Research Association
P.O. Box 31
Ashton, MD 20861 USA
301-774-5414

Document Management Industries
Association
433 E. Monroe Ave.
Alexandria, VA 22301 USA
703-836-6232
www.dmia.org

Healthcare Information and
Management Systems Society
230 E. Ohio St.
Suite 500
Chicago, IL 60611-3269 USA
312-664-4467
www.himss.org

International Disk Drive Equipment
and Materials Association
470 Lakeside Dr.
Sunnyvale, CA 94085 USA
408-991-9430
www.idema.org

PRISM International: Professional
Records & Information Services
Management
131 US 70
West Garner, NC 27529-3905 USA
919-881-0677
www.prismintl.org

Society of American Archivists
527 S. Wells St., Fifth Floor
Chicago, IL 60607 USA
312-922-0140
www.archivists.org

American Public Works Association
2345 Grand Boulevard
Suite 500
Kansas City, MO 64108-2641 USA
www.APWA.net

Business Disaster Preparedness
Council, The Lee County
Government
P.O. Box 398
Fort Myers, FL 33902-0398 USA
239-332-2737
www.leegov.com

Business Recovery Planners
Assoc. of WI
5910 Mineral Point Road
Madison, WI 53705-4456 USA
608-231-7502

Canadian Emergency Preparedness
Association
6715 Henry Street
Chilliwack, BC V2R 2C2 Canada
www.cepa-acpc.ca

Center for Domestic Preparedness
P.O. Box 5100
Anniston, AL 36205-5100 USA
866-213-9553
cdp.dhs.gov

IMAGE Society
P.O. Box 6221
Chandler, AZ 85246-6221 USA
www.public.asu.edu/~image

Medical Records Institute
425 Boylston St.
Fourth Floor
Boston, MA 02116 USA
617-964-3923
www.medrecinst.com

Professional Records and
Information Services Management
International
131 US 70
West Garner, NC 27529 USA
919-771-0657
www.prismintl.org

Association of Contingency
Planners National HQ
7044 South 13th Street
Oak Creek, WI 53154-1429 USA
414-768-8000
www.acp-international.com

Business Network of Emergency
Resources, Inc.
11 Hanover Square
Suite #501
New York, NY 10005 USA
212-599-1599
www.bnetinc.org/home.html

Business Recovery Planners
Association of SE WI
1957 South 81st Street
West Allis, WI 53219-1011 USA
414-543-8100

Caribbean Disaster Emergency
Response Agency
Manor Lodge Building #1
Lodge Hill USA
246-425-0386
www.cdera.org

Church World Service Emergency
Response Program
475 Riverside Dr.
Suite 700
New York, NY 10115 USA

Institute of Certified Records
Managers
5818 Molloy Rd.
Syracuse, NY 13211 USA
315-234-1904
www.icrm.org

National Association of
Government Archives and Records
Administrators
90 State St.
Suite 1009
Albany, NY 12207 USA
518-463-8644
www.nagara.org

Society for Information
Management
401 North Michigan Avenue
Chicago, IL 60611-4267 USA
312-527-6734
www.simnet.org

America Prepared Campaign
45 Rockfeller Plaza
New York, NY 10111 USA
212-332-6302
www.americaprepared.org

Business Continuity Planners
Association
P.O. Box 75930
St. Paul, MN 55175-0930 USA
651-223-9801
www.bcpa.org

Business Recovery Managers
Association
P.O. Box 2184
San Francisco, CA 94126 USA
925-355-8660
www.brma.com

Canadian Centre for Emergency
Preparedness
860 Harrington Ct.
Burlington, ON L7N 3N4 Canada
www.ccep.ca

Center for Domestic Preparedness
P.O. Box 5100
Anniston, AL 36205-5100 USA
866-213-9553
cdp.dhs.gov

Contingency Planning Exchange
11 Hanover Square
Suite 501
New York, NY 10005 USA
www.cpeworld.org

Disaster Forum Association
11215 Jasper Avenue
Suite 437
Edmonton, AB T5K 0L5 Canada

Disaster Recovery Info. Exchange
P.O. Box 27035
Kitchener, ON N2M 5P2 Canada
www.drie-swo.org/

Disaster Recovery Institute
2175 Sheppard Avenue East
Suite 310
Toronto ON M2J 1W8 Canada
888-728-3742
www.dri.ca

Great Lakes Business Recovery
Group
1647 Dancer Drive
Suite 102
Rochester Hills, MI 48307-3312
USA
248-650-9900

International Association of
Emergency Managers
201 Park Washington Court
Falls Church, VA 22046-4527 USA
www.iaem.com

Mid-America Contingency
Planning Forum
P.O. Box 38112
St. Louis, MO 63138 USA
314-466-3509
www.brookstech.net/mcpf/
index.html

Midwest Contingency Planners
P.O. Box 1632
Indianapolis, IN 46206-1632 USA
765-778-8758

New England Disaster Recovery
Information Exchange
P.O. Box 52120
Boston, MA 02205 USA
www.nedrix.com

Disaster Preparedness and
Emergency Response Association
P.O. Box 797
Longmont, CA 80502 USA
970-532-3362
www.disasters.org

Disaster Recovery Information
Exchange
157 Adelaide Street West
P.O. Box 247
Toronto, M5H 4E7 Canada
647-299-9743
www.drie.org

DR Information E-Change Group
CVS Pharmacy, Inc.
1 CVS Drive #1
Woonsocket, RI 02895 USA
401-765-1500

Great Plains Contingency Planners
P.O. Box 33
Omaha, NE 68101-0033 USA
www.greatplainscontingencyplann
ers.com

International Disaster Recovery
Association
Shrewsbury, MA USA
www.idra.com

Mid-Atlantic Disaster Recovery
Association Inc.
4500 Paint Branch Parkway
College Park, MD 20743 USA
703-456-5744
www.madra.org

National Emergency Management
Association
P.O. Box 11910
Lexington, KY 40578 USA
www.nemaweb.org

Northeast Florida Association
of Contingency Planners
P.O. Box 23556
Jacksonville, FL 32256 USA
www.acp-international.com/
neflorida

Disaster Recovery Info
Exchange-Ottawa
P.O. Box 20518
390 Rideau Street
Ottawa, K1N 1A3 Canada
613-238-2909
www.drieottawa.org

Disaster Recovery Information
Exchange (West)
POB 1557 Station M
Calgary, T2P 3B9 Canada
www.drie.org/west/

DRI International
1400 I St. NW Suite 1050
Washington, D.C. 20005 USA
202-962-3979
www.drii.org

Health Canada, Centre for
Emergency Preparedness and
Response
100 Colonnade Road, PL6201A
Ottawa, ON K1A 0K9 Canada
www.phac-aspc.gc.ca/new_e.html

ISA International Society of
Arborculture
P.O. Box 3129
Champaign, IL 61826-3129 USA
217-355-9411
www.isa-arbor.com

Mid-Island Emergency
Coordinators and Managers
#210 660 Primrose Street
Qualicum Beach, BC V9K 1S7
Canada
www.qualicumbeach.com

National Fire Protection Agency
One Batterymarch Park
Quincy, MA 02169-7471 USA
617-984-7275
www.nfpa.org

Northeast States Emergency
Consortium
1 West Water Street
Suite 205
Wakefield, MA 01880-1301 USA
www.nesec.org

Ontario Association of Emergency
Managers
P.O. Box 67043
2150 Burnhamthorpe Road W.
Mississauga, ON L5L 5V4 Canada
www.oaem.ca

Partnership for Emergency
Planning
c/o Sprint
MS:KSOPHM0106-1B402
6480 Sprint Parkway-Truman C
Overland Park, KS 66251 USA
913-315-8224
www.pepkc.org

PSEG
80 Park Plaza L1a
Newark, NJ 07102 USA
www.pseg.com

Public Risk Management
Association
500 Montgomery Street
Suite 750
Alexandria, VA 22314 USA
703-528-7701
www.primacentral.org

Puerto Rico Info Security
Emergency Mgmt. Assoc.
John Robles and Associates
P.O. Box 29715
San Juan, PR 00929-0715 USA
787-647-3961

Terrorist Incident Response
Association
234 West Dixie Avenue
Marietta, GA 30060 USA
678-640-9743
www.tiraonline.org

The American Civil Defense
Association
11576 South State Street
Suite #502
Draper, UT 84020 USA
801-501-0077
www.tacda.org

The Center for Biosecurity of
UPMC
The Pier IV Building
621 East Pratt Street
Suite 210
Baltimore, MD 21202 USA
443-573-3304
www.upmc-biosecurity.org

Air and Waste Management
Association
One Gateway Center
Third Floor
420 Ft. Duquesne Blvd.
Pittsburgh, PA 15222-1435 USA
412-232-3444
www.awma.org

Alliance of Indiana Rural Water
5715-A Churchman Ave.
Indianapolis, IN 46203 USA
317-789-4200
www.inh2o.org

American Academy of
Environmental Engineers
130 Holiday Ct.
Suite 100
Annapolis, MD 21401 USA
410-266-3311
www.aaee.net

American Academy of Sanitarians
1568 Le Grand Circle
Lawrenceville, GA 30043-8191
USA
678-407-1051
www.sanitarians.org

American Society of Sanitary
Engineering
901 Canterbury Road
Suite A
Westlake, OH 44145 USA
440-835-3040
www.asse-plumbing.org

American Water Resources
Association
P.O. Box 1626
Middleburg, VA 20118-1626 USA
540-687-8390
www.awra.org

American Water Works
Association-Chesapeake Section
2981 Beacon Dr.
Manchester, MD 21102 USA
410-374-0318
www.csawwa.org

Asphalt Recycling and Reclaiming
Association
Three Church Circle
Suite 250
Annapolis, MD 21401-1902 USA
410-267-0023
www.arra.org

Association for Preservation
Technology International
4513 Lincoln Ave.
Suite 213
Lisle, IL 60532 USA
630-968-6400
www.apti.org

Association for the Environmental
Health of Soils
150 Fearing St.
Amherst, MA 01002 USA
413-549-5170
www.aehs.com

Association of Conservation
Engineers
c/o Missouri Department of
Conservation
P.O. Box 180
Jefferson City, MO 65102 USA
573-52241153739
http://conservationengineers.org

Association of Consulting Foresters
of America
312 Montgomery St.
Suite 208
Alexandria, VA 22314 USA
703-548-0990
www.acf-foresters.org

Association of Environmental
Authorities
2333 Whitehorse-Mercerville Rd.
Suite Three
Mercerville, NJ 08619-1978 USA
609-584-1877
www.aeanj.org

Association of Metropolitan Water
Agencies
1620 I St. NW #500
Washington, D.C. 20006 USA
202-331-2820
www.amwa.net

Association of State Wetland
Managers
P.O. Box 269
Berne NY 12023-9746 USA
518-872-1804
www.aswm.org

Ecological Society of America
1707 H St. NW
Suite 400
Washington, D.C. 20006 USA
202-833-8773
www.esa.org

Environmental Industry
Associations
4301 Connecticut Ave. NW
Suite 300
Washington, D.C. 20008-2304
USA
202-244-4700
www.envasns.org

Federal Water Quality Association
P.O. Box 14303
Washington, D.C. 20044-4303
USA
202-566-2582
www.fwqa.org

Forest Resources Association
600 Jefferson Plaza
Suite 350
Rockville, MD 20852 USA
301-838-9385
www.forestresources.org

Groundwater Management
Districts Association
P.O. Box 905
Colby, KS 67701-0905 USA
785-462-3915
www.gmdausa.org

Association of Environmental
Engineering and Science Professors
2303 Naples Court
Champaign, IL 61822 USA
217-398-6969
www.aeesp.org

Association of State Dam Safety
Officials
450 Old Vine Street
2nd Floor
Lexington, KY 40507-1544 USA
www.damsafety.org

Association of Water Technologies
8201 Greensboro Dr.
Suite 300
McLean VA 22102 USA
703-610-9012
www.awt.org

Environmental Assessment
Association
USA
320-763-4320
www.iami.org/eaa.cfm

Environmental Information
Association
6935 Wisconsin Ave.
Suite 306
Chevy Chase, MD 20815-6112
USA
301-961-4999
www.eia-usa.org

Federation of Environmental
Technologists
9451 N. 107th St.
Milwaukee, WI 53224-1105 USA
414-354-0070
www.fetinc.org

Forestry Conservation
Communications Association
P.O. Box 3217
Gettysburg, PA 17325 USA
717-338-1505
www.fcca-usa.org

Institute of Environmental Sciences
and Technology
5005 Newport Dr.
Suite 506
Rolling Meadows, IL 60008-3841
USA
847-255-1561
www.iest.org

Association of Fish and Wildlife
Agencies
444 North Capitol St. NW
Suite 275
Washington, D.C. 20001 USA
202-624-7890
www.fishwildlife.org

Association of State Floodplain
Managers
2809 Fish Hatchery Road
Suite 204
Madison, WI 53713 USA
608-274-0123
www.floods.org/home/

California Association of Sanitation
Agencies
925 L St.
Suite 1400
Sacramento, CA 95814 USA
916-446-0388
www.casaweb.org

Environmental Design Research
Association
P.O. Box 7146
Edmond, OK 73083-7146 USA
405-330-4863
www.edra.org

Environmental Mutagen Society
1821 Michael Faraday Dr.
Suite 300
Reston, VA 20190 USA
703-438-8220
www.ems-us.org

Floodplain Mangement Association
P.O. Box 712080
Santee, CA 92072-2080 USA
619-204-4380
www.floodplain.org

Ground Water Protection Council
13308 N. MacArthur Blvd.
Oklahoma City, OK 73142 USA
405-516-4972
www.gwpc.org

International Association of
Wildland Fire
P.O. Box 261
Hot Springs, SD 57747-0261 USA
206-600-5113
www.iawfonline.org

International Erosion Control
Association
3001 S. Lincoln Ave.
Suite 8
Steamboat Springs, CO 80487 USA
970-879-3010
www.ieca.org

International Water Resources
Association
Southern Illinois University
4535 Faner Hall
Carbondale, IL 62901-4516 USA
618-453-5138
www.iwra.siu.edu

Municipal Waste Management
Association
1620 I St. NW
Suite 300
Washington, D.C. 20006 USA
202-293-7330
www.usmayors.org/uscm/mwma

National Alliance of Preservation
Commissions
P.O. Box 1605
Athens, GA 30603 USA
706-542-4731
www.uga.edu/sed/pso/
programs/napc/napc.htm

National Association for
Environmental Management
1612 K. St. NW
Suite 1102
Washington, D.C. 20006 USA
202-986-6616
www.naem.org

National Association of Clean
Water Agencies
1816 Jefferson Place NW
Washington, D.C. 20036-2505 USA
202-833-2672
www.nacwa.org

National Association of
Conservation Districts
509 Capitol Ct. NE
Washington, D.C. 20002 USA
202-547-6223
www.nacdnet.org

National Association of
Environmental Professionals
389 Main St. Suite 202
Malden, MA 02148 USA
781-397-8870
www.naep.org

National Association of Flood and
Stormwater Management Agencies
1301 K St. NW Suite 800 East
Washington, D.C. 20005 USA
202-218-4122
www.nafsma.org

National Association of Service and
Conservation Corps
666 11th St. NW
Suite 1000
Washington, D.C. 20001-4542
USA
202-737-6272
www.nascc.org

National Association of Sewer
Service Companies
1314 Bedford Ave.
Suite 201
Baltimore, MD 21208 USA
410-486-3500
www.nassco.org

National Association of Water
Companies
1725 K St. NW
Suite 200
Washington, D.C. 20006 USA
202-833-8383
www.nawc.org

National Environmental Health
Association
720 S. Colorado Blvd.
North Tower
Suite 1000-N
Denver, CO 80246-1926 USA
303-756-9090
www.neha.org

National Environmental, Safety
and Health Training Association
P.O. Box 10321
Phoenix, AZ 85064-0321 USA
602-956-6099
www.neshta.org

National Ground Water Association
601 Dempsey Road
Westerville, OH 43081-8978 USA
614-898-7791
www.ngwa.org

National Parks Conservation
Association
1300 19th St. NW Suite 300
Washington, D.C. 20036 USA
202-223-6722
www.eparks.org

National Rural Water Association
2915 S. 13th St.
Duncan, OK 73533 USA
580-252-0629
www.nrwa.org

National Solid Wastes Management
Association
4301 Connecticut Ave. NW
Suite 300
Washington, D.C. 20008 USA
202-244-4700
www.nswma.org

National Solid Wastes Management
Association-Northeast Region
290 Turnpike Rd.
PMB #407
Westborough, MA 01581-2843
USA

National Water Resources
Association
3800 N. Fairfax Drive
Suite Four
Arlington, VA 22203 USA
703-524-1544
www.nwra.org

National Waterways Conference
4650 Washington Blvd., #608
Arlington, VA 22201 USA
703-243-4090
www.waterways.org

National Wildlife Rehabilitators
Association
2625 Clearwater Road
Suite 110
St. Cloud, MN 56301 USA
320-230-9920
www.nwrawildlife.org

New England Water Environment
Association
100 Tower Office Park
Suite K
Woburn, MA 01801 USA
781-939-0908
www.newea.org

New England Water Works
Association
125 Hopping Brook Rd.
Holliston, MA 01746 USA
508-893-7979
www.newwa.org

NORA: An Association of
Responsible Recyclers
5965 Amber Ridge Road
Haymarket, VA 20169 USA
703-753-4277
www.noranews.org

Ontario Environment Industry
Association
2175 Sheppard Avenue East
Suite 310
Toronto, ON M2J 1W8 Canada
www.ceia.on.ca

River Management Society
P.O. Box 9048
Missoula, MT 59807 USA
406-549-0514
www.river-management.org

Society for Human Ecology
College of the Atlantic 105 Eden St.
Bar Harbor, ME 04609 USA
207-288-5015
www.societyforhumanecology.org

Solid Waste Association of North
America-Alabama Chapter
P.O. Box 240757
Montgomery, AL 36124 USA
334-260-7970

Spill Control Association of
America
2105 Laurel Bush Road
Suite 200
Bel Air, MH 21015 USA
443-640-1085
www.scaa-spill.org

Washington Association of Sewer
and Water Districts
2800 S. 192nd St.
Suite 104
Sea Tac, WA 98188-5166 USA
206-246-1299
www.waswd.org

Water Environment Federation
601 Wythe St.
Alexandria, VA 22314-1994 USA
703-684-2400
www.wef.org

AIM Global
125 Warrrendale Road
Suite 100
Warrendale, PA 15086 USA
724-934-4470
www.aimglobal.org

North American Association for
Environmental Education
2000 P St. NW
Suite 540
Washington, D.C. 20036 USA
202-419-0412
www.naaee.org

Pacific Northwest Waterways
Association
9115 S.W. Oleson Rd.
Suite 101
Portland, OR 97232 USA
503-234-8556
www.pnwa.net

Society for Ecological Restoration
285 W. 18th St., #1
Tucson, AZ 85701 USA
520-622-5485
www.ser.org

Society of Environmental
Toxicology and Chemistry
1010 N. 12th Ave.
Pensacola, FL 32501-3367 USA
850-469-1500
www.setac.org

Solid Waste Association of North
America-California Chapters
1414 K St. Suite 320
Sacramento, CA 95814 USA
916-446-4656
www.swanacal-leg.org

United States Committee on
Irrigation and Drainage
1616 17th St.
Suite 483
Denver, CO 80202 USA
303-628-5430
www.uscid.org

Washington Refuse & Recycling
Association
4160 Sixth Ave. SE
Suite 205
Lacey, WA 98503 USA
360-943-8859
www.wrra.org

Water Quality Association
4151 Naperville Road
Lisle, IL 60532 USA
630-505-0160
www.wqa.org

American Fire Sprinkler
Association
9696 Skillman St. Suite 300
Dallas, TX 75243 USA
214-349-5965 x118
www.firesprinkler.com

Northeast Resource Recovery
Association
2101 Dover Rd.
Epsom, NH 03234 USA
603-736-4401
www.recyclewithus.org

Pacific Water Quality Association
17300 17th St.
Suite J-266
Tustin, CA 92780-7918 USA
760-644-7348
www.pwqa.org

Society for Environmental
Geochemistry and Health
4698 S. Forrest Ave.
Springfield, MO 65810 USA
417-885-1166
www.segh.net

Soil and Water Conservation
Society
945 S.W. Ankeny Road
Ankeny, IA 50023 USA
515-289-2331
www.swcs.org

Southeast Desalting Association
2409 S.E. Dixie Hwy.
Stuart, FL 34996 USA
772-781-7698
www.southeastdesalting.com

Virginia Association of Soil and
Water Conservation Districts
7308 Hanover Green Dr.
Suite 100
Mechanicsville, VA 23111 USA
804-559-0324
www.vaswcd.org

Washington State Recycling
Association
6100 Southcenter Blvd.
Ste 180
Tukwila, WA 98188 USA
206-244-0311
www.wsra.net

Battery Council International
401 N. Michigan Ave.
Suite 2200
Chicago, IL 60611-4267 USA
312-644-6610
www.batterycouncil.org

Cleaning Equipment Trade
Association
968 Lake St., South
Suite 202
Forest Lake, MN 55025 USA
651-982-0010
www.ceta.org

Fire and Emergency Manufacturers
and Services Association
P.O. Box 147
Lynnfield, MA 01940-0147 USA
781-334-2771
www.femsa.org

International Foodservice
Manufacturers Association
Two Prudential Plaza
180 N. Stetson Ave. Suite 4400
Chicago, IL 60601 USA
312-540-4400
www.ifmaworld.com

National Fire Sprinkler Association
P.O. Box 1000
Patterson, NY 12563 USA
845-878-4200
www.nfsa.org

Portable Sanitation Association
International
7800 Metro Parkway
Suite 104
Bloomington, MN 55425 USA
952-854-8300
www.psai.org

United Association of Equipment
Leasing
78120 Calle Estado
Suite 201
La Quinta, CA 92253 USA
760-564-2227
www.uael.org

Air Conditioning and Plumbing
Contractors Association of Florida
P.O. Box 180458
Casselberry, FL 32718-0458 USA
407-260-1313

Air Movement and Control
Association International
30 W. University Dr.
Arlington Heights, IL 60004-1893
USA
847-394-0150
www.amca.org

Commercial Food Equipment
Service Association
2211 W. Meadowview Road
Suite 20
Greensboro, NC 27407 USA
336-346-4700
www.cfesa.com

Fire Apparatus Manufacturers'
Association
P.O. Box 397
Lynnfield, MA 01940-0397 USA
781-334-2911
www.fama.org

International Oxygen
Manufacturers Association
1255 23rd St. NW Suite 200
Washington, D.C. 20037-1174
USA
202-521-9300
www.iomaweb.org

National Mobility Equipment
Dealers Association
3327 W. Bearss Ave.
Tampa, FL 33618 USA
813-264-2697
www.nmeda.org

Production Equipment Rental
Association
P.O. Box 55515
Sherman Oaks, CA 91413-0515
USA
818-906-2467
www.productionequipment.com

Waste Equipment Technology
Association
4301 Connecticut Ave. NW
Suite 300
Washington, D.C. 20008-2403
USA
202-244-4700
www.wastec.org

Air Conditioning and Refrigeration
Institute
4100 N. Fairfax Dr.
Suite 200
Arlington, VA 22203 USA
703-524-8800
www.ari.org

Alabama Roofing, Sheet Metal,
Heating and Air Conditioning
Contractors Association
P.O. Box 381236
Birmingham, AL 35238 USA
205-981-0086

Eastern Association of Equipment
Lessors
62 William St.
Fourth Floor
New York, NY 10005-1545 USA
212-809-1602
www.eael.org

Independent Office Products and
Furniture Dealers Association
301 N. Fairfax St.
Suite 200
Alexandria, VA 22314-2696 USA
703-549-9040
www.iopfda.org/

National Association of Fire
Equipment Distributors
104 S. Michigan Ave.
Suite 300
Chicago, IL 60603 USA
312-263-8100
www.nafed.org

North American Association of
Food Equipment Manufacturers
161 N. Clark St.
Suite 2020
Chicago, IL 60601 USA
312-821-0201
www.nafem.org

Safety Equipment Distributors
Association
105 Laurel Bush Road
Suite 200
Bel Air, MD 21015 USA
443-640-1065
www.safetycentral.org

Air Conditioning Contractors of
America
2800 Shirlington Road
Suite 300
Arlington, VA 22206 USA
703-575-4477
www.acca.org

American Construction Inspectors
Association
12995 Sixth St.
Suite 69
Yucaipa, CA 92399 USA
909-795-3939
www.acia-rci.org

American Fence Association
800 Roosevelt Rd.
Bldg. C
Suite 312
Glen Ellyn, IL 60137 USA
630-942-6598
www.americanfenceassociation.
com

American Filtration and
Separations Society
7608 Emerson Ave. South
Richfield, MN 55423 USA
612-861-1277
www.afssociety.org

American Floorcovering Alliance
210 W. Cuyler St.
Dalton, GA 30720-8209 USA
706-278-4101
www.americanfloor.org

American Institute of Architects
1735 New York Ave. NW
Washington, D.C. 20006-5292
USA
202-626-7300
www.aia.org

American Institute of Building
Design
2505 Main St.
Suite 209B
Stratford, CT 06615 USA
www.aibd.org

American Institute of Constructors
P.O. Box 26334
Alexandria, VA 22314 USA
703-683-4999
http://aicnet.org

American Public Power Association
2301 M St. NW
Suite 300
Washington, D.C. 20037-1484
USA
202-467-2952
www.appanet.org

American Society of Concrete
Contractors
2025 S. Brentwood Blvd.
St. Louis, MO 63144 USA
314-962-0210
www.ascconline.org

American Society of Farm
Managers and Rural Appraisers
950 S. Cherry St.
Suite 508
Denver, CO 80246-2664 USA
303-692-1211
www.asfmra.org

Associated Builders and
Contractors
4250 N. Fairfax Dr.
Suite 900
Arlington, VA 22203 USA
703-812-2000
www.abc.org

Associated Underground
Contractors
P.O. Box 1640
Okemos, MI 48805-1640 USA
517-347-8336
www.aucmi.org

Association for Facilities
Engineering
8160 Corporate Park Drive
Suite 125
Cincinnati, OH 45242 USA
513-489-2473
www.afe.org

Association for Facilities
Engineering
8160 Corporate Park Drive
Suite 125
Cincinnati, OH 45242 USA
513-489-2473
www.afe.org

Association of Construction
Inspectors
1224 N. Nokomis NE
Alexandria, MN 56308 USA
320-763-7525
www.iami.org/aci.html

Association of Energy Service
Companies
10200 Richmond Ave.
Suite 253
Houston, TX 77042 USA
713-781-0758
www.aesc.net

Building Owners & Managers
Association
1101 15th Street NorthWest
Suite 800
Washington, D.C. 20005 USA
202-408-2662
www.boma.org

Building Owners and Managers
Association International
1201 New York Ave. NW
Suite 300
Washington, D.C. 20005 USA
202-408-2662
www.boma.org

Building Owners and Managers
Institute International
1521 Ritchie Hwy.
Arnold, MD 21012 USA
410-974-1410
www.bomi-edu.org

Building Service Contractors
Association International
401 North Michigan Avenue
22nd Floor
Chicago, IL 60611-4267 USA
312-321-5167
www.bscai.org

Building Systems Councils of the
National Association of
HomeBuilders
1201 15th St., N.W.
Washington, D.C. 20005-2800
USA
202-266-8576
www.nahb.com

Construction Specifications Institute
99 Canal Center Plaza
Suite 300
Alexandria, VA 22314-1791 USA
www.csinet.org

Construction Suppliers Association
11205 Alpharetta Hwy.
Suite F-1
Roswell, GA 30076 USA
770-751-6373
www.gosca.com

Eastern Building Material Dealers
Association
908 N. Second St.
Harrisburg, PA 17102 USA
717-441-6045
www.ebmda.org

Eastern Contractors Association
Six Airline Dr.
Albany, NY 12205-1095 USA
518-869-0961
www.ecainc.org

Electrical Generating Systems
Association
1650 S. Dixie Hwy.
Suite 500
Boca Raton, FL 33432-7462 USA
561-750-5575
www.egsa.org

Engineering Contractors
Association
8310 Florence Ave.
Downey, CA 90240 USA
www.ecaonline.net

Florida Building Material
Association
P.O. Box 65
Mount Dora, FL 32756-0065 USA
352-383-0366
www.fbma.org

Independent Electrical Contractors
4401 Ford Ave.
Suite 1100
Alexandria, VA 22302 USA
703-549-7351
www.ieci.org

International Association of
Refrigerated Warehouses
1500 King Street
Suite 201
Alexandria, VA 22314-2730 USA
703-373-4300
www.iarw.org

Mason Contractors Association
1429 S. Big Bend Blvd.
Suite A
St. Louis, MO 63117-2203 USA
314-645-1966
www.mcastle.com

Mobile Air Conditioning Society
Worldwide
P.O. Box 88
Lansdale, PA 19446 USA
215-631-7020
www.macsw.org

Multihazard Mitigation Council of
the National Institute of Building
Science
1090 Vermont Avenue North West
Suite 700
Washington, D.C. 20005-4905
USA
202-289-7800
www.nibs.org/mmc/mmchome.html

Energy and Environmental
Building Association
10740 Lyndale Ave., South
Suite 10W
Bloomington, MN 55420 USA
952-881-1098
www.eeba.org

Federal Facilities Council
500 Fifth St. NW
Washington, D.C. 20001 USA
202-334-3374
http://www7.nationalacademies
.org/ffc

Heavy Construction Contractors
Association
10756-B Ambassador Dr.
Suite 201
Manassas, VA 20109 USA
703-392-7410
www.hcca.net

Inland Northwest Associated
General Contractors of America
P.O. Box 3266
Spokane, WA 99220-3266 USA
509-535-0391
www.northwestagc.net

International Concrete Repair
Institute
3166 S. River Road
Suite 132
Des Plaines, IL 60018 USA
847-827-0830
www.icri.org

Metal Building Contractors and
Erectors Association
P.O. Box 499
Shawnee Mission, KS 66201 USA
913-432-3800
www.mbcea.org

Mobile Industrial Caterers'
Association International
304 W. Liberty St.
Suite 201
Louisville, KY 40202 USA
502-583-3783 x219
www.mobilecaterers.com

National Academy of Building
Inspection Engineers
P.O. Box 522158
Salt Lake City, UT 84152 USA
www.nabie.org

Engineering and Utility Contractors
Association
17 Crow Canyon Ct.
Suite 100
San Ramon, CA 94583 USA
925-855-7900
www.euca.com

Floor Covering Installation
Contractors Association
7439 Millwood Dr.
West Bloomfield, MI 48322-1234
USA
248-661-5015
www.fcica.com

Heavy Constructors Association
3101 Broadway
Suite 780
Kansas City, MO 64111 USA
816-753-6443
www.heavyconstructors.org

International Association of Heat
and Frost Insulators and Asbestos
Workers
9602 M. L. King Jr. Hwy
Lanham, MD 20706 USA
301-731-9101
www.insulators.org/index.htm

International Facility Management
Association
1 East Greenway Plaza
Suite 1100
Houston, TX 77046 USA
713-623-4362
www.ifma.org

Midwest Roofing Contractors
Association
4840 Bob Billings Pkwy.
Lawrence, KS 66049-3876 USA
785-843-4888
www.mrca.org

Modular Building Institute
413 Park St.
Charlottesville, VA 22902-4737
USA
434-296-3288
www.mbinet.org

National Air Duct Cleaners
Association
1518 K Street North West
Suite 503
Washington, D.C. 20005 USA
202-737-2926
www.nadca.com

National Air Filtration Association
P.O. Box 68639
Virginia Beach, VA 23471 USA
757-313-7400
www.nafahq.org

National Association of
Miscellaneous, Ornamental and
Architectural Products
Contractors
10382 Main St.
Box 280
Suite 200
Fairfax, VA 22038 USA
703-591-1870

National Association of State
Facilities Administrators
2760 Research Park Dr.
P.O. Box 11910
Lexington, KY 40578-1910 USA
859-244-8181
www.nasfa.net

National Business Owners
Association
P.O. Box 111
Stuart, VA 24076 USA
276-251-7500
www.nboa.org

National Electrical Contractors
Association
Three Bethesda Metro Center
Suite 1100
Bethesda, MD 20814-5330 USA
301-657-3110
www.necanet.org

National Institute of Building
Sciences
1090 Vermont Ave. NW
Suite 700
Washington, D.C. 20005-4905
USA
202-289-7800
www.nibs.org

National Precast Concrete
Association
10333 N. Meridian St.
Suite 272
Indianapolis, IN 462901 USA
317-571-9500
www.precast.org

National Association of County
Engineers
25 Massachusetts Ave. NW
Washington, D.C. 20001 USA
202-393-5041
www.countyengineers.org

National Association of Power
Engineers
One Springfield St.
Chicopee, MA 01013-2624 USA
413-592-6273
www.powerengineers.com

National Association of
Waterproofing & Structural
Repair Contractors
8015 Corporate Drive
Suite A
Baltimore, MD 21236 USA
410-931-3332
www.nawsrc.org

National Conference of States on
Building Codes and Standards
505 Huntmar Park Dr.
Suite 210
Herndon, VA 20170 USA
703-481-2035
www.ncsbcs.org

National Electronic Distributors
Association
1111 Alderman Dr.
Suite 400
Alpharetta, GA 30005-4175 USA
678-393-9990
www.nedaassoc.org

National Leased Housing
Association
1818 N St. NW
Suite 405
Washington, D.C. 20036 USA
202-785-8888
www.hudnlha.com

National Property Management
Association
1102 Pinehurst Road
Dunedin, FL 34698-5427 USA
727-736-3788
www.npma.org

National Association of Electrical
Distributors
1100 Corporate Square Dr.
Suite 100
St. Louis, MO 63132 USA
314-991-9000
www.naed.org

National Association of Residential
Property Managers
638 Independence Pkwy.
Suite 100
Chesapeake, VA 23320 USA
512-381-6091
www.narpm.org

National Association of
Waterproofing and Structural
Repair Contractors
8015 Corporate Dr.
Suite A
Baltimore, MD 21236 USA
410-931-3332
www.nawsrc.org

National Demolition Association
16 N. Franklin St.
Suite 203
Doylestown, PA 18901-3536 USA
215-348-4949
www.demolitionassociation.com

National Hydropower Association
One Massachusetts Ave.
Suite 850
Washington, D.C. 20001 USA
202-682-1700
www.hydro.org

National Pest Management
Association
9300 Lee Hwy.
Suite 301
Fairfax, VA 22031 USA
703-352-6762
www.pestworld.org

National Roofing Contractors
Association
O'Hare International Center
10255 W. Higgins Road
Suite 600
Rosemont, IL 60018-5607 USA
847-299-9070
www.nrca.net

National Utility Contractors
Association
4301 N. Fairfax Dr.
Suite 360
Arlington, VA 22203 USA
703-358-9300
www.nuca.com

Plumbing-Heating-Cooling
Contractors-National Association
180 S. Washington St.
Falls Church, VA 22042 USA
703-237-8100
www.phccweb.org

Scaffold Industry Association
P.O. Box 20574
Phoenix, AZ 85036 USA
602-257-1144
www.scaffold.org

Western Building Material
Association
P.O. Box 1699
Olympia, WA 98507 USA
360-943-3054
www.wbma.org

American Academy of
Environmental Medicine
7701 E. Kellogg Dr.
Suite 625
Wichita, KS 67207-1705 USA
316-684-5500
www.aaem.com

American College of Contingency
Planners
701 Lee St.
Suite 600
Des Plaines, IL 60016 USA
847-759-8601

American Hospital
Association-Northeast Region
Five New England Executive Park
Burlington, MA 01803-5006 USA
781-272-8456
www.aha.org

American Psychological Association
750 First St. NE
Washington, D.C. 20002-4242
USA
202-336-5500
www.apa.org

National Utility Contractors
Association of Arizona
11811 N. Vatum Blvd.
Suite 3031
Phoenix, AZ 85028 USA
602-953-7665
www.nucaaz.org

Power Sources Manufacturers
Association
P.O. Box 418
Mendham, NJ 07945-0418 USA
973-543-9660
www.psma.com

Southern Building Material
Association
P.O. Box 18667
Charlotte, NC 28218 USA
704-376-1503
www.southernbuilder.org

Western States Roofing
Contractors Association
1098 Foster City Blvd., #206
Foster City, CA 94404 USA
650-570-5441
www.wsrca.com

American Academy of Experts in
Traumatic Stress
368 Veterans Memorial Highway
Commack, NY 11725-4322 USA
631-543-2217
www.aaets.org

American Conference of
Governmental Industrial
Hygienists
1330 Kemper Meadow Dr.
Suite 600
Cincinnati, OH 45240-1634 USA
513-742-2020
www.acgih.org

American Industrial Hygiene
Association
2700 Prosperity Avenue
Suite 250
Fairfax, VA 22031 USA
703-849-8888
www.aiha.org

American Public Health
Association
800 I St. NW
Washington, D.C. 20001 USA
202-777-2742
www.apha.org

Northwest Wall and Ceiling Bureau
1032-A N.E. 65th St.
Seattle, WA 98115-6609 USA
206-524-4243
www.nwcb.org

Precast/Prestressed Concrete
Institute
209 W. Jackson Blvd.
Suite 500
Chicago, IL 60606 USA
312-786-0300
www.pci.org

United Union of Roofers,
Waterproofers and Allied Workers
1660 L St. NW
Suite 800
Washington, D.C. 20036-5646 USA
202-463-7663
www.unionroofers.com

American Ambulance Association
8201 Greensboro Dr.
Suite 300
McLean, VA 22102 USA
703-610-9018
www.the-aaa.org

American Health Quality
Association
1155 21st St. NW
Suite 202
Washington, D.C. 20036 USA
202-331-5790
www.ahqa.org

American Mosquito Control
Association
15000 Commerce Pkwy.
Suite C
Mt. Laurel, NJ 08054 USA
856-439-9222
www.mosquito.org

American School Counselor
Association
1101 King St.
Suite 625
Alexandria, VA 22314 USA
703-683-2722
www.schoolcounselor.org

American School Health
Association
7263 State Route 43
Box 708
Kent, OH 44240 USA
330-678-1601
www.ashaweb.org

American Sexually Transmitted
Diseases Association
P.O. Box 13827
Research Triangle Park, NC 27709
USA
919-361-8400
www.ashastd.org/

American Society for Healthcare
Environmental Services
One N. Franklin St.
Chicago, IL 60606 USA
312-422-3860
www.ashes.org

American Society for Healthcare
Risk Management
One N. Franklin St.
Chicago, IL 60606 USA
312-422-3982
www.ashrm.org

American Society of Tropical
Medicine and Hygiene
60 Revere Dr., Suite 500
Northbrook, IL 60062-1577 USA
847-480-9592
www.astmh.org

American Trauma Society
7611 South Osborne Road
Suite 202
Upper Marlboro, MD 20772 USA
301-574-4300
www.AMTRAUMA.org

Association for Professionals in
Infection Control and Epidemiology
1275 K St. NW
Suite 1000
Washington, D.C. 20005-4006
USA
202-789-1890
www.apic.org

Association of Occupational and
Environmental Clinics
1010 Vermont Ave. NW
Suite 513
Washington, D.C. 20005 USA
202-347-4976
www.aoec.org

Association of Ohio Philanthropic
Homes, Housing and Services for
the Aging
855 S. Wall St.
Columbus, OH 43206-1921 USA
614-444-2882
www.aopha.org

Association of State & Territorial
Health Officials
1275 K Street North West
Suite 800
Washington, D.C. 20005-4006
USA
202-371-9090
www.astho.org

Disaster Response Network
750 1st Street NorthEast
Washington, D.C. 20002 USA
www.apa.org

Disease Management Association
of America
701 Pennsylvania Ave. NW
Suite 700
Washington, D.C. 20004 USA
202-3575980
www.dmaa.org

Global Health Council
1111 19th St. NW
Suite 1120
Washington, D.C. 20036 USA
202-833-5900
www.globalhealth.org

Human Factors and Ergonomics
Society
P.O. Box 1369
Santa Monica, CA 90406-1369
USA
310-394-1811
www.hfes.org

Infectious Diseases Society of
America
66 Canal Center Plaza
Suite 600
Alexandria, VA 22314 USA
703-299-0200
www.idsociety.org

Mental Health America
2000 N. Beauregard St.
Sixth Floor
Alexandria, VA 22311 USA
703-838-7523
www.nmha.org

National Association of County
and City Health Officials
1100 17th Street North West
Second Floor
Washington, D.C. 20036 USA
202-783-5550
www.naccho.org

National Association of Health
Data Organizations
448 East 400 South
Suite 301
Salt Lake City, UT 84111 USA
801-587-9104
www.nahdo.org

National Association of Local
Boards of Health
1840 E. Gypsy Lane Rd.
Bowling Green, OH 43402 USA
419-353-7714
www.nalboh.org

National Board for Certified
Counselors
Three Terrace Way
Greensboro, NC 27403 USA
336-547-0607
www.nbcc.org

National EMS Pilots Association
526 King St.
Suite 415
Alexandria, VA 22314-3143 USA
703-836-8930
www.nemspa.org

National Health Association
P.O. Box 30630
Tampa, FL 33630-3630 USA
813-855-6607
www.anhs.org

National Health Council
1730 M St. NW
Suite 500
Washington, D.C. 20036 USA
202-785-3910
www.nhcouncil.org

National Rehabilitation Association
633 S. Washington St.
Alexandria, VA 22314-4109 USA
703-836-0850

National Rehabilitation Counseling
Association
P.O. Box 4480
Manassas, VA 20108 USA
703-361-2077
http://nrca-net.org

Nevada State Medical Association
3660 Baker Lane
Suite 101
Reno, NV 89509 USA
775-825-6788
http://nsmadocs.org

North American Association For
Ambulatory Care
4019 Quayle Briar Dr.
Valrico, FL 33594 USA
www.nafac.com

Society for Public Health Education
750 First St. NE
Suite 910
Washington, D.C. 20002 USA
202-408-9804
www.sophe.org

American Academy of Safety
Education
Safety Science & Technology Dept.
Central Missouri State Univ.
Warrensburg, MO 64093 USA
660-543-4972

American Planning Association
122 South Michigan Avenue
Suite 1600
Chicago, IL 60603-6111 USA
312-431-9100
www.planning.org

American Society for Public
Administration
1301 Pennsylvania Ave. NW,
Suite 840
Washington, D.C. 20004 USA
202-393-7878
www.aspanet.org

Art Libraries Society of North
America
232-329 March Road
Box 11
Ottawa, ON K2K 2E1 Canada
613-599-3074
www.arlisna.org

National Wellness Institute
1300 College Court
P.O. Box 827
Stevens Point, WI 54481-0827
USA
715-342-2969
www.nationalwellness.org

New England Pest Management
Association
USA
603-228-1231
www.nepma.org

Society for Academic Emergency
Medicine
901 N. Washington Ave.
Lansing, MI 48906-5137 USA
517-485-5484
www.saem.org

Society of Forensic Toxicologists
P.O. Box 5543
Mesa, AZ 85211-5543 USA
480-839-9106
www.soft-tox.org

American Institute of Certified
Planners
1776 Massachusetts Ave. NW
Washington, D.C. 20036-1904
USA
202-872-0611
www.planning.org

American Planning
Association-National Capital
Area Chapter
P.O. Box 508
Washington, D.C. 20044-0508
USA
301-495-4576
www.ncac-apa.org

Archaeological Institute of America
656 Beacon St.
Boston, MA 02215-2006 USA
617-353-9361
www.archaeological.org

Association for Federal
Information Resources
Management
P.O. Box 2851
Washington, D.C. 20013 USA
www.affirm.org

Nebraska Health Care Association
3900 N.W. 12th St.
Suite 100
Lincoln, NE 68521-3015 USA
402-435-3551
www.nehca.org

New England Public Health
Association
30 Dwight Drive
Middlefield, CT 06455 USA
860-349-8995

Society for Occupational and
Environmental Health
6728 Old McLean Village Dr.
McLean, VA 22101 USA
703-556-9222
www.soeh.org

American National Standards
Institute
1819 L St. NW
Sixth Floor
Washington, D.C. 20036 USA
202-293-8020
www.ansi.org

American Society for
Environmental History
119 Pine St.
Suite 301
Seattle, WA 98101 USA
206-343-0226
www.aseh.net

Archivists and Librarians in the
History of the Health Sciences
1216 Fifth Ave.
New York, NY 10029 USA
212-822-7321
www.library.ucla.edu/
libraries/biomed/alhhs

Association of Collegiate Schools
of Planning
6311 Mallard Trace
Tallahassee, FL 32312 USA
850-385-2054
www.acsp.org

Association of Research Libraries
21 Dupont Circle NW
Suite 800
Washington, D.C. 20036 USA
202-296-2296
www.arl.org

Atlantic Oceanographic and
Meteorological Laboratory
4301 Rickenbacker Causeway
Miami, FL 33149 USA
305-361-4450
www.aoml.noaa.gov

Canadian Environmental Auditing
Association
1-6820 Kitimat Road
Mississauga, ON L5N 5M3 Canada
www.ceaa-acve.ca

Chemical Manufacturers
Association
1300 Wilson Boulevard
Arlington, VA 22209-2323 USA
703-741-5000

ComCARE Alliance
1701 K Street North West
12th Floor
Suite 400
Washington, D.C. 20006 USA
202-429-0574

Congressional Hazards Caucus
4220 King Street
Alexandria, VA 22302 USA
703-379-2480
www.hazardscaucus.org

Disability Preparedness Center
1010 Wisconsin Avenue NorthWest
Washington, D.C. 20007 USA
202-338-7158

Earthquakes Canada (East)
Natural Resources Canada
7 Observatory Crescent
Ottawa, ON K1A 0Y3 Canada
613-995-0600 French
www.seismo.nrcan.gc.ca

Earthquakes Canada (West)
P.O. Box 6000
9860 West Saanich Road
Sidney, BC V8L 4B2 Canada
www.pgc.nrcan.gc.ca

Global Disaster Information
Network
26128 Talamore Drive
Suite 201
South Riding, VA 20152 USA
202-647-5070
www.gdin.org

Health Sciences Libraries
Consortium
3600 Market St.
Suite 550
Philadelphia, PA 19104-2646 USA
215-222-1532
www.hslc.org

International Hurricane Protection
Assciation World Office
2501 Floral Road
Lantana, FL 33462 USA

Library and Information
Technology Association
50 E. Huron St.
Chicago, IL 60611-2795 USA
312-2804269
www.lita.org

National Association of Counties
25 Massachusetts Ave. NW
Suite 500
Washington, D.C. 20001 USA
202-942-4221
www.naco.org

National Association of Flood and
Storm Water Management Agencies
1301 K Street NorthWest
Suite 800 East
Washington, D.C. 20005 USA
www.nafsma.org

National Voluntary Organizations
Active in Disaster
P.O. Box 151973
Alexandria, VA 22315 USA
703-339-5596
www.nvoad.org

Public Library Association
50 E. Huron St.
Chicago, IL 60611 USA
312-280-5752
www.pla.org

Responder Knowledge Base
5765-F Burke Centre Parkway
PMB 331
Burke, VA 22015 USA
703-641-3731
www.rkb.mipt.org

Society of Environmental
Journalists
P.O. Box 2492
Jenkintown, PA 19046 USA
215-884-8174
www.sej.org

Southern California Earthquake
Center
University of Southern California
3651 Trousdale Parkway
Los Angeles, CA 90089 USA
www.scec.org

Special Libraries Association
331 S. Patrick St.
Alexandria, VA 22314 USA
703-647-4900
www.sla.org

The Infrastructure Security
Partnership
101 Constitution Avenue
Northwest
Suite 375 East
Washington, D.C. 20001 USA
202-789-7855
www.tisp.org

The National Environmental
Services Center
West Virginia University
P.O. Box 6064
Morgantown, WV 26506 USA

Alabama Association of Insurance
and Financial Advisors
2361 Fairlane Dr.
Suite A-1
Montgomery, AL 36116 USA
334-271-4900

Alliance of Insurance Agents and
Brokers
1768 Arrow Hwy. Suite 105
La Verne, CA 91750-5332 USA
909-392-0836
www.agentsalliance.com

American Academy of Actuaries
1100 17th St. NW
Suite 700
Washington, D.C. 20036-4601 USA
202-223-8196
www.actuary.org

American Insurance Association
1130 Connecticut Ave. NW
Suite 1000
Washington, D.C. 20036 USA
202-828-7100
www.aiadc.org

American Society of Appraisers
555 Herndon Parkway
Suite 125
Herndon, VA 20170 USA
703-733-2110
www.appraisers.org

Casualty Actuarial Society
4350 N. Fairfax Dr.
Suite 250
Arlington, VA 22203 USA
703-276-3100
www.casact.org

Fiduciary and Risk Management
Association
P.O. Box 48297
Athens, GA 30604 USA
704-365-3344
www.thefirma.org

Institute of Business Appraisers
P.O. Box 17410
Plantation, FL 33318 USA
954-584-1144
www.go-iba.org

Insurance Loss Control Association
P.O. Box 2075
Columbus, OH 43216-2075 USA
614-221-9950
www.insurancelosscontrol.org

Liability Insurance Research Bureau
3025 Highland Pkwy.
Suite 800
Downers Grove, IL 60515-1291
USA
630-724-2252
www.lirb.org

National Association of Fire
Investigators
857 Tallevast Rd.
Sarasota, FL 34243 USA
941-359-2800
www.nafi.org

American Insurance Association-
Southeast Regional Office
5565 Glenridge Connector
Suite 425
Atlanta, GA 30342 USA
404-261-8834

Appraisal Institute
550 W. Van Buren St.
Suite 1000
Chicago, IL 60607 USA
312-335-4100
www.appraisalinstitute.org

Certified Risk Managers
International
P.O. Box 27027
Austin, TX 78755-2027 USA
www.thenationalalliance.com

Global Association of Risk
Professionals
100 Pavonia Ave., Suite 405
Jersey City, NJ 07310 USA
201-222-0054
www.garp.com

Insurance Consumer Affairs
Exchange
P.O. Box 746
Lake Zurich, IL 60047 USA
847-991-8454
www.icae.com

International Association of
Assessing Officers
314 W. 10th St.
Kansas City, MO 64105 USA
816-701-8100
www.iaao.org

National Association of
Catastrophe Adjusters
P.O. Box 821864
North Richland Hills, TX 76182
USA
817-498-3466
www.nacatadj.org

National Association of Jewelry
Appraisers
P.O. Box 18
Rego Park, NY 11374-0018 USA
718-896-1536
www.najappraisers.com

American Risk & Insurance
Association
716 Providence Road
Malvern, PA 19355-3402 USA
610-640-1997
www.aria.org

Associated Risk Managers
c/o ARM Partners
Gallagher Center
Two Pierce Place
Itasca, IL 60143 USA
630-285-4186
www.armiweb.com

Claims Support Professional
Association
6451 North Federal Highway
Suite 121
Fort Lauderdale, FL 33308 USA
www.claimssupport.com/
firstgen.htm

Independent Insurance Agents and
Brokers of America
127 S. Peyton St.
Alexandria, VA 22314-2803 USA
703-706-5443
www.independentagent.com

Insurance Information Institute
110 William Street
New York, NY 10038 USA
212-346-5500
www.iii.org

International Society of Appraisers
1131 S.W. Seventh St.
Suite 105
Renton, WA 98055 USA
206-241-0359
www.isa-appraisers.org

National Association of Certified
Valuation Analysts
1111 Brickyard Rd.
Suite 200
Salt Lake City, UT 84106-5401
USA
801-486-0600
www.navca.com

National Association of Master
Appraisers
303 W. Cypress St.
San Antonio, TX 78212 USA
www.masterappraisers.org

National Association of
Professional Insurance Agents
400 N. Washington St.
Alexandria, VA 22314-2312 USA
703-518-1340
www.pianet.com

National Association of Public
Insurance Adjusters
21165 Whitfield Place #105
Potomac Falls, VA 20165 USA
703-433-9217
www.napia.com

National Insurance Crime Bureau
1111 E. Touhy Ave.
Suite 400
Des Plaines, IL 60018 USA
847-544-7000
www.nicb.org

National Risk Retention
Association
4248 Park Glen Road
Minneapolis, MN 55416 USA
952-928-4656
www.nrra-usa.org

Professional Insurance Agents
Association
25 Chamberlain Street
P.O. Box 997
Glenmont, NY 12077-0997 USA
www.piaonline.com

Property Casualty Insurers
Association of America
2600 River Rd.
Des Plaines, IL 60018-3286 USA
847-297-7800
www.pciaa.net

Property Insurance Association
of Louisiana
P.O. Box 60730
New Orleans, LA 70160 USA
504-831-6930
www.pial.org

Property Loss Research Bureau
3025 Highland Parkway Suite 800
Downers Grove, IL 60515-1291
USA
630-724-2200
www.plrb.org

Public Agency Risk Managers
Association
P.O. Box 6810
San Jose, CA 95150 USA
408-356-4627
www.parma.com

Public Entity Risk Institute
11350 Random Hills Road
Suite 210
Fairfax, VA 22030 USA
703-352-1846
www.riskinstitute.org

Risk and Insurance Management
Society
655 Third Ave.
New York, NY 10017 USA
www.rims.org

RMA—The Risk Management
Association
1801 Market St.
Suite 300
Philadelphia, PA 19103 USA
215-446-4000
www.rmahq.org

Society for Risk Analysis
1313 Dolley Madison Boulevard
Suite 402
McLean, VA 22101 USA
703-790-1745
www.sra.org

Society of Actuaries
475 N. Martingale Road
Suite 600
Schaumburg, IL 60173-2226 USA
847-706-3541
www.soa.org

Society of Cost Estimating and
Analysis
101 S. Whiting St.
Suite 201
Alexandria, VA 22304 USA
703-751-8069
www.sceaonline.net

Society of Risk Management
Consultants
330 South Executive Drive
Suite 301
Brookfield, WI 53203-4275 USA
www.srmcsociety.org

Southwestern Insurance
Information Service
8303 N. Mopac, Bldg. B
Suite B-231
Austin, TX 78759 USA
512-795-8214
www.siisinfo.org

The Risk Management Association
One Liberty Place
1650 Market Street
Suite 2300
Philadelphia, PA 19103 USA
215-446-4100
www.rmahq.org

University Risk Management and
Insurance Association
P.O. Box 1027
Bloomington, IN 47402 USA
812-855-6833
www.urmia.org

Utah Association of Independent
Insurance Agents
4885 South 900 East
Suite 302
Salt Lake City, UT 84117-5746
USA
801-269-1200
www.uaiia.org

Vermont Insurance Agents
Association
P.O. Box 1387
Montpelier, VT 05601 USA
802-229-5884
www.viaa.org

Virginia Association of Insurance
and Financial Advisors
3108 Parham Rd.
Suite 100-A
Richmond, VA 23294-4415 USA
804-747-6020
www.naifanet.com/virginia

Western Insurance Agents
Association
11190 Sun Center Dr.
Suite 100
Rancho Cordova, CA 95670 USA
916-443-4221
www.wiaagroup.org

Windstorm Insurance Network, Inc.
2929 Langley Avenue
Suite 203
Pensacola, FL 32504 USA
850-473-0601
www.windnetwork.com

Academy of Certified Hazardous
Materials Managers, Inc.
P.O. Box 1216
Rockville, MD 20849 USA
301-916-3306
www.achmm.org

Acadiana Safety Association
1126 Coolidge Blvd., Complex Four
Lafayette, LA 70503 USA
337-234-4640
www.acadianasafety.org

American Biological Safety
Association
1202 Allanson Rd.
Mundelein, IL 60060-3808 USA
847-949-1517
www.absa.org

American Pyrotechnics Association
P.O. Box 30438
Bethesda, MD 20824-0438 USA
301-907-8181
www.americanpyro.com

American Red Cross BICEPP
Program
3747 Euclid Avenue
Cleveland, OH 44115 USA
216-431-3311

American Red Cross National
Headquarters
2025 East Street North West
Washington, D.C. 20006 USA
www.redcross.org/services/disaster

American Society of Safety
Engineers
1800 East Oakton Street
Des Plaines, IL 60018-2100 USA
847-699-2929
www.asse.org

American Traffic Safety Services
Association
15 Riverside Pkwy.
Suite 100
Fredericksburg, VA 22406 USA
540-368-1701
www.atssa.com

Automatic Fire Alarm Association
P.O. Box 951807
Lake Mary FL 32795-1807 USA
407-833-9133
www.afaa.org

Aviation Safety Institute
P.O. Box 690
Worthington, OH 43085-0690
USA
614-793-1679
www.aviationsafetyinstitute.com

Board of Certified Safety
Professionals
208 Burwash Ave.
Savoy, IL 61874-9510 USA
217-359-9263
www.bcsp.org

Campus Safety, Health and
Environmental Management
Association
1121 Spring Lake Dr.
Itasca, IL 60143 USA
630-775-2227
www.cshema.org

Evergreen Safety Council
401 Pontius Ave. N
Seattle, WA 98109-5423 USA
206-382-4090
www.esc.org

Home Safety Council
1725 Eye Street North West
Suite 300
Washington, D.C. 20006 USA
202-349-1100
www.homesafetycouncil.org

Industrial Accident Prevention
Association
207 Queens Quay West
Toronto, ON M5J 2Y3 Canada
www.iapa.ca

Institute for Business and Home
Safety
4775 East Fowler Avenue
Tampa, FL 33617 USA
813-286-3400
www.ibhs.org

Institute of Hazardous Materials
Management
11900 Parklawn Drive
Suite 450
Rockville, MD 20852-2676 USA
www.ihmm.org

International Association of Bomb
Technicians and Investigators
P.O. Box 160
Goldvein, VA 22720-0160 USA
540-752-4533
www.iabti.org

International Association of
Fire Chiefs
4025 Fair Ridge Drive
Suite 300
Fairfax, VA 22033-2868 USA
703-273-0911
www.iafc.org

International Association of Fire
Fighters
1750 New York Ave. NW
Washington, D.C. 20006-5395 USA
202-737-8484
www.iaff.org

International Association of
Industrial Accident Boards and
Commissions
5610 Medical Circle
Suite 24
Madison, WI 53711 USA
608-663-6355
www.iaiabc.org

International Fire Marshals
Association
One Batterymarch Park
Quincy, MA 02169 USA
617-984-7424
www.nfpa.org/ifma

International Safety Equipment
Association
1901 North Moore Street
Arlington, VA 22209-1762 USA
703-525-1695
www.safetyequipment.org

International Society of Air Safety
Investigators
107 E. Holly Ave.
Suite 11
Sterling, VA 20164-5405 USA
703-430-9668
www.isasi.org

National Safety Council
1121 Spring Lake Drive
Itasca, IL 60143-3201
USA630-285-1121
www.nsc.org

Neighborhood Safety Net
2060 D Suite 669
Avenid De Los Arboles
Thousand Oaks, CA 91362 USA
www.neighborhoodsafety.net

System Safety Society
P.O. Box 70
Unionville, VA 22567-0070 USA
540-854-8630
www.system-safety.org

Women in the Fire Service
P.O. Box 5446
Madison, WI 53705 USA
608-233-4768
www.wfsi.org

American Federation of Police and
Concerned Citizens
6350 Horizon Dr.
Titusville, FL 32780 USA
321-264-0911
www.aphf.org

Homeland Security Industries
Association
666 11th St. NW
Suite 315
Washington, D.C. 20001 USA
202-331-3096
www.hsianet.org

International Association of
Professional Security Consultants
525 SW 5th Street Ste A
Des Moines, IA 50309-4501 USA
515-282-8192
www.iapsc.org

International Society of Fire Service
Instructors
2425 Hwy. 49 East
Pleasant View, TN 37146 USA
www.isfsi.org

National Safety Management
Society
P.O. Box 4460
Walnut Creek, CA 94596-0460
USA
925-944-7094
www.nsms.us

Safety and Loss Prevention
Management Council
950 North Glebe Road
Suite 210
Arlington, VA 22203-4181 USA
703-838-1861
http://www.truckline.com/
Federation/Councils/slpmc/

The National Environmental,
Safety & Health Training
Association—NESHTA
P.O. Box 10321
Phoenix, AZ 85064-0321 USA
602-956-6099
http://neshta.org

American Society for Amusement
Park Security and Safety
Riverside Park, P.O. Box 307
Agawam, MA 01001 USA
413-786-9390

International Association of
Campus Law Enforcement
Administrators
342 N. Main St.
West Hartford, CT 06117-2507
USA
860-586-7517
www.iaclea.org

International Guards Union
of America
Route 8 Box 32-14
Amarillo, TX 79118-9427 USA
806-622-2424
www.amaonline.com/igua

National Association for Search
and Rescue
P.O. Box 232020
Centreville, VA 20120-2020 USA
703-222-6277
www.nasar.org

National Volunteer Fire Council
1050 17th St. NW
Suite 490
Washington, D.C. 20036 USA
202-887-5700
www.nvfc.org

Society of Fire Protection Engineers
7315 Wisconsin Ave.
Suite 1225W
Bethesda, MD 20814 USA
301-718-2910
www.sfpe.org

United States Lifesaving Association
P.O. Box 322
Avon-By-the-Sea, NJ 07717 USA
714-968-9360
www.usla.org

American Association of Homeland
Security Professionals
1717 K Street North West
Suite 600
Washington, D.C. 20036 USA
www.aahsp.org

Canadian Society for Industrial
Security Inc.
141 Bentley Avenue Unit B
Ottawa, ON K2E 6T7 Canada
800-461-7748
www.csis-scsi.org

International Association of Chiefs
of Police
515 N. Washington St.
Suite 400
Alexandria, VA 22314-2340 USA
703-836-6767
www.theiacp.org

International Security Management
Association
P.O. Box 623
Buffalo, IA 52728 USA
563-381-4008
www.isma.com

International Security Officers,
Police, and Guards Union
411 Hempstead Ave.
Suite 101
West Hempstead, NY 11552-2333
USA
718-836-3508

National Association of Police
Organizations
750 First St. NE
Suite 920
Washington, D.C. 20002-4241
USA
202-842-4420
www.napo.org

National Burglar and Fire Alarm
Association
2300 Vally View Ln.
Suite 230
Irving, TX 75062 USA
214-260-5970
www.alarm.org

National Guard Association of
the U.S.
One Massachusetts Ave. NW
Washington, D.C. 20001 USA
202-789-0031
www.ngaus.org

United Federation of Police &
Security Officers
540 N. State Rd. Box 76
Briarcliff Manor, NY 10510-0076
USA
914-941-4103

American Animal Hospital
Association
12575 W. Bayaud Ave.
Lakewood, CO 80228 USA
303-986-2800
www.aahanet.org

American Concrete Pavement
Association—Northeast Chapter
800 N. Third St.
Suite 201
Harrisburg, PA 17102 USA
717-441-3506
www.ne.pavement.com

American Public Works Association
2345 Grand Boulevard
Suite 500
Kansas City, MO 64108-2641 USA
816-472-6100
www.apwa.net

National Alarm Association of
America
P.O. Box 3409
Dayton, OH 45401-3409 USA
www.naaa.org

National Association of School
Safety and Law Enforcement
Officers
P.O. Box 3147
Sowego, NY 13126 USA
315-529-4858
www.nassleo.org

National Council of Investigation
and Security Services
7501 Sparrows Point Blvd.
Baltimore, MD 21219-1927 USA
www.nciss.org

National Police and Security
Officers Association of America
P.O. Box 663
South Plainfield, NJ 07080-0663
USA
http://npoaa.tripod.com

American Association of Crop
Insurers
One Massachusetts Ave. NW
Suite 800
Washington, D.C. 20001-1401
USA
202-789-4100
www.cropinsurers.org

American Health Lawyers
Association
1025 Connecticut Ave. NW
Suite 600
Washington, D.C. 20036-5405
USA
202-833-1100
www.healthlawyers.org

American Road and
Transportation Builders
Association
1219 28th St. NW
Washington, D.C. 20007-3712
USA
202-289-4434
www.artba.org

National Association of Chiefs of
Police
6350 Horizon Dr.
Titusville, FL 32780 USA
321-264-0911
www.aphf.org

National Border Patrol Council
P.O. Box 678
Campo, CA 91906-0678 USA
619-478-5145
www.nbpc.net

National Defense Industrial
Association
2111 Wilson Boulevard
Suite 400
Arlington, VA 22201-3061 USA
703-522-1820
www.ndia.org

The National Burglar & Fire Alarm
Association
2300 Valley View Lane Suite 230
Irving, TX 75062 USA
214-260-5970
www.alarm.org

American Academy of Forensic
Sciences
410 N. 21st St.
Colorado Springs, CO 80904 USA
719-636-1100
www.aafs.org

American Association of Port
Authorities
1010 Duke St.
Alexandria, VA 22314-3589 USA
703-684-5700
www.aapa-ports.org

American Public Transportation
Association
1666 K St. NW
Suite 1100
Washington, D.C. 20006 USA
202-496-4852
www.apta.com

American Salvage Pool Association
2100 Roswell Road
Suite 200C—PMB 709
Marietta GA 30062 USA
678-560-6678
www.aspa.com

American Society of Directors of
Volunteer Services
One N. Franklin, 27th Floor
Chicago, IL 60606 USA
312-422-3939
www.asdvs.org

American Veterinary Medical
Association
1931 North Meacham Road
Suite 100
Schaumburg, IL 60173-4360 USA
847-925-8070
www.avma.org

Animal Transportation Association
1111 East Loop North
Houston, TX 77029 USA
713-532-2177
www.aata-animaltransport.org

Antique Appraisal Association
of America
11361 Garden Grove Blvd.
Garden Grove, CA 92843 USA
714-530-7090

Association for Business
Simulation and Experiential
Learning
Wayne State Univ.
Marketing Dept.
Detroit, MI 48202-3930 USA
313-577-4551
www.absel.org

Association of Diving Contractors
International
5206 F.M. 1960 W.
Suite 202
Houston, TX 77069 USA
281-893-8388
www.adc-int.org

Association of Transportation
Professionals
Three Church Circle, PMB 250
Annapolis, MD 21401 USA
410-267-0023
www.atlp.org

Automotive Fleet and Leasing
Association
1000 Westgate Dr.
St. Paul, MN 55114 USA
651-2906274
www.aflaonline.org

BEMA—The Baking Industry
Suppliers Association
7101 College Blvd.
Suite 1505
Overland Park, KS 66210 USA
913-338-1300
www.bema.org

Business Forms Management
Association
319 S.W. Washington
Suite 710
Portland, OR 97204-2618 USA
503-227-3393
www.bfma.org

Catholic Relief Services
209 West Fayette St.
Baltimore, MD 21201-3443 USA
800-736-3467

Consumer Specialty Products
Association
900 17th St. NW
Suite 300
Washington, D.C. 20006-2111 USA
202-872-8110
www.cspa.org

Council on the Safe Transportation
of Hazardous Articles
7803 Hill House Court
Fairfax Station, VA 22039 USA
703-451-4031
www.costha.com

Electric Power Research Institute
(EPRI)
3412 Hillview Avenue
Palo Alto, CA 94304-1395 USA
www.epri.com

Electrical Safety Foundation
International
1300 North 17th Street Suite 1752
Rosslyn, VA 22209 USA
703-841-3229
www.electrical-safety.org

Energy Security Council
5555 San Felipe St.
Suite 100
Houston, TX 77056-3634 USA
713-296-1893
www.energysecuritycouncil.org

Evidence Photographers
International Council
600 Main St.
Honesdale, PA 18431 USA
570-253-5450
www.epicheadquarters@verizon.net

Independent Armored Car
Operators Association
102 E. Ave. J
Lancaster, CA 93535 USA
661-726-9864
www.iacoa.com

Independent Pet and Animal
Transportation Association
International
745 Winding Trail
Holly Lake Ranch, TX 75765 USA
903-769-2267
www.ipata.com

Institute of Nuclear Materials
Management
60 Revere Dr.
Suite 500
Northbrook, IL 60062 USA
847-480-9573
www.inmm.org

International Association for Food
Protection
6200 Aurora Ave.
Suite 200W
Des Moines, IA 50322-2864 USA
515-276-3344
www.foodprotection.org

International Association for
Impact Assessment
1330 23rd St. South
Suite C
Fargo, ND 58103 USA
701-297-7912
www.iaia.org

International Association of
Accident Reconstruction Specialists
P.O. Box 534
Grand Ledge, MI 48837-0534
USA
517-622-3135
www.iaars.org

International Association of Arson
Investigators
12770 Boenker Road
Bridgeton, MO 63044 USA
314-739-4224
www.firearson.com

International Association of
Dive Rescue Specialists
201 N. Link Lane
Ft. Collins, CO 80524 USA
970-482-1562
www.iadrs.org

International Bottled Water
Association
1700 Diagonal Road
Suite 650
Alexandria, VA 22314 USA
703-683-5213
www.bottledwater.org

International Foodservice
Distributors Association
201 Park Washington Ct.
Falls Church, VA 22046-4621 USA
703-532-9400
www.ifdaonline.org

International Safe Transit
Association
1400 Abbott Rd., Suite 160
East Lansing, MI 48823-1900 USA
517-333-3437
www.ista.org

National Association of Federal
Veterinarians
1910 Sunderland Pl. NW
Washington, D.C. 20036-1608
USA
202-223-4878
www.nafv.org

National Association of Trailer
Manufacturers
1320 S.W. Topeka Blvd.
Topeka, KS 66612-1817 USA
785-272-4433
www.natm.com

National Freight Transportation
Association
P.O. Box 1321
Exton, PA 19341 USA
610-363-7747
www.nftahq.org

National Institute of Packaging,
Handling and Logistics Engineers
177 Fairson Court
Lewisburg, PA 17837-6844 USA
570-528-6475
www.niphle.com

International Association of
Financial Crimes Investigators
873 Embarcadero Dr.
Suite Five
El Dorado Hills, CA 95762 USA
916-939-5000
www.iafci.org

International Bridge, Tunnel and
Turnpike Association
1146 19th St. NW
Suite 800
Washington, D.C. 200363725
USA
202-659-4620
www.ibtta.org

International Packaged Ice
Association
P.O. Box 1199
Tampa, FL 33601-1199 USA
813-258-1690
www.packagedice.com

Investment Recovery Association
638 W. 39th St.
Kansas City, MO 64111 USA
816-561-5323
www.invrecovery.org

National Association of Fleet
Administrators
100 Wood Ave. South
Suite 310
Iselin, NJ 08830-2709 USA
732-494-8100
www.nafa.org

National Association of
Wastewater Transporters
336 Chestnut Ln.
Ambler, PA 19002 USA
215-643-6798
www.nawt.org

National Frozen and Refrigerated
Foods Association
4755 Linglestown Road
Suite 300 P.O. Box 6069
Harrisburg, PA 17112-0069 USA
717-657-8601
www.nfraweb.org

National Training and Simulation
Association
2111 Wilson Blvd.
Suite 400
Arlington, VA 22201-3061 USA
703-247-2569
www.trainingsystems.org

International Association of Food
Industry Suppliers
1451 Dolley Madison Blvd.
McLean, VA 22101-3850 USA
703-761-2600
www.iafis.org

International Fire Photographers
Association
143 40th St.
New Orleans, LA 70124 USA
504-482-9616

International Refrigerated
Transportation Association
1500 King St.
Alexandria, VA 22314 USA
703-373-4300
www.irta.org

National Air Transportation
Association
4226 King St.
Alexandria, VA 22302 USA
703-845-9000
www.nata.aero/

National Association of Industrial
Technology
3300 Washtenaw Ave.
Suite 220
Ann Arbor, MI 48104-4200 USA
734-677-0720
www.nait.org

National Food and Energy Council
P.O. Box 309
Wilmington, OH 45177-0309 USA
937-383-0001
www.nfec.org

National Funeral Directors
Association
13625 Bishops Drive
Brookfield, WI 53005-6607 USA
262-789-1880
www.nfda.org

Northeast Bottled Water
Association
20 Eastwood Dr.
Plainville, CT 06062 USA
860-517-9808
www.nebwa.org

Nuclear Energy Institute
1776 I St. NW
Suite 400
Washington, D.C. 20006-3708
USA
202-739-8021
www.nei.org

Security Industry Association
635 Slaters Lane
Suite 110
Alexandria, VA 22314 USA
703-683-2075
www.siaonline.org

Telecommunications Industry
Association
2500 Wilson Blvd.
Suite 300
Arlington, VA 22201-3834 USA
703-907-7734
www.tiaonline.org

Transportation & Logistics Council
120 Main St.
Huntington, NY 11743-0630 USA
631-549-8984
www.tlcouncil.org

Truck Renting and Leasing
Association
675 N. Washington St.
Suite 410
Alexandria, VA 22314 USA
703-299-9120
www.trala.org

Wireless Communications
Association International
1333 H St. NW
Suite 700W
Washington, D.C. 20005 USA
202-452-7823
www.wcai.com

Pennsylvania Asphalt Pavement
Association
3540 N. Progress Ave.
Suite 206
Harrisburg, PA 17110-9637 USA
717-657-1881
www.pahotmix.org

Southwest Movers Association
700 E. 11th St.
Austin, TX 78701 USA
512-476-0107
www.southwestmovers.org

The Humane Society of the
United States
2100 L Street North West
Washington, D.C. 20037 USA
202-452-1100
www.hsus.org

Transportation Institute
5201 Auth Way
Camp Springs, MD 20746 USA
301-423-3335
www.trans-inst.org

Wildlife Disease Association
P.O. Box 1897
Lawrence, KS 66044-8897 USA
785-843-1221
www.wildlifedisease.org/

Secondary Materials and Recycled
Textiles Association
7910 Woodmont Ave.
Suite 1130
Bethesda, MD 20814-3015 USA
301-656-1077
www.smartasn.org

Taxicab, Limousine and Paratransit
Association
3849 Farragut Ave.
Kensington, MD 20895 USA
301-946-5701
www.tlpa.org

Towing and Recovery Association
of America
2121 Eisenhower Ave.
Suite 200
Alexandria, VA 22314-4686 USA
703-684-7713
www.towserver.org

Tree Care Industry Association
3 Perimeter Road
Unit 1
Manchester, NH 03103 USA
603-314-5380
www.tcia.org

Wildlife Management Institute
1146 19th St. NW
Suite 700
Washington, D.C. 20036 USA
202-371-1808
www.wildlifemanagementinstitute.org

Table 10.2 Disaster Recovery Consultants

American BioRecovery Association
P.O. Box 828
Ipswich, MA 01938
987-356-4606
www.americanbiorecovery.com

OptiMetrics, Inc.
3115 Professional Drive
Ann Arbor, MI 48104-5131
734-973-1177
www.optimetrics.org

Agility Recovery Solutions
Suite 350E
2101 Rexford Road
Charlotte, NC 28211
704-341-8700
www2.agilityrecovery.com

Appropriate Systems, LLC
68 Greenrale Avenue
Wayne, NJ 07470
973-904-1547
www.a-systems.biz

Avalution Consulting
800-941-0381
www.avalutionconsulting.com

Bel Esprit Partners
11160 Anderson Lakes Parkway
Suite 208
Eden Prairie, MN 55344
952-223-5404
www.belesprit-inc.com

BSI Management Systems
6205 Airport Road
Mississauga, Ontario L4V 1E1
416-620-9991
www.bsiamerica.com

Bioterroism Preparedness and
Response Planning
Centers for Disease Control and
Prevention
1600 Clifton Road
Atlanta, GA 30333
404-639-3431
www.bt.cdc.gov

Response Biomedical Corp.
8081 Lougheed Highway
Burnaby, British Columbia V5J 5J1
604-681-4101
www.responsebio.com

AccessPoint Business Recovery
Services
1103 Victory Drive
Port Moody, British Columbia
V3H 1K3
604-250-4829
www.businessrecovery.ws

Align Business Group
1302 Ventura Canyon Dr.
Katy, TX 77494
713-410-5489
www.alignbg.com

Asempra Technologies
640 West California Avenue
Sunnyvale, CA 94086-3624
408-215-5800
www.asempra.com

Balardo Group
90 Adelaide Street West
Suite 401
Toronto, Ontario M5H 3V9
416-657-2322
www.balardogroup.com

BRProactive Inc.
1141 East Bennett Avenue
Glendora, CA 91741
626-852-0412
www.brproactive.com

Business & Government
Continuity Services
P.O. Box 1706
Oklahoma City, OK 73101
405-737-8348
www.businesscontinuity.info

Knight-Star Enterprises, Inc.
HUB/Woman-Owned Business
P.O. Box 7873
Waco, TX 76714
254-776-1996
www.knightstar.us

University of Pittsburgh Medical
Center
Clinicians' Biosecurity Network
200 Lothrop St.
Pittsburgh, PA 15213-2582
443-573-3304
www.upmc-cbn.org/dmz/about_net-
work.html

Adjusters International
25800 Northwestern Highway
Suite 885
Southfield, MI 48075-8403
248-352-2100
www.globemwai.com

Anacomp
15378 Avenue of Science
San Diego, CA 92128
858-716-3400
www.anacomp.com

Assurity River Group
P.O. Box 24506
Minneapolis, MN 55108-5265
612-435-2170
www.assurityriver.com

Barney F Pelant & Associates
243 Harvard Lane
P.O. Box 6204
Bloomingdale, IL 60108-2141
630-894-6989
www.bfpelantassoc.com

BSI Management Systems
12110 Sunset Hills Road
Reston, VA 20190
703-464-1931
www.bsiamerica.com

Business Continuity Institute
P.O. Box 4474
Worcester WR6 5YA
44-08706038783
www.thebci.org

Business Continuity Planning Asia
#03-15 Keypoint
371 Beach Road
Singapore 199597
656-325-2080
www.bcpasia.com

Business Recovery Consultants Inc.
27 Westwood Boulevard
Hockessin, DE 19707-2059
302-234-8178
www.brcrecovery.com

CEM Associates Inc.
2218 Little John Trail
Newton, NC 28658
828-310-2859
www.cemassociates.org

CommandGlobal
14902 Preston Road
Suite 404-343
Dallas, TX 75254
214-550-8888
www.commandglobal.com

Concurrent Technologies
150 Allen Road
Liberty Corner, NJ 07938
800-345-3895
www.concurrenttechnologies.com

Contingency Now Inc.
6032 Buffalo Avenue
Van Nuys, CA 91401-3024
913-484-5317 310-686-9094
www.contingencynow.com

Continuity First
P.O. Box 28796
Richmond, VA 23228
804-559-6623
www.continuityfirst.com

Continuity Solutions, Inc.
6649 North High Street
Worthington, OH 43085
614-885-5001
www.csigroup.cc

Corporate Risk Solutions Inc.
Suite 450
8725 Rosehill Road
Lenexa, KS 66215
913-422-0410
www.corprisk.net

CPM Global Assurance Newsletter
Suite 777
3141 Fairview Park Drive
Falls Church, VA 22042
215-348-1084
www.contingencyplanning.com

Business Guard Inc.
Suite 900
601 Penn. Avenue Northwest
Washington, D.C. 20004-2601
202-497-3800
www.businessguardinc.com

CAPS Business Recovery Services
2 Enterprise Drive
Shelton, CT 06484
203-925-3900
www.capsbrs.com

CGI-AMS
Enterprise Security Practice
4050 Legato Road
Fairfax, VA 22033
703-267-8000
www.ams.com/security

Comprehensive Solutions
250 North Sunny Slope Road
Suite 300
Brookfield, WI 53005
262-785-8101
www.comp-soln.com

Consonus
301 Gregson Drive
Cary, NC 27511
919-379-8000
www.consonus.com

Contingency Planners Inc.
2510 Frederick Drive
Conway, AR 72034-9698
501-329-0958
www.contingency-planners.com

Continuity Research
2 Hambleton Court
Baltimore, MD 21208
www.continuityresearch.com

ContinuityLink
3300 Liebert
Montreal, Quebec QC H1L 5S3
514-572-4517
www.continuitylink

CoSentry
Suite 132
1001 Fort Crook Road North
Bellevue, NE 68005
402-492-7800
www.cosentry.com

CRI Network Inc
Suite 401 901 Gordon Street
Victoria, British Columbia
V8L 2K6
250-889-5030
www.crinetwork.com

Business Protection Systems
Suite 220
5041 LaMart Drive
Riverside, CA 92507
951-341-5050
www.businessprotection.com

CDM
#210 14420 Albemarle Point Place
Chantilly, VA 20151
703-968-0900
www.cdm.com

CIMCO Communications Inc.
Suite 700
1901 South Meyers Road
Oakbrook Terrace, IL 60181
877-691-8080
www.cimco.net

Computer Solutions Inc.
Suite 2
6 Commerce Street
Branchburg, NJ 08876-6041
908-823-3200
www.internetcsi.com

Consortium of Business Continuity
Professionals, Inc. (CBCPi) Tools
Suite 115-399
10580 North McCarran Boulevard
Reno, NV 89503-1895
877-621-2227
www.cbcpinc.com

ContingenZ Corporation
4th Floor 227 Fowling Street
Playa Del Rey, CA 90293
310-306-0166
www.contingenz.com

Continuity Shield
1 Yonge Street Suite 1801
Toronto, Ontario M5E 1W7
416-483-0464
www.continuityshield.com

Continuous Solutions Inc.
1314 Madeira Southeast
Albuquerque, NM 87113
505-228-0864
www.continuoussolutions.com

CPACS LLC
590 Danbury Road
Ridgefield, CT 06877
203-431-8720
www.cpacsweb.com

Criterion Strategies Inc.
580 Broadway
Suite 305
New York, NY 10012
212-343-1134
www.criterionstrategies.com

DAMICON LLC
13 Jackson Road
Burlington, MA 01803
781-789-8238
www.damicon.com

Deucalion
7456 South West Baseline Road
#119
Hillsboro, OR 97123
403-543-4695
www.deucalion.net

Disaster Preparedness of
North Texas
2820 E. University Dr.
Suite #174
Denton, TX 76209
940-365-3442

Disaster Survival Planning
Network (DSPN)
5352 Plata Rosa Court
Camarillo, CA 93012
www.dspnetwork.com

EBC Partners Inc.
Suite 108
6753 Thomasville Road
Tallahassee, FL 32312
850-894-4043
www.ebcpartners.com

EMS Solutions
260 Whitney Street
San Francisco, CA 94131
415-643-4300
www.ems-solutions.com

Excelliant
1201 Lee Branch Lane
Birmingham, AL 35242
205-313-9180
www.excelliant.com

Global Security Systems, LLC
308 East Pearl Street Suite 202
Jackson, MS 39201
337-237-7757
www.gssnet.us

Hitachi Data Systems
750 Central Expressway
Santa Clara, CA 95050
408-970-1000
www.hds.com

DaVinci Technology Corporation
(DaVinciTek)
89 Headquarters Plaza
North Tower, 14th Floor
Morristown, NJ 07960
973-993-4860
www.davincitek.com

Disaster Management Inc.
1531 Southeast Sunshine Avenue
Port St. Lucie, FL 34952-6011
772-335-9750
www.disastermgt.com

Disaster Recovery Consultants,
LLC
41 Lenox Court
Montville, NJ 07045
908-328-7719

Don J. Brooks Holdings Ltd.
695 Proudfoot Lane Suite 718
London, Ontario N6H 4Y7
519-657-5472
www.b2bcontinuity.com

eBRP Solutions
7895 Tranmere Drive
Unit 25
Mississauga, Ontario L5S 1W9
905-677-0404
www.ebrp.net

EnSafe Inc.
5724 Summer Trees Drive
Memphis, TN 38134
901-372-7962
www.ensafe.com

FirstMerit Corp.
6625 West Snowville Road
Brecksville, OH 44141-3209
440-838-4044
www.contingencyplanningguide.com

Hannah-Watrous Continuity
Strategies
P.O. Box 54
60 West Main Street
Chester, CT 06412-1345
860-227-5046
www.hanwat.com

HZX Computer Systems
Consultants—BCP Services
41 Palomino Crescent
Toronto, Ontario M2K 1W2
416-221-6603

Decisive Technologies Inc.
38 Auriga Drive
Suite 200
Ottawa, Ontario K2E 8A5
613-482-2649
www.decisive.ca

Disaster Masters
146-23 61st Road
Flushing, NY 11367-1203
718-939-5800
www.theplan.com/disaster.htm

Disaster Recovery System (DRS)
1732 Remson Avenue
Merrick, NY 11566
516-623-2038
www.drsbytamp.com

DSP Network
5352 Plata Rosa Court
Camarillo, CA 93012
www.dspnetwork.com

Emotional Continuity Management
Inner Directions LLP
2815 Van Giesen
Richland, WA 99354
509-942-0443
www.emotionalcontinuity.com

Ethix Consulting, LLC
202 Berkley Drive Suite 210
Harrisburg, PA 17112
717-651-1520
www.ethixconsulting.com

Gaintner Bandler Reed & Peters
PLC
2198 East Camelback Road
Suite 205
Phoenix, AZ 85016
602-381-0381

Hillmann Group, LLC
1600 Route 22 East
Union, NJ 07083
908-688-7800
www.hillmanngroup.com

IBM Business Continuity and
Recovery Services
1 New Orchard Road
Armonk, NY 10504-1722
877-426-6006
www.us.ibm.com

IBM Business Resilience and
Continuity Services
300 Long Meadow Road
Sterling Forest, NY 10979
www.ibm.com/services/resilience

IBM Global Services
10 North Martingale Road
Schaumburg, IL 60173
www.ibm.com

Inflow Inc.
Suite E5 550 East 84th Avenue
Thornton, CO 80229-5338
www.inflow.com

InfoSENTRY Services Inc.
P.O. Box 28048
2 Hannover Square, Suite 2330
Raleigh, NC 27601
919-838-8570
www.infosentry.com

Interstate Restoration Co.
1661 Kersley Circle
Heathrow, FL 32746
404-307-8238
www.interstaterestoration.com

iTeam Consulting
Calle 22 5-88 Oficina 301
Bogotá 0001
571-282-6587
www.iteamconsulting.co.za

Jack Henry & Associates Inc.
663 West Highway 60
P.O. Box 807
Monett, MO 65708
417-235-6652
www.jackhenry.com

Jannaway & Associates
102 Robinson Avenue
Toronto, Ontario M1L 3T3
416-569-3274

Jonathan Ward & Associates
3600 Yonge Street
Suite 625
Toronto, Ontario M4N 3R8
www.wardassociates.ca

JS&A Consulting
2510 Frederick Drive
Conway, AZ 72034-9698
501-329-0958
www.contingency-planners.com

KETCHConsulting
P.O. Box 641
Waverly, PA 18471
570-563-0868
www.ketchconsulting.com

Keyworth Fire and Safety
Consultants, Inc.
1645 Von Braun Trail
Elk Grove Village, IL 60007-3148
630-776-0284
www.keyworthfireandsafety.com

KL Security Enterprises LLC
2842 East 860 North
P.O. Box 165
Stockland, IL 60967
866-867-0306
www.klsecurity.com

Knight-Star Enterprises, Inc.
HUB/Woman-Owned Business
P.O. Box 7873
Waco, TX 76714
254-776-1996
www.knightstar.us

KPMG, LLP
Risk Advisory Services
345 Park Avenue
New York, NY 10154
212-872-4380
www.kpmg.com

LBL Technology Partners
2501 Wayzata Boulevard
Minneapolis, MN 55405
612-381-8939
http:/www.drplan@iblco.com

Lunngroup Consulting Inc
255 Newport Dr Metrotown RPO
Port Moody, British Columbia
V3H SH1
604-780-4816
www.lunngroup.com

M Corby & Associates Inc.
255 Park Avenue 8th Floor
Worcester, MA 01609-1953
www.mcorby.com

Mainstay Consulting Group, LLC
20 Calle Pastadero
San Clemente, CA 92672
949-369-1111
www.mainstayconsulting.com

Marsh Inc.
1166 Avenue of the Americas
New York, NY 10036-2774
www.marshriskconsulting.com

Meredith Management Group
Station Square Three Suite 202
37 North Valley Road
Paoli, PA 19301
610-725-8286
www.mmg-ems.com

MLC & Associates Inc.
Suite 265 3525 Hyland Avenue
Costa Mesa, CA 92626
949-222-1202
www.mlcandassociates.com

Montague Risk Management
9 Underhill Avenue
Locust Valley, NY 11560
516-676-9234
www.montaguetm.com

National Security Research
2231 Crystal Drive Suite 500
Arlington, VA 22202
703-647-2200
www.nsrinc.com

Network Technology Group Inc.
7127 Florida Boulevard
Baton Rouge, LA 70806
225-214-3800
www.ntg.com

NW Recovery Specialties LLC.
9309 South 223rd Place
Kent, WA 98031
206-786-7431
nwrecoveryspecialists.com

Oppenheim Consulting, LLC
4 Adams Court
Plainsboro, NJ 08536-2324
609-721-1874
www.oppenheimconsulting.com

Optimus Consulting
22 Technology Parkway South
Norcross, GA 30092
770-447-1951
www.optimussolutions.com/oc

Pekarcik Global Industries &
Ventures
695 Coral Court
Los Altos Hills, CA 94024
650-814-1625
www.pekarcikglobalventures.com

Penta Associates Ltd.
Suite 501 11 Hanover Square
New York, NY 10005
212-344-2010
www.pentaassoc.com

Performance Technology Group
1615 Knecht Avenue
Baltimore, MD 21227
410-347-5200
www.ptgcorp.com

PreEmpt Inc.
3787 Upper Denton Road
Weatherford, TX 76085-8360
817-596-5018
www.preemptinc.com

ProfitStars, A Jack Henry Company
1025 Central Expressway South
Dallas, TX 75013
913-307-1324
www.profitstars.com

Recovery Specialties LLC
Suite 102 5731 Mustang Drive
Simi Valley, CA 93063-6312
805-581-3227
recoveryspecialties.com

River Bend Business Continuity
One Omega Drive
Stamford, CT 06907
203-978-7444
www.riverbend1.com

Scivantage
10 Exchange Place, 13th Floor
Jersey City, NJ 07302-4931
646-452-0050
www.scivantage.com

SEH
Butler Square Building
Suite 710C
100 North 6th Street
Minneapolis, MN 55403-1515
612-216-1889
www.sehinc.com

StoneHenge Partners Inc.
401 South Boston Suite 400
Tulsa, OK 74103
918-971-1999
www.stonehenge.or

Suite 400
1950 Spectrum Circle
Marietta, GA 30067
877-304-2917
www.monarchresiliency.com

The Courton Group
134 Fort Greene Place
Brooklyn, NY 11217
718-855-1440
www.courtongroup.com

Phonetic Systems
5th Floor 300 Concord Road
Billerica, MA 01821
978-439-3600
www.pitango.com

PriceWaterhouseCoopers LLC
Suite 3000 Royal Trust Tower
Toronto-Dominion Centre
77 King Street West
Toronto, Ontario M5K 1G8
416-863-1133
www.pwc.com/ca

Protiviti Inc.
5720 Stoneridge Drive
Pleasanton, CA 94588
415-402-3600
www.protiviti.com

Redmond Worldwide, Inc
6637 Bergen Place
Brooklyn, NY 11220
718-745-0582
www.redmondworldwide.com

RP Risk Advisors
5 Lyons Mall, Suite 322
Basking Ridge, NJ 07920
908-310-6381
rpriskadvisors.com

SECTOR, Inc.
2 MetroTech Center
Brooklyn, NY 11201
212-383-2000
www.sectorinc.com

Sentryx
Suite 207 2476 Argentia Road
Mississauga, Ontario L5N 6M1
www.sentryx.com

Strategic BCP
102 Ava Court
Plymouth Meeting, PA 19462
610-275-4227
www.strategicbcp.com

TAMP Systems
1732 Remson Avenue
Merrick, NY 11566
516-623-2038
www.drsbytamp.com

The Gimbal Group, Inc.
2111 Wilson Boulevard Suite 700
Arlington, VA 22201
703-351-5054
www.gimbal.com

Pitney Bowes
27 Waterview Drive
Shelton, CT 06484-4705
203-922-5810
www.pb.com

Professional Loss Adjusters Inc.
343 Washington Street
Newton, MA 02458
617-850-0477
www.pla-us.com

R&A Crisis Management Services
650 South River Road Suite 713
Des Plaines, IL 60016-8344
847-827-4267
www.raconsulting.net

Resiliency Solutions Inc.
9208 Valaretta Drive
Gretna, NV 68028
www.resiliencysolutions

SBP Consulting Services Inc
97 Dovercourt Road
Toronto, Ontario M6J 3C2
416-723-7953

Seek Safety Services, Inc.
8409 Pickwick Lane #271
Dallas, TX 75225
214-668-2285
www.seeksafetyservices.com

STERIS Corporation
5960 Heisley Road
Mentor, OH 44060
440-354-2600
www.steris.com

Structured Technical Services, LLC
2850 South West Cedar Hills
Boulevard #330
Beaverton, OR 97005-1393
503-449-7703
www.structuredtechnical.com

TDG, Inc
82 Sorrentino Way
Mays Landing, NJ 08330
703-623-2702
www.tdginc.com

The Howe Partnership
Suite #800
#2 St. Clair Ave. East
Toronto, Ontario M4T 2T5
416-721-1053

The Radian Group LLC
Suite 201
40 Shuman Boulevard
Naperville, IL 60563
630-305-7100
www.theradiangroup.com

Turnbull Consulting Inc.
P.O. Box 475
Wallace, NC 28466-0475
910-285-8606
www.turnbullconsulting.us

URS Corporation
200 Orchard Ridge Dr.
Suite 101
Gaithersburg, MD 20878
301-258-6554
www.urscorp.com

Usg Inc.
7831 East Bush Lake Road
Suite 100
Minneapolis, MN 55416
952-835-2349
www.usg-inc.com

Vistastor (Peakdata LLC)
6309 Monarch Park Place
Niwot, CO 80503
303-652-2630
www.peakdata.com

Wester & Associates
22 Laurel Drive
Corte Madera, CA 94925
415-945-9327

CAPS Business Recovery Services
2 Enterprise Drive
Shelton, CT 06484
203-925-3900
www.capsbrs.com

Disaster Management Inc.
1531 Southeast Sunshine Avenue
Port St. Lucie, FL 34952-6011
772-335-9750
www.disastermgt.com

Trackis, Inc.
2211 Centerbrook Ln.
Katy, TX 77450
832-435-9604
www.trackis.com

TwoSeven, Inc.
P.O. Box 11064
Norfolk, VA 23501
804-339-5890
www.twoseven.com

USG Inc.
550 West Adams Street
Chicago, IL 60661-3676
312-436-4000
www.usg.com

Virtela Communications Inc.
5680 Greenwood Plaza Boulevard
Greenwood Village, CO 80111
720-475-4000
www.virtela.net

Waypoint Advisory
P .O. Box 984
Concordville, PA 19331
610-358-1202
www.waypointadvisory.com

Wester & Associates
172 Joicey Boulevard
Toronto, Ontario M5M 2V2
416-489-9327

Attainium Corp.
14540 John Marshall Highway
Suite 103
Gainesville, VA 20155
571-248-8200
www.attainium.net

Comprehensive Solutions
250 North Sunny Slope Road
Suite 300
Brookfield, WI 53005
262-785-8101
www.comp-soln.com

EBC Partners Inc.
Suite 108
6753 Thomasville Road
Tallahassee, FL 32312
850-894-4043
www.ebcpartners.com

Transformyx Inc.
8510 Quarters Lake Road
Baton Rouge, LA 70809
225-761-0088
www.transformyx.com

United Security Group
5775 Wayzata Boulevard
Suite 700
Minneapolis, MN 55416
952-582-2955
www.usg-inc.com

USG Inc.
Suite 100
7831 East Bush Lake Road
Edina, MN 55439-3117
952-806-0911
www.contingencyplanningguide.com

Virtual Corporation
98 Route 46
Suite 12
Village Green Annex
Budd Lake, NJ 07828
973-426-1444
www.virtual-corp.net

WebEx Communications
3979 Freedom Circle
Santa Clara, CA 95054
877-509-3239
www.webex.com

Zonecast Inc.
210 2021 Peyton Avenue
Burbank, CA 91504
818-232-7567
www.zonecast.com

Business Continuity Planners, Inc
(BCPI)
12685 Dorsett Road #126
Maryland Heights, MO 63043
314-541-4913
www.bus-cont-plan.com

DaVinci Technology Corporation
(DaVinciTek)
89 Headquarters Plaza
North Tower, 14th Floor
Morristown, NJ 07960
973-993-4860
www.davincitek.com

Global Security Systems, LLC
308 East Pearl Street
Suite 202
Jackson, MS 39201
337-237-7757
www.gssnet.us

IBM Global Services
10 North Martingale Road
Schaumburg, IL 60173
www.ibm.com

Montague Risk Management
9 Underhill Avenue
Locust Valley, NY 11560
516-676-9234
www.montaguetm.com

Price Hollingsworth Co.
1580 Louis Avenue
Elk Grove Village, IL 60007
847-718-9460
www.pricehollingsworth.com

RP Risk Advisors
5 Lyons Mall, Suite 322
Basking Ridge, NJ 07920
908-310-6381
rpriskadvisors.com

TAMP Systems
1732 Remson Avenue
Merrick, NY 11566
516-623-2038
www.drsbytamp.com

The Revere Group
Suite #325
325 North LaSalle Street
Chicago, IL 60610
312-873-3400
www.reveregroup.com

Associated Records and
Information Services
P.O. Box 937
Caddo Mills, TX 75135-0937
903-527-2156
www.associatedrecords.com

CAPS Business Recovery Services
2 Enterprise Drive
Shelton, CT 06484
203-925-3900
www.capsbrs.com

Continuity Solutions, Inc.
6649 North High Street
Worthington, OH 43085
614-885-5001
www.csigroup.cc

KETCHConsulting
P.O. Box 641
Waverly, PA 18471
570-563-0868
www.ketchconsulting.com

Optimus Consulting
22 Technology Parkway South
Norcross, GA 30092
770-447-1951
www.optimussolutions.com/oc

ProfitStars, A Jack Henry
Company
1025 Central Expressway South
Dallas, TX 75013
913-307-1324
www.profitstars.com

Strategic BCP
102 Ava Court
Plymouth Meeting, PA 19462
610-275-4227
www.strategicbcp.com

The Courton Group
134 Fort Greene Place
Brooklyn, NY 11217
718-855-1440
www.courtongroup.com

bigbyte.cc
123 Central Avenue Northwest
Albuquerque, NM 87102
505-255-5422
www.bigbyte.cc

Comprehensive Solutions
250 North Sunny Slope Road
Suite 300
Brookfield, WI 53005
262-785-8101
www.comp-soln.com

Copper Harbor Consulting Inc.
12 Grant Street
Needham, MA 02492
781-400-1305
www.copperharborconsulting.com

MLC & Associates Inc.
Suite 265
3525 Hyland Avenue
Costa Mesa, CA 92626
949-222-1202
www.mlcandassociates.com

PreEmpt Inc.
211 Foxbury Drive
Euless, TX 76040
www.preemptinc.com

Protiviti Inc.
5720 Stoneridge Drive
Pleasanton, CA 94588
415-402-3600
www.protiviti.com

Strohl Systems
631 Park Avenue
King of Prussia, PA 19406
610-768-4135
www.strohlsystems.com

The Gimbal Group, Inc.
2111 Wilson Boulevard
Suite 700
Arlington, VA 22201
703-351-5054
www.gimbal.com

Adjusters International
25800 Northwestern Highway
Suite 85
Southfield, MI 48075-8403
248-352-2100
www.globemwai.com

Caps Business Recovery Planning
1 Enterprise Drive
Shelton, CT 06484
203-925-3900
www.capsbrs.com

Concergent LLC.
Suite T4
245 North Waco
Wichita, KS 67202-1132
316-613-4747
www.contingencyplanningguide.com

Crisis Management International
Inc.
Suite 420
8 Piedmont Center Northeast
Atlanta, GA 30305
404-841-3400
www.cmiatl.com

DaVinci Technology Corporation
(DaVinciTek)
89 Headquarters Plaza
North Tower, 14th Floor
Morristown, NJ 07960
973-993-4860
www.davincitek.com

Eyeview Recovery Consulting
8923 West Sunset Drive
Wonder Lake, IL 60097
815-728-7031
www.eyeviewrecovery.com

KETCHConsulting
P.O. Box 641
Waverly, PA 18471
570-563-0868
www.ketchconsulting.com

Montague Risk Management
9 Underhill Avenue
Locust Valley, NY 11560
516-676-9234
www.montaguetm.com

Pitney Bowes Inc.
1 Elmcroft Road
Stamford CT 06926-0700
203-356-5000
www.pb.com

Recovery-Plus Planning Products
& Services
3509 Harvest Court
Suite 4B
Island Lake, IL 60042-9536
847-487-5700
www.recovery-plus.com

Sage Business Associates Inc.
6156 Powell Road
Parker, CO 80134
303-841-4467

The Courton Group
134 Fort Greene Place
Brooklyn, NY 11217
718-855-1440
www.courtongroup.com

Datalink
8170 Upland Circle
Chanhassen, MN 55317
952-944-3462
www.datalink.com

Disaster Recovery Services Pty Ltd
(Australia) Tools
Suite 285
Sea Bridge House
377 Kent Street NSW
Sydney NSW 2000
61-296369331

HP Business Recovery Services
3000 Hanover Street
Palo Alto, CA 94304-1185
650-857-1501
www.hp.com

La Socit Prudent Inc.
2075 rue Victoria bureau 113
St-Lambert, Quebec J4S 1H1
450-672-7966
www.prudent.qc.ca

Optimus Consulting
22 Technology Parkway South
Norcross, GA 30092
770-447-1951
www.optimussolutions.com/oc

ProfitStars, A Jack Henry
Company
1025 Central Expressway South
Dallas, TX 75013
913-307-1324
www.profitstars.com

RP Risk Advisors
5 Lyons Mall
Suite 322
Basking Ridge, NJ 07920
908-310-6381
rpriskadvisors.com

SEM3 Solutions
19 Jackson Drive
Raynham, MA 02767
508-717-7208
www.sem3solutions.com

The Revere Group
Suite #325 325 North
LaSalle Street
Chicago, IL 60610
312-873-3400
www.reveregroup.com

Evolving Solutions
3989 County Road 116
Hamel, MN 55340-9341
763-516-6500
www.evolvingsol.com

EBC Partners Inc.
Suite 108
6753 Thomasville Road
Tallahassee, FL 32312
850-894-4043
www.ebcpartners.com

IBM Global Services
10 North Martingale Road
Schaumburg, IL 60173
www.ibm.com

Mikron Consulting
1073 Oakwood Center
Chanburgh, IL 60193
847-909-9516
www.mikronconsulting.com

Phoenix Consulting Services
Suite K-285
4450 California Avenue
Bakersfield, CA 93309
661-396-8336

Protiviti Inc.
5720 Stoneridge Drive
Pleasanton, CA 94588
415-402-3600
www.protiviti.com

RSM McGladrey Inc.
801 Nicollet Avenue
11th Floor
West Tower
Suite 700
Minneapolis, MN 55402
612-573-8750
www.rsmmcgladrey.com

Siegel Rich Division-Rothstein,
Kass and Company, P.C.
1350 Avenue of the Americas
15th Floor
New York, NY 10019
212-997-0500
www.rkco.com

GeoAge
Suite 9
3740 St. Johns Bluff Road
Jacksonville, FL 32224
904-565-9855
www.geoage.com

ISSI DATA
22122 20th Avenue South
East
#152
Bothell, WA 98201
426-483-4801
www.issidata.com

PentaSafe Security
Technologies Inc.
#1800 1233 West Loop South
Houston, TX 77027-9106
713-860-9390
www.netiq.com

TTR Group Inc: Security Services
P.O. Box 7185
Colorado Springs, CO 80933-7185
719-265-8378
www.ttrg.org

Alertnow
4000 Westchase Boulevard
Suite 190
Raleigh, NC 27609
919-841-0175
www.alertnowusa.com

Contingency Management
Consultants
4 Shawnee Lane
Orinda, CA 94563-3217
925-254-1663
www.businesscontinuity.com

Landslide Observatory-
JCET/UMBC
Suite 320
5523 Research Park Drive
Baltimore, MD 21228
410-455-6362
www.jcet.umbc.edu

Optimus Consulting
22 Technology Parkway South
Norcross, GA 30092
770-447-1951
www.optimussolutions.com/oc

TSC Consulting Inc.
5030 Champion Blvd. G6-205
Boca Raton, FL 33496
561-829-8118
www.thesecurecomputer.com

4Sight Group
Suite 134-233
4001 Kennett Pike
Wilmington, DE 19807-2315
www.4sightgroup.com

KL Security Enterprises LLC
2842 East 860 North
P.O. Box 165
Stockland, IL 60967
866-867-0306
www.klsecurity.com

ProfitStars, A Jack Henry
Company
1025 Central Expressway South
Dallas, TX 75013
913-307-1324
www.profitstars.com

VirtuIT Systems
Suite 163
119 Rockland Center
Nanuet, NY 10954
877-847-8848
www.virtuitsystems.com

Computersite Engineering Inc.
2904 Rodeo Park Drive East
Building 125
Santa Fe, NM 87506
505-982-8300
www.upsite.com

Emotional Continuity
Management
Inner Directions LLP 2815 Van
Giesen
Richland, WA 99354
509-942-0443
www.emotionalcontinuity.com

Montague Risk Management
9 Underhill Avenue
Locust Valley, NY 11560
516-676-9234
www.montaguetm.com

RecoveryPlanner
2 Enterprise Drive
Shelton, CT 06484
203-925-3950
www.recoveryplanner.com

ACS Image Solutions
102 Business Park Drive
Suite I
Ridgeland, MS 39157
601-977-4000

Patrina Corporation: 3rd Party Data
Management
Two Wall Street
New York, NY 10005
212-233-1155
www.patrina.com

The Courton Group
134 Fort Greene Place
Brooklyn, NY 11217
718-855-1440
www.courtongroup.com

Consonus All Ways On Data
Centers
180 East 100 South Questar
Building
Salt Lake City, UT 84111
888-452-8000
www.consonus.com

IBM Global Services
10 North Martingale Road
Schaumburg, IL 60173
www.ibm.com

Oppenheimer Wolff & Donnelly
LLP
45 South 7th Street Plaza VII,
Suite 3300
Minneapolis, MN 55402
612-607-7000
www.oppenheimer.com

Recovery-Plus Availability Planning
Suite 4B 3509 Harvest Court
Island Lake, IL 60042-9536
847-487-5700
www.contingencyplanningguide.com

2bcool Enterprises Inc.
503-333 2 Avenue NE
Calgary, Alberta T2E 0E5
403-276-3855
www.2bcool.ca

Adjusters International
25800 Northwestern Highway
Suite 885
Southfield, MI 48075-8403
248-352-2100
www.globemwai.com

Advanced Process Solutions Inc.
2416 Basil Drive
Raleigh, NC 27612
919-844-1625
www.aps4you.com

AMTI
2900 Sabre Street
Suite 800
Virginia Beach, VA 23452
757-431-8597
www.amti.net

BearingPoint Inc.
1676 International Drive
McLean, VA 22030
703-747-3000
www.bearingpoint.com

Business Contingency Group
Suite 333
18034 Ventura Boulevard
Encino, CA 91316
www.businesscontingencygroup.com

Chase Environmental Group Inc
P.O. Box AB
Centralia, IL 62801
618-533-6740

Clean Harbors Environmental
Services, Inc.
42 Longwater Drive
P.O. Box 9149
Norwell, ME 02061-9149
781-792-5000
www.cleanharbors.com

Contingency Planning Solutions
Inc.
518 South Westland Drive
Appleton, WI 54914
920-734-0241
www.contingencyplans.com

DataCenterManager.com
P.O. Box 805975
Chicago, IL 60680
312-451-1052
www.datacentermanager.com

DPS Management Consultants
2320 Gravel Drive
Fort Worth, TX 76118-6950
817-284-7711
www.dpsconsultants.net

Earthquake Solutions
122 East Walnut Avenue
Suite A
Monrovia, CA 91016
626-256-7900
www.earthquakesolutions.com

Agility Recovery Solutions
7621 Little Avenue Suite 218
Charlotte, NC 28226
704-341-8700
www.agilityrecovery.com

Arlington Associates, LLC
770 Arlington Circle
Novato, CA 94947-4976
415-883-0884

Booz-Allen & Hamilton Inc.
8283 Greensboro Drive
McLean, VA 22102-3838
703-902-5000
www.boozallen.com

Business Continuity Services Inc.
 207 West Ash Street
Lombard, IL 60148
630-629-6327
www.businesscontinuitysvcs.com

Chubb Group of Insurance
Companies
55 Water Street
New York, NY 10041
212-612-4000
www.chubb.com

Comprehensive Solutions
250 North Sunny Slope Road
Suite 300
Brookfield, WI 53005
262-785-8101
www.comp-soln.com

Curtis 1000
Suite 500
1725 Breckinridge Parkway
Duluth, GA 30096
678-380-9095
www.curtis1000.com

Disaster Recovery System (DRS)
1732 Remson Avenue
Merrick, NY 11566
516-623-2038
www.drsbytamp.com

Dreamcatcher Disaster Resilience
LLC
1647 Dancer Drive
Rochester Hills, MI 48307
248-650-9900
www.dreamcatcher-dr.com

EBC Partners Inc.
Suite 108
6753 Thomasville Road
Tallahassee, FL 32312
850-894-4043
www.ebcpartners.com

Agility Recovery Solutions
2281 N. Sheridan Way
Mississauga, Ontario L5K 2S3
www.agilityrecovery.com

AXCESS Disaster Consulting Group
P.O. Box 91825
West Vancouver British Columbia,
V7V 4S1
604-657-6760
www.strategis.ca.gc

Burney Industries
6702 Linwood Avenue
Shreveport, LA 71106
318-861-4327

CGI
1130 Sherbrooke St. West 7th Floor
Montreal, Quebec H3A 2M8
514-841-3200
www.cgi.com

Clas Consulting LLC
26 Needham Street
Norfolk, MA 02056-1624
508-613-2171
www.clasconsulting.com

Consolidated Risk Management
1717 East 9th Street
Suite 1125
Cleveland, OH 44114-2804
216-623-1777
www.risk-manage.com

DAMICON LLC
13 Jackson Road
Burlington, MA 01803
781-789-8238
www.damicon.com

Disaster Resource Mgmt
11700 Mountain Park Road
Roswell, GA 30075
770-605-6477

Eagle Rock Alliance Ltd.
80 Main Street 3rd Floor
West Orange, NJ 07052
973-325-9900
www.eaglerockalliance.com

EGP & Associates Inc.
Suite 395
3625 Brookside Parkway
Alpharetta, GA 30022
678-904-8730
www.egpinc.com

Elliot Consulting Services
Suite 326
7853 Gunn Highway
Tampa, FL 33626
813-792-8833
www.elliot-consulting.com

Emotional Continuity
Management
Inner Directions LLP
2815 Van Giesen
Richland, WA 99354
509-942-0443
www.emotionalcontinuity.com

Equivus Consulting
#361 4238 North Arlington
Heights Road
Arlington Heights, IL 60004
866-378-4887
www.equivus.com

First General Enterprises Inc.
Suite 121
6451 North Federal Highway
Fort Lauderdale, FL 33304-3115
954-537-5556
www.firstgeneralservices.com

Grant Thornton LLP
Suite 900
18400 Von Karman Avenue
Irvine, CA 92612
949-553-1600
www.gt.com

Harland Financial Solutions
312 Plum Street
Cincinnati, OH 45202
513-381-9400
www.intrieve.com

Idea Integration
5251 DTC Parkway
Suite 1045
Greenwood Village, CO 80111
303-824-5600
www.idea.com

Intrado
1601 Dry Creek Drive
Longmont, CO 80503
720-494-5800
www.intrado.com

Janus Associates
9th Floor
River Plaza
9 West Broad Street
Stamford, CT 06902
203-251-0200
www.janusassociates.com

Emergency Management Solutions,
LLC
P.O. Box 77151
Colorado Springs, CO 80970-7151
719-359-6762

Enterprise Connections
Suite 600
1800 Century Park East
Los Angeles, CA 90067
310-229-5744
www.enterpriseconnections.com

Ernst & Young LLP
1401 McKinney Street Suite 1200
Houston, TX 77010
713-750-8147
www.ey.com

First Response Restoration
Network
1919 South Michigan Street
South Bend, IN 46613
574-288-0500
www.firstresponsedrs.com

Greenley & Associates
Incorporated
5 Corvus Court
Ottawa, Ontario K2E 7Z4
613-247-0342
www.greenley.ca

Homeland Restoration Network
Suite 301
3585 Highway 317
Suwanee, GA 30024
www.homelandrestoration.net

IEM Inc.
8555 United Plaza Boulevard
Suite 100
Baton Rouge, LA
225-952-8191
www.ieminc.com

J&H Marsh & McLennan Inc.
212 Carnegie Center
Princeton, NJ 08543
609-520-2900
www.marsh-financial.com/
risk.html

JM Hickey and Associates
9050 South Hamilton Avenue
Chicago, IL 60620-6102
773-239-3310

Emergency Visions
2110 Spring Hill Court
Smyrna, GA 30080
770-436-2474
www.emergencyvisions.com

Enterprise Risk Worldwide, Inc
551 Fifth Avenue
Suite 3025
New York, NY 10176
212-599-1878
www.enterpriseriskworldwide.com

E-ternity Business Continuity
Consultants Inc.
Suite 100
2110 Matheson Boulevard East
Mississauga, Ontario L4W 5E1
416-410-2142
www.e-ternity.ca

Gerard Group International LLC
164 Westford Road Suite 15
Tyngsborough, MA 01879
978-649-4575
www.gerardgroup.com

HAN Consulting
Suite 243
4960 Almaden Expressway
San Jose, CA 95118
408-234-9936
www.han-consulting.com

IBM Global Services
10 North Martingale Road
Schaumburg, IL 60173
www.ibm.com

International Dynamics Research
Corporation
Suite 4
2480 Briarcliff Road Northeast
Atlanta, GA 30329
770-723-9785

Janco Associates Inc.
11 Eagle Landing Court
Park City, UT 84060
435-940-9300 x101
www.e-janco.com

KETCHConsulting
P.O. Box 641
Waverly, PA 18471
570-563-0868
www.ketchconsulting.com

Knight-Star Enterprises, Inc.
HUB/Woman-Owned Business
P.O. Box 7873
Waco TX 76714
254-776-1996
www.knightstar.us

Montague Risk Management
9 Underhill Avenue
Locust Valley, NY 11560
516-676-9234
www.montaguetm.com

Network Services
Sunset Place
Hawthorne, NY 10532
202-470-3250
www.na-access.com

Organizational Resilience
International LLC
85 Warren Street
Concord, NH 03301
603-369-3481
www.oriconsulting.com

Phoenix Consulting Services
Suite K-285
4450 California Avenue
Bakersfield, CA 93309
661-396-8336

Pitney Bowes Inc.
1 Elmcroft Road
Stamford, CT 06926-0700
203-356-5000
www.pb.com

ProfitStars, A Jack Henry
Company
1025 Central Expressway South
Dallas, TX 75013
913-307-1324
www.profitstars.com

SafetyMate Corp.
1642 Mcgaw Avenue
Irvine, CA 92614-5632
949-252-1570

www.safetyservicesinc.com

Science Applications International
Corporation (SAIC)
301 Laboratory Road
P.O. Box 2501
Oak Ridge, TN 37831
865-482-9031
www.saic.com

Lucien G. Canton CEM LLC
783 45th Avenue
San Francisco CA 94121
415-221-2562

Multi Risk International
6955, boul. Taschereau, 210
Brossard, Quebec J4Z 1A7
450-443-2500
www.multirisques.net

Oppenheimer Wolff & Donnelly
LLP
45 South 7th Street Plaza VII
Suite 3300
Minneapolis, MN 55402
612-607-7000
www.oppenheimer.com

PBS&J
Suite 350 7406 Fullerton Street
Jacksonville, FL 32256
904-363-6100
www.pbsj.com

Phoenix Continuity Solutions, LLC
25800 Northwestern Highway
Suite L-60
Southfield, MI 48075
248-263-3855
www.pcsphoenix.com

Pre-Emergency Planning, LLC
513A North Highway 51
Poynette, WI 53955
608-635-2903
www.pre-emergency.com

Reynolds Bone & Griesbeck
5100 Wheelis Drive
Suite 300
Memphis, TN 38117
901-682-2431
www.rbgcpa.com

Sage Business Associates Inc.
6156 Powell Road
Parker, CO 80134
303-841-4467

Seek Safety Services, Inc.
8409 Pickwick Lane #271
Dallas, TX 75225
214-668-2285
www.seeksafetyservices.com

McWains Chelsea Inc
736 Speedwell Avenue
Morris Plains NJ 07950
973-993-5700
www.mcwains.com

National Institute for Occupational
Safety and Health
200 Independence Ave, SW Hubert
H. Humphery Bldg. 715H
Washington, D.C. 20201
202-260-9727
www.cdc.gov/niosh

Optimus Consulting
22 Technology Parkway South
Norcross, GA 30092
770-447-1951
www.optimussolutions.com/oc

Pearces 2 Consulting Corporation
5730 Sunshine Falls Lane
North Vancouver, British Columbia
V7G 2T9
604-929-4560
www.pearces2.com

Phoenix Disaster Services
10221 Desert Sands Street
Suite 111
San Antonio, TX 78216-3944
210-541-0505
www.pdstx.com

Professional Loss Adjusters Inc.
343 Washington Street
Newton, MA 02458
617-850-0477
www.pla-us.com

RP Risk Advisors
5 Lyons Mall
Suite 322
Basking Ridge, NJ 07920
908-310-6381
rpriskadvisors.com

Scanlon Associates Inc.
117 Aylmer Ave.
Ottawa, Ontario K1S 2X8
613-730-9239

SEH
Butler Square Building
Suite 710C
100 North 6th Street
Minneapolis, MN 55403-1515
612-216-1889
www.sehinc.com

SRA International Inc.
4300 Fair Lakes Court
Fairfax, VA 22033
703-803-1500
www.sra.com

Strohl Systems
631 Park Avenue
King of Prussia, PA 19406
610-768-4135
www.strohlsystems.com

The Center for Continuity &
Recovery Advice (CCRA)
P.O. Box 291214
Davie, FL 33329
954-383-4031
www.recoveryadvice.com

The Revere Group
Suite #325
325 North LaSalle Street
Chicago, IL 60610
312-873-3400
www.reveregroup.com

TRC
712 South Wheeling Avenue
Tulsa, OK 74104
918-585-1990
www.trcdisastersolutions.com

USG Recovery Centers
7831 East Bush Lake Road
Suite 100
Edina, MN 55439
612-874-6500
www.usgrecoverycenter.com

Blue Heron Consulting Corp.
90 Airpark Drive
Suite 200
Rochester, NY 14624
585-464-8035
www.blueheron-consulting.com

Professional Emergency
Management Ltd. (PROEM)
71 Cheyanne Meadows Way
Calgary, Alberta T3R 1B6
403-560-9456
www.proem.com

The National Environmental
Services Center
West Virginia University
P.O. Box 6064
Morgantown, WV 26506

Stephens Associates, Inc
157 Broad Street
Suite 202
Red Bank, NJ 07701
732-842-1903
www.stevensassocinc.com

TAMP Systems
1732 Remson Avenue
Merrick, NY 11566
516-623-2038
www.drsbytamp.com

The Courton Group
134 Fort Greene Place
Brooklyn, NY 11217
718-855-1440
www.courtongroup.com

The Steele Foundation
Suite 2450
101 California Street
San Francisco, CA 94111
415-781-4300
www.steelefoundation.com

USG Inc.
7801 East Bush Lake Road
Suite 150
Edina, MN 55439
952-835-2349
www.usgerp.com

Vigilant Services Group
1556 Sunshine Tree Boulevard
Longwood, FL 32779
407-786-4194
www.vigilantservicesgroup

JALCO Services, Inc
521 Helena Avenue
Wyckoff, NJ 07481
201-847-2019
www.jalcoservices.com

SEH Inc.
6418 Normandy Lane #100
Madison, WI 53719
608-270-5364
www.sehinc.com

Stratford Solutions
2988 Monmouth Rd.
Cleveland Heights, OH 44118
216-932-5690
www.stratfordsolutions.com

TelLAWCom Labs, Inc.
100 Ovilla Oaks Dr.
Suite 200
Ovilla, TX 75154
214-888-1300
www.tellawcomlabs.com

The Gartner Group
Suite 600
5950 Canoga Avenue
Woodland Hills, CA 91367
818-710-8855

The Systems Audit Group Inc.
25 Ellison Road
Newton, MA 02459-1434
617-332-3496
www.risk-help.com

Usg Inc.
7831 East Bush Lake Road
Suite 100
Minneapolis, MN 55416
952-835-2349
www.usg-inc.com

Montague Risk Management
9 Underhill Avenue
Locust Valley, NY 11560
516-676-9234
www.montaguetm.com

Templar Titan Inc.
Suite 400
402 West Broadway
San Diego, CA 92101
800-779-0322
www.templartitan.com

Adjusters International
25800 Northwestern Highway
Suite 885
Southfield, MI 48075-8403
248-352-2100
www.globemwai.com

Agnes Huff Communications
Group LLC
Suite 100
Howard Hughes Center
6601 Center Drive West
Los Angeles, CA 90045
310-641-2525
www.ahuffgroup.com

Allan Bonner Communications
Management
393 King St. W.
Suite #701
Toronto, Ontario M5V 3G8
416-961-3620
www.allanbonner.com

Asesores en Emergencias y Desastres
Paseo Jurica 105-25PB Jurica,
Quertaro/Mexico
Queretaro, TX 76100
442-218-4424
www.asemde.com

Barron Emergency Consulting
33 Gaskins Road
Milton, MA 02186
617-298-8265
www.barronemergencyconsulting.com

Bernstein Crisis Management
Corp.
1013 Orange Avenue
Monrovia, CA 91016
626-305-9277
www.bernsteincrisismanagement.com

C4CS, LLC
5625 Hempstead Road
Suite101
Pittsburgh, PA 15217
412-708-0940
www.c4cs.com

Childress Duffy Goldblatt Ltd.
Suite # 2200
515 North State Street
Chicago, IL 60610
312-494-0200
www.childresslawyers.com

Comprehensive Solutions
250 North Sunny Slope Road
Suite 300
Brookfield, WI 53005
262-785-8101
www.comp-soln.com

COPE Solutions Inc
3274 Rosedale Rd North
Smiths Falls, Ontario K7A 4S7
613-223-1128
www.copesolutions.com

Crisis Care Network
2855 44th Street South West
Suite 360
Grandville, MI 49418
www.crisiscare.com

Crisis Consulting Group
P.O. Box 84153
Fairbanks, AK 99708
848-333-1376
www.fwep.org

Crisis Management International
Inc.
Suite 420
8 Piedmont Center Northeast
Atlanta, GA 30305
404-841-3400
www.cmiatl.com

Crisis Response Planning
Corporation
464 Gowland Crescent
Milton, Ontario L9T 4E5
905-876-0229
www.crpc.com

Criterion Strategies Inc
580 Broadway
Suite 305
New York, NY 10012
212-343-1134
www.criterionstrategies.com

Disaster Management Inc.
1531 Southeast Sunshine Avenue
Port St. Lucie, FL 34952-6011
772-335-9750
www.disastermgt.com

Dynamic Options Group
#320 5830 Northwest Expressway
Oklahoma City, OK 73132-5239
05-476-5976
www.contingencyplanningguide.com

Eagle Watch International
3292 Thompson Bridge Road
Suite 333
Gainesville, GA 30506
770-287-7694
www.eaglewatchinvestigations.com

EBC Partners Inc.
Suite 108
6753 Thomasville Road
Tallahassee, FL 32312
850-894-4043
www.ebcpartners.com

EnviroMED Inc.
Suite 300
4400 East Broadway Boulevard
Tucson, AZ 85711
520-881-1000

Envision—Planning Solutions Inc.
131 Scenic Hill Close NW
Calgary, Alberta T3L 1R1
403-241-8883

ERMS Corporation
Suite 200
2916 South Sheridan Way
Oakville, Ontario L6J 7J8
905-829-8216
www.ermscorp.com

FIRECON
P.O. Box 231
East Earl, PA 17519
717-354-2411
www.firecon.com

FirstCall Network Inc.
5423 Galeria Drive
Baton Rouge, LA 70816
225-295-8123
www.firstcall.net

Frontline Corporate
Communications Inc.
650 Riverbend Drive
Kitchener, Ontario N2K 3S2
www.fcc.onthefrontlines.com

Global Impact Inc.
21 Gardiner Dr.
Bradford, Ontario
416-791-6109
www.riskandthreat.net

Global Security Systems, LLC
308 East Pearl Street
Suite 202
Jackson, MS 39201
337-237-7757
www.gssnet.us

Hazmat DQE
8112 Woodland Drive
Indianapolis, IN 46278
www.dqeready.com

Hour-Zero Crisis Consulting Ltd.
9914 - 86 Ave.
Edmonton, Alberta T6E 2L7
780-439-0999
www.hour-zero.com

Institute for Crisis Management
Suite 140
950 Breckenridge Lane
Louisville, KY 40207
502-891-2507
www.crisisconsultant.com

KI Canada Ltd.
707 Alness Street #206
Toronto, Ontario M3J 2H8
416-661-1818
www.kicanada.com

Montague Risk Management
9 Underhill Avenue
Locust Valley, NY 11560
516-676-9234
www.montaguetm.com

Pitney Bowes Inc.
1 Elmcroft Road
Stamford, CT 06926-0700
203-356-5000
www.pb.com

ProfitStars, A Jack Henry Company
1025 Central Expressway South
Dallas, TX 75013
913-307-1324
www.profitstars.com

R&A Crisis Management Services
650 South River Road
Suite 713
Des Plaines, IL 60016-8344
847-827-4267
www.raconsulting.net

RP Risk Advisors
5 Lyons Mall, Suite 322
Basking Ridge, NJ 07920
908-310-6381
rpriskadvisors.com

Terra Firma Enterprises
181 Westminster Avenue
Ventura, CA 93003
805-642-5232
www.terrafirmaenterprises.com

The Redfern Group
14450 T.C. Jester Blvd.
Suite 205
Houston, TX 77014
281-866-9451
www.redferncpr.com

IBM Global Services
10 North Martingale Road
Schaumburg, IL 60173
www.ibm.com

International Critical Incident
Stress Foundation
3290 Pine Orchard Lane
Suite 106
Ellicott City, MD 21042
410-750-9600
www.icisf.org

Levick Strategic Communications
1900 M Street NW Suite 400
Washington, D.C. 20036
202-973-1300
www.levick.com

Oppenheimer Wolff &
Donnelly LLP
45 South 7th Street Plaza VII
Suite 3300
Minneapolis, MN 55402
612-607-7000
www.oppenheimer.com

Prepared Response
Suite 1525
600 University Street
Seattle, WA 98101-4155
206-223-5544
www.preparedresponse.com

Project Time & Cost,Inc. (PT&C)
2727 Paces Ferry Road
Suite 1-1200
Atlanta, GA 30339
770-444-9799

R.D. Zande & Associates, Inc.
1500 Lake Shore Drive
Suite 100
Columbus, OH 43204
614-486-4383
www.zande.com

Strategic Teaching Associates, Inc.
4158 Forestbrook Drive
Liverpool, NY 13090
315-622-5924
www.drpwithdrtom.com

The Courton Group
134 Fort Greene Place
Brooklyn, NY 11217
718-855-1440
www.courtongroup.com

URS Corporation
200 Orchard Ridge Dr.
Suite 101
Gaithersburg, MD 20878
301-258-6554
www.urscorp.com

Incident Mitigation, Inc.
17340 West 12 Mile Road Suite 101
Southfield, MI 48076
248-552-0821
www.incidentmitigation.com

KETCHConsulting
P.O. Box 641
Waverly, PA 18471
570-563-0868
www.ketchconsulting.com

Lucien G. Canton CEM LLC
783 45th Avenue
San Francisco, CA 94121
415-221-2562

Optimus Consulting
22 Technology Parkway South
Norcross, GA 30092
770-447-1951
www.optimussolutions.com/oc

Pricewaterhouse Coopers LLP
300 Madison Avenue
24th Floor
New York, NY 10017
646-471-4000
www.pwc.com/us

Protiviti Inc.
5720 Stoneridge Drive
Pleasanton, CA 94588
415-402-3600
www.protiviti.com

Raido Response
1109 First Avenue
Suite 212
Seattle, WA 98101
206-628-9156
www.raidoresponse.com

TAMP Systems
1732 Remson Avenue
Merrick, NY 11566
516-623-2038
www.drsbytamp.com

The Mindszenthy & Roberts Corp.
115 Marlborough Place
Toronto, Ontario M5R1X5
416-924-2425
www.mrcom.com

Abtron Corp.
40 Underhill Boulevard
Suite 10C
Syosset, NY 11791
516-364-4678

Alternative Environmental
Solutions Inc.
2217 Liberty Street
Monroe, LA 71201
318-388-4833

Atlantis Waterproofing &
Mold Control
18238 Showalter Road
Hagerstown, MD 21742
800-572-2001

Braun Intertec Corp.
11001 Hampshire Avenue South
Minneapolis, MN 55438
952-995-2000
www.brauncorp.com

CDM
1 Cambridge Place
50 Hampshire Street
Cambridge, MA 02139
617-452-6000
www.cdm.com

CH2M Hill
9191 South Jamaica Street
Englewood, CO 80112
303-771-0900
www.ch2m.com

Emergency Flood Services, Inc.
370 Franklin Drive
Suite A
Dacula, GA 30019-3408
770-963-7443
www.contractors.com/efsi

ENSR: Environmental
Consultants and Engineers
2 Technology Park Drive
Forge Village
Westford, MA 01886
978-589-3000
www.ensr.com

Environmental Resources
Management (ERM)
350 Eagleview Boulevard
Suite 200
Exton, PA 19341
610-524-3500
www.erm.com

ACM Environmental Inc.
26598 US 20 West
South Bend, IN 46628
574-234-8435
www.acmenv.com

AMI&More, LLC
265 West Highway 54
Suite 198 PMB
Durham, NC 27713
919-450-0265
www.amiandmore.com

Baker Tanks, Inc.
35173 Highway 30
Geismar, LA 70734
225-677-8763
www.bakertanks.com

Bureau Veritas
45525 Grand River Avenue
Suite 200
Novi, MI 48374
248-344-2661
www.us.bureauveritas.com

CDW Consultants, Inc.
40 Speen Street
Suite 301
Framingham, MA 01701
508-875-2657
www.cdwconsultants.com

EBC Partners Inc.
Suite 108
6753 Thomasville Road
Tallahassee, FL 32312
850-894-4043
www.ebcpartners.com

EMG
11011 McCormick Road
Hunt Valley, MD 21031
www.emgcorp.com

Environmental & Occupational
Risk Management
283 East Java Drive
Sunnyvale, CA 94089
408-822-8100
www.eorm.com

Fehr-Graham Associates
Suite 200
221 East Main Street
Freeport, IL 61032
815-235-7643
www.fehr-graham.com

Air Quality Control
636 Olanta Highway
Florence, SC 29541
888-580-4379
www.aqcsc.com

Assured Indoor Air Qulaity
6616 Forest Park Road
Dallas, TX 75235
214-855-0222

Bercha Group
P.O. Box 61105
Calgary, Alberta T2N 4S6
403-270-2221
www.berchagroup.com

Calvin, Giordano & Associates
Suite 600
1800 Eller Drive
Ft. Lauderdale, FL 33316
954-921-7781
www.calvin-giordano.com

Cemtech International
10816 Highway 21
Suite D
P.O. Box 914
Hillsboro MO 63050
636-789-2803
www.cemtech-international.com

Ecology & Environment Inc
368 Pleasant View Drive
Lancaster, NY 14086
716-684-8060
www.ene.com

Engineering Professionals &
Constructors, Inc.
601 Front Street
Lisle, IL 60532
708-906-8818
epcltd@comcast.net

Environmental Management &
Engineering Inc.
Suite 1
5242 Bolsa Avenue
Huntington Beach, CA 92649
714-379-1096
www.emeiaq.com

First Environmental Nationwide
5223 Riverside Drive
Macon, GA 31210
478-477-2323
www.firstenvironmental.com

Geotech Computer Systems, Inc.
6535 South Dayton Street
Suite 2100
Englewood, CO 80111
303-740-1999
www.geotech.com

Haz-Matters Inc.
6801 North Highway 79
Suite #1
Black Hawk, SD 57718
605-343-4898
www.hazmatters.com

K. S. Ware & Associates, LLC
54 Lindsley Avenue
Nashville, TN 37210
615-255-9702
www.kswarellc.com

P W Grosser Consulting, Inc
630 Johnson Avenue
Suite 7
Bohemia, NY 11716
631-589-6353
www.pwgrosser.com

Philip Environmental Services
Corp.
210 West Sand Bank Road
Columbia, IL 62236
618-281-7020
www.contactpsc.com

Roux Associates Inc.
209 Shafter Street
Islandia, NY 11749
631-232-2600
www.rouxinc.com

Shaw Environmental &
Infrastructure Incorporated
2790 Mosside Boulevard
Monroeville, PA 15146
412-372-7701
www.shawgrp.com

Superior Environmental
Corporation
1128 Franklin Street
Marne, MI 49435
616-667-4000
www.superiorenvironmental.com

Trammell Consulting
6163 Stillwell-Beckett Road
Oxford, OH 45056
513-680-2574
trammellconsulting.com

Golder Associates Corp.
3730 Chamblee Tucker Road
Atlanta, GA 30341
770-496-1893
www.golder.com

HWS Consulting Group Inc.
825 J Street
P.O. Box 80358
Lincoln, NE 68501-0358
402-479-2200
www.hws.com

MacDonald-Bedford LLC
2900 Main Street
Suite 200
Alameda, CA 94501
510-521-4020
www.macdonaldbedford.com

PBS Engineering & Environmental
4412 South West Corbett Avenue
Portland, OR 97239
503-417-7586

Professional Service Industries Inc.
(PSI)
1901 S.Meyers Blvd
Suite 400
Oakbrook Terrace, IL 60181

Safety & Environmental
Management Planning Inc.
235 Antigua Drive
Lafayette, LA 70503
337-981-5391
www.semp.com

Shaw Environmental Inc.
88C Elm Street
Hopkinton, MA 01748-1656
508-497-6139
www.shawgrp.com

Terracon Environmental Inc.
16000 College Boulevard
Lenexa, KS 66219
913-599-6886

Trinity Consultants
2311 W 22nd Street
Suite 315
Oak Brook, IL 60523
630-574-9400
www.trinityconsultants.com

Green Environmental Inc.
52 Accord Park Drive
Norwell, MA 02061
617-479-0550
www.greenenvironmental.com

Integrated Environmental Services
1445 Marietta Boulevard
Atlanta, GA 30318
404-352-2001
www.iescylinders.com

O'Connor Associates Environmental
Inc.
318 11th Ave SE
Ste 200
Calgary, Alberta T2G 0Y2
403-294-4200
www.oconnor-associates.com

PerkinElmer
45 William Street
Wellesley, MA 02481
781-237-5100
www.perkinelmer.com

Quality Inspection Service
7420 Stanford Avenue
La Mesa, CA 91941
619-466-2581

SCI Inc
16345 Anetietam Avenue
Baton Rouge, LA 70817
225-270-0701

Shaw Group
4171 Essen Lane
Baton Rouge, LA 70809
225-932-2500
www.shawgrp.com

Tetra Tech Inc.
30 3475 East Foothill Boulevard
Pasadena, CA 91107
626-683-0066
www.tetratech.com

URS Corporation
200 Orchard Ridge Dr.
Suite 101
Gaithersburg, MD 20878
301-258-6554
www.urscorp.com

Washington Group International
720 Park Boulevard
P.O. Box 73
Boise, ID 83729
208-386-5000
www.wgint.com

XL Insurance
520 Eagleview Boulevard
Exton, PA 19341
610-458-0570
www.xlenvironmental.com

Certified Firestop LLC
1868 Forsyth Avenue c/o 101
Monroe, LA 71201
318-665-2060
www.certifiedfirestop.com

Firepoint Technologies Inc
27-180 Wilkinson Road
Brampton, Ontario L6T 4W8
905-874-9400
www.firepoint.cc

Fyrsafe Engineering Inc.
Suite 108
1225 Carnegie Street
Rolling Meadows, IL 60008-1032
847-392-1111

Haberill Contracting
40 Lynch Lane
Everett, Ontario L0M1J0
705-435-4219

Harrington Group Inc.
Suite 310
3055 Breckinridge Boulevard
Duluth, GA 30096
770-564-3505
www.hgi-fire.com

KL Security Enterprises LLC
2842 East 860 North
P.O. Box 165
Stockland, IL 60967
866-867-0306
www.klsecurity.com

Loss Control Association
Incorporated
172 Middletown Boulevard
Suite 204B
Langhorne, PA 19047
215-750-6841
www.losscontrolassociates.com

MATRIX Risk Consultants Inc
3130 S Tech Boulevard
Miamisburg, OH 45342-4882
937-886-0000
www.matrixrc.com

Rally Fire Protection Services
P.O. Box 14
Wheaton, IL 60189-0014
630-690-4200

Risk, Reliability, and Safety
Engineering
2525 South Shore Blvd.
Suite 206
League City, TX 77573
281-334-4220
www.rrseng.com

Rolf Jensen & Assoc. Inc
600 West Fulton Street
Suite 500
Chicago, IL 60661
312-879-7200
www.rjainc.com

Rollinger Engineering Inc.
2000 South Dairy Ashford Street
Suite 455
Houston, TX 77077-5719
281-558-5000
www.reifirepro.com

SimplexGrinnell LLC
6423 Shelby View Drive
Memphis, TN 38134-7614
901-386-0532
www.contingencyplanningguide.com

The Deatherage Companies
1805 North 16th Street
Broken Arrow, OK 74012-9339
918-355-2344

TriData Corporation
1000 Wilson Boulevard
Arlington, VA 22209-3927
703-351-8308

Assurity River Group
Suite 260
550 Main Street
New Brighton, MN 55112
651-259-6880
www.assurityriver.com

Coffing Corporation
5336 LeSourdsville West
Chester Road
Hamilton, OH 45011
513-755-8866
www.coffingco.com

DaVinci Technology Corporation
(DaVinciTek)
89 Headquarters Plaza North
Tower, 14th Floor
Morristown, NJ 07960
973-993-4860
www.davincitek.com

EBC Partners Inc.
Suite 108
6753 Thomasville Road
Tallahassee, FL 32312
850-894-4043
www.ebcpartners.com

Forsythe Solutions Group
7770 Frontage Road
Skokie, IL 60077
847-213-7000
www.forsythesolutions.com

Hospitech Solutions
150 River Road
Suite G4B
Montville, NJ 07045
973-263-9800
www.hospitechsolutions.com

IBM Global Services
10 North Martingale Road
Schaumburg, IL 60173
www.ibm.com

InfoSENTRY Services Inc.
P.O. Box 28048
2 Hannover Square
Suite 2330
Raleigh, NC 27601
919-838-8570
www.infosentry.com

MLC & Associates Inc.
Suite 265
3525 Hyland Avenue
Costa Mesa, CA 92626
949-222-1202
www.mlcandassociates.com

SECTOR, Inc.
2 MetroTech Center
Brooklyn, NY 11201
212-383-2000
www.sectorinc.com

lThe Courton Group
134 Fort Greene Place
Brooklyn, NY 11217
718-855-1440
www.courtongroup.com

Bocada, Inc.
10500 North East 8th Street
Bellevue, WA 98004
425-818-4400
www.bocada.com

DaVinci Technology Corporation
(DaVinciTek)
89 Headquarters Plaza North
Tower, 14th Floor
Morristown, NJ 07960
973-993-4860
www.davincitek.com

Infosys International
110 Terminal Drive
Plainview, NY 11803
516-576-9494
www.infosysinternational.com

Murray Associates
P.O. Box 668
Oldwick, NJ 08858-0668
908-832-7900
www.spybusters.com

R. Grossman & Associates
4058 Spruce Avenue
Egg Harbor Township, NJ
08234-5807
609-383-3456
www.tech-answers.com

Science Applications International
Corporation Justice and Security
Solutions
8301 Greensboro Drive
McLean, VA 22102
703-676-6046
www.saic.com

ProfitStars, A Jack Henry
Company
1025 Central Expressway South
Dallas, TX 75013
913-307-1324
www.profitstars.com

Sigma Business Solutions Inc.
55 York Street
Suite 1100
Toronto, Ontario M5J 1R7
416-594-1991
www.sigma-sbs.com

CERIAS: Center for Information
Assurance & Security
Purdue University
656 Oval Drive
West LaFayette, IN 47907-2086
765-494-7841
www.cerias.purdue.edu

Guardium Citypoint
230 Third Avenue
Waltham, MA 02451
781-487-9400
www.guardium.com

Ingenuity-Business by Design
3512 Balsam Boulevard
Port Orchard, WA 98366
360-731-0522

PC SYSWARE Inc.
57 Squires Avenue
Toronto, Ontario M4B 2R6
416-951-0110
www.securesmb.ca

RP Risk Advisors
5 Lyons Mall
Suite 322
Basking Ridge, NJ 07920
908-310-6381
rpriskadvisors.com

SecureWorks
P.O. Box 95007
Atlanta, GA 30347
404-327-6339
www.secureworks.com

RP Risk Advisors
5 Lyons Mall
Suite 322
Basking Ridge, NJ 07920
908-310-6381
rpriskadvisors.com

SRA International Inc.
4300 Fair Lakes Court
Fairfax, VA 22033
703-803-1500
www.sra.com

Anteon
3211 Jermantown Road
Suite 700
Fairfax, VA 22030
703-246-0200
www.anteon.com

CERT Coordination Center
Software Engineering Institute
Carnegie Mellon University
Pittsburgh, PA 15213-3890
412-268-7090
www.cert.org

IBM Global Services
10 North Martingale Road
Schaumburg, IL 60173
www.ibm.com

M Corby & Associates Inc.
8th Floor 255 Park Avenue
Worcester, MA 01609-1953
508-792-4320
www.mcorby.com

ProfitStars, A Jack Henry Company
1025 Central Expressway South
Dallas, TX 75013
913-307-1324
www.profitstars.com

Sarcom, Inc.
8337-A Green Meadows Dr. N
Lewis Center, OH 43035
614-845-1300
www.sarcom.com

Securify
20425 Stevens Creek Boulevard
Cupertino, CA 95014
408-343-4300
www.securify.com

TechGuard Security
St. Louis West County Office
743 Spirit 40 Park Drive
Suite 206
Chesterfield, MO 63005
636-519-4848
www.techguardsecurity.com

Childress Duffy Goldblatt Ltd.
Suite # 2200
515 North State Street
Chicago IL 60610
312-494-0200
www.childresslawyers.com

Fraser Investigative Engineering
Services, P.C.
7670 Chase Road
Lima, NY 14485
585-582-2533
www.fraseries.com

Helmsman Management Services
Inc.
9 Riverside Road
Weston, MA 02493
617-243-7985
www.helmsmantpa.com

LWG Consulting
3455 Commercial Avenue
Northbrook, IL 60062
847-559-3000
www.lwgconsulting.com

Murray Associates Eavesdropping
Detection
P.O. Box 668
Oldwick, NJ 08858-0668
908-832-7900
www.spybusters.com

Risk Consultants Inc.
P.O. Box 490850
Atlanta, GA 30349
770-964-1226
www.riskcon.com

Spybusters.com
P.O. Box 668
Oldwick, NJ 08858
908-832-7900
www.spybusters.com

The Courton Group
134 Fort Greene Place
Brooklyn, NY 11217
718-855-1440
www.courtongroup.com

Cannon Cochran Management
Services Inc.
2 East Main Street
Danville, IL 61832
217-446-1089
www.ccmsi.com

CTL Group
5400 Old Orchard Road
Skokie, IL 60077-1030
847-965-6500
www.ctlgroup.com

Gallagher Bassett Services Inc.
(National HQ)
2 Pierce Place
Itasca, IL 60143
630-773-3800

JMG Consultants, Inc.
330 Jervis Avenue
Copiague, NY 11726
631-495-1051
www.ienga.com

Madsen Kneppers & Associates
Inc.
Suite 310
1855 Olympic Boulevard
Walnut Creek, CA 94596
925-934-3235
www.mkainc.com

Overland Solutions Inc.
Suite 400
11880 College Boulevard
Overland Park, KS 66210
913-451-3222
www.olsi.net

Ruf & Associates, LLC
510 East 770 North
Orem, UT 84097
801-764-9100
www.rufassociates.com

URS Corporation
200 Orchard Ridge Dr.
Suite 101
Gaithersburg, MD 20878
301-258-6554
www.urscorp.com

TriAxis Storage Solutions Inc.
15 Midstate Office Park
Auburn, MA 01501
508-721-9691
www.triaxisinc.com

Cap Index, Inc.
The Commons at Lincoln Center
150 John Robert Thomas Dr.
Exton, PA 19341
610-903-3000
www.capindex.com

Flagship Reconstruction
#100
Arlington, TX 76014
817-695-1339
www.flagship-services.com

Gilbane CAT-Response
7 Jackson Walkway
Providence, RI 02903
401-456-5530
www.gilbaneco.com

Loss Solutions Group
500 Libson Road
Canterbury, CT 06331
860-237-8041
www.losssolutionsgroup.com

Meadowbrook Insurance Group
Inc.
26255 American Drive
Southfield, MI 48034
248-358-1100
www.meadowbrook.com

Regional Reporting Inc
40 Fulton Street 20th Floor
New York, NY 10038
212-964-5973
www.regionalreporting.com

Schirmer Engineering Corp.
Suite 200
707 Lake Cook Road
Deerfield, IL 60015
847-272-8340
www.schirmerengineering.com

Forbes Calamity Prevention
Pte Ltd.
75C Duxton Road
Singapore 089534
63243091
www.calamityprevention.com

Access Northeast
34 Saint Martin Drive
Marlboro, MA 01752
508-281-7600
www.axsne.com

Deloitte & Touche LLP
10 Westport Road
Wilton, CT 06897
203-761-3000
www.deloitte.com

Montague Risk Management
9 Underhill Avenue
Locust Valley, NY 11560
516-676-9234
www.montaguetm.com

ServePath
SF2 Data Center & Office
360 Spear Street 2nd Floor
San Francisco CA 94105
415-869-7000
www.servepath.com

Lightedge Solutions Inc.
#407 511 Eleventh Avenue South
Minneapolis, MN 55415-1536
612-252-2318
www.lightedge.com

ADVA Optical Networking
Suite 705
1 International Boulevard
Mahwah, NJ 07495
201-258-8300
www.advaoptical.com

Desamd USA
Suite 620
511 East John Carpenter Fairway
Irving TX 75062
877 - DESAMD2
www.desamd.com

ProfitStars, A Jack Henry
Company
1025 Central Expressway South
Dallas, TX 75013
913-307-1324
www.profitstars.com

The Courton Group
134 Fort Greene Place
Brooklyn, NY 11217
718-855-1440
www.courtongroup.com

Center for Infectious Disease
Research & Policy
Academic Health Center-University
of Minnesota
420 Delaware Street Southeast,
MMC 263
Minneapolis, MN 55455
612-626-6770
www.cidrap.umn.edu

b4Ci Inc.
100 Ovilla Oaks Drive
Suite 200
Ovilla, TX 75115
214-888-1300
www.rewireit.com

IBM Global Services
10 North Martingale Road
Schaumburg, IL 60173
www.ibm.com

SECTOR, Inc.
2 MetroTech Center
Brooklyn, NY 11201
212-383-2000
www.sectorinc.com

The Patricia Bennett Group
259-A E Browning Road
Bellmawr, NJ 08031
856-931-1604
www.bennettgrp.com

KETCHConsulting
P.O. Box 641
Waverly, PA 18471
570-563-0868
www.ketchconsulting.com

Montague Risk Management
9 Underhill Avenue
Locust Valley, NY 11560
516-676-9234
www.montaguetm.com

Samkoff Consulting Group, LLC
Harrisburg, PA 17110
717-545-3632

Center for Infectious Disease
Research & Policy
Academic Health Center-University
of Minnesota
420 Delaware Street Southeast,
MMC 263
Minneapolis, MN 55455
612-626-6770
www.cidrap.umn.edu

TAMP Systems
1732 Remson Avenue
Merrick, NY 11566
516-623-2038
www.drsbytamp.com

Crisis Communications
Toronto, Ontario
416-261-9828
www.topstory.ca

Focus Group Risk & Strategic
Communications Consultants
29 Welgate Road
Medford, MA 02155
781-483-3054
www.focusgroupconsulting.com

Hennes Communications
2841 Berkshire Road
Cleveland, OH 44118
216-321-7774
www.hennescommunications.com

URS Corporation
200 Orchard Ridge Dr. Suite 101
Gaithersburg, MD 20878
301-258-6554
www.urscorp.com

CYNREDE Inc.
Suite 108
23152 Verdugo Drive
Laguna Hills, CA 92653
800-258-1225
www.cynrede.com

SECTOR, Inc.
2 MetroTech Center
Brooklyn, NY 11201
212-383-2000
www.sectorinc.com

Vital Records Control of Florida
11901 Amedicus Lane
Fort Myers, FL 33907
239-337-4030
www.vrcoffl.com

Acordia
24 East Greenway Plaza
Suite 1100
Houston, TX 77046
713-507-9476
www.acordia.com

Compaudit Services
266 Harristown Road
Suite106
GlenRock, NJ 07452
201-689-4040
www.nationalrisk.com

Frontline Corporate
Communications Inc.
650 Riverbend Drive
Kitchener, Ontario N2K 3S2
www.fcc.onthefrontlines.com

PR Direct
36 King St., East
Toronto, Ontario M5C 2L9
416-507-2028
www.prdirect.ca

Weber Shandwick
101 Main Street Suite 8
Cambridge, MA 02142
617-661-7900
www.webershandwick.com

Advanced Records Management
Services, Inc.
249 North Street
Danvers, MA 01923
www.armsrecords.com

Pitney Bowes Inc.
1 Elmcroft Road
Stamford, CT 06926-0700
203-356-5000
www.pb.com

Vault Management Incorporated
1805 West Detroit
Broken Arrow, OK 74012
918-258-7781
www.vault-tulsa.com

Vital Records Control of South
Carolina
255 Eagle Drive
North Charleston, SC 29445
843-566-7650
www.vrcofsc.com

Williams Records Management
1925 East Vernon Avenue
Los Angeles, CA 90058
323-234-3453
www.williamsrecords.com

Claims Administrative Services Inc.
501 Shelley Drive
2nd Floor
Tyler, TX 75701
903-509-8484
www.cas-services.com

Emergency Management
International (USA) Inc.
Suite 331
1022 Lincoln Street
Rhinelander, WI 54501-3618
715-362-7577
www.emi-usa.com

Godec Randall & Associates
3944 North 14th Street
Phoenix, AZ 85014
602-266-5556
www.godecrandall.com

Ten United
375 N Front Street Suite 400
Columbus, OH 43215
614-221-7667
www.tenunited.com

William Russell & Associates, Inc.
305 West Masonic View Avenue
Alexandria, VA 22301-2418
703-739-6277

Archive Document Storage Inc.
345 10th Street
Jersey City, NJ 07302
201-716-7900
www.archivedocumentstorage.com

RP Risk Advisors
5 Lyons Mall
Suite 322
Basking Ridge, NJ 07920
908-310-6381
rpriskadvisors.com

Vital Records Control of Arkansas
Suite 14
1401 Murphy Drive
Maumelle, AR 72113
501-374-7775 and 501-374-7177
www.vrcofar.com

Vital Records Control of Tennessee
4011 East Raines Road
Memphis, TN 38118
901-363-6555
www.vrcoftn.com

COMCO Safety Consulting Inc.
4141 Norse Way
Long Beach, CA 90808
562-425-4886
www.comcosafety.com

Executive Environmental Services
Corp.
507 Mission Street
South Pasadena, CA 91030
626-441-7050
www.execenv.com

Fire & Safety Specialist Inc.
P.O. Box 9713
College Station, TX 77842
979-690-7559

KEMPER Insurance Companies
One Kemper Drive
Long Grove, IL 60049
847-320-3237
www.kemperinsurance.com

MEMIC Safety Services
261 Commercial Street
P.O. Box 11409
Portland, ME 04104
207-791-3300
www.memic.com

North American Risk Services
P.O. Box 945055
Maitland, FL 32794-5044
www.narisk.com

Pooler Consultants Ltd.
321 Upland Drive
Lafayette, LA 70506
337-984-1601

SEH
Butler Square Building
Suite 710C
100 North 6th Street
Minneapolis, MN 55403-1515
612-216-1889
www.sehinc.com

Factor Security & Continuity
Suite 100
801 West Main Street
Boise, ID 83702
208-639-5993
www.factorsecurity.com

Honeywell Access Systems
135 West Forest Hill Avenue
Oak Creek, WI 53154
414-766-1700
www.honeywellaccess.com

HSA Engineers & Scientists
4019 East Fowler Avenue
Tampa, FL 33617
813-971-3882
www.hsa-env.com

Liberty International Risk Services
175 Berkeley Street MS 01E
Boston, MA 02116
617-574-5555
www.libertyinternational.com

National Floor Safety Institute
P.O. Box 92607
Southlake, TX 76092
817-749-1700
www.nfsi.org

Octagon Risk Services
Suite 645
2101 Webster Street
Oakland, CA 94612
510-302-3000
www.octagonrs.com

Railroad Safety Consultants, Inc.
P. O. Box 22746
Lake Buena Vista, FL 32830
407-319-4819
www.railroadsafetyconsultants.com

Specialty Risk Services
16th Floor 225 Asylum Street
Hartford, CT 06103
860-520-2599

Allied Security
3606 Horizon Drive
King of Prussia, PA 19406
www.alliedsecurity.com

Gerard Group International LLC
164 Westford Road
Suite 15
Tyngsborough, MA 01879
978-649-4575
www.gerardgroup.com

Initial Security
3355 Cherry Ridge Street
Suite 200
San Antonio, TX 78230-4818
210-349-6321
www.initialsecurity.com

Inservco Insurance
2 North 2nd Street
Harrisburg, PA 17101
412-247-4565
www.inservco.net

Madsen, Kneppers & Associates
Inc.
#C 3020 Holcomb Bridge
Road Northwest
Norcross, GA 30071
770-446-9606
www.mkainc.com

Nelson Architectural Engineers Inc.
Suite 220
2740 Dallas Parkway
Plano, TX 75093
469-429-9000
www.nae-us.com

Olney Safety Consultants
P.O. Box 31
Ashton, MD 20861

SAIC
10260 Campus Point Drive
San Diego, CA 92121
858-826-6000
www.saic.com

U.S. HealthWorks-Preventive
Services Division
1600 Genessee Street
Suite 700
Kansas City, MO 64102
678-713-6038

Competitive Insights Inc
24 Valencia Street
Ottawa, Ontario K2G 6T1
613-843-9944
www.competitiveinsightsinc.com

HMA Consulting
2929 Briarpark Drive
Suite 325
Houston, TX 77042
www.hmaconsulting.com

IT Matters Inc.
412, 1000 8th Avenue SW
Calgary, Alberta T2P 3M7
403-503-0772
www.itmatters.ca

KETCHConsulting
P.O. Box 641
Waverly, PA 18471
570-563-0868
www.ketchconsulting.com

Phoenix Genesis Corp
(The PGC Consortium)
Suite 401 355 I Street Southwest
Washington, D.C. 20024
202-737-8030
www.pgcsecurity.com

Protiviti Inc.
5720 Stoneridge Drive
Pleasanton, CA 94588
415-402-3600
www.protiviti.com

Securitas AB
P.O. Box 12307
Lindhagensplan 70
Stockholm SE-102 28
+46 8 657 74 00
www.securitas.com

Special Protection Inc.
P.O. Box 286
Hermiston, OR 97838
541-567-7566

TRW
12011 Sunset Hills Road
Reston, VA 20190-3262
www.contingencyplanningguide.com

Amcom Software Inc.
10400 Yellow Circle Drive
Eden Prairie, MN 55343
800-852-8935
www.amcomsoft.com

CapRock Communications
4400 S Sam Houston Parkway E
Houston, TX 77048-5902
832-668-2300
www.caprock.com

Integrity Communications
130 Route 209
Port Jervis, NY 12771-5130
845-649-5387
www.integritycd.com

SECTOR, Inc.
2 MetroTech Center
Brooklyn, NY 11201
212-383-2000
www.sectorinc.com

MERIT Security
P.O. Box 7236
Redwood City, CA 94063
650-366-0100
www.meritsecurity.com

Precision Security Consulting
Suite 129
5929L Jeanne D'Arc Blvd.
Ottawa, Ontario K1C 7K2
613-863-2993

Safety & Risk Control Services Inc
395 Main Street
Suite 4
Metuchen, NJ 08840
732-906-2244
www.safetyrisk.com

Securitas Treasury Ireland Ltd
Ground Floor Plaza
1 Custom House Plaza IFSC
Dublin 1
+353 1 435 0100
www.securitas.com

Strategic Technology Group
100 Medway Road
Suite 3000
Milford, MA 01757
508-473-4949
www.drthermos.com

Walker International LLC
P.O. Box 4311
Manchester, NH 03108
603-930-4141
www.walkerintl.com

Ascendent Telecom
Suite 204
17141 Ventura Boulevard
Encino, CA 91316-4036
818-728-2021
www.ascendentsystems.com

Federal Engineering Inc.
10600 Arrowhead Drive
Fairfax, VA 22030
703-359-8200
www.fedeng.com

PanAmSat Corporation
20 Westport Road
Wilton, CT 06897
203-210-8000
www.panamsat.com

The Koxlien Group
800 Wisconsin Street
Unit 103
Eau Claire WI 54703
715-831-5581
www.koxlien.com

Network Systems Architects
Corporation
14 Page Terrace
Stoughton, MA 02072
781-297-5300
www.nsaservices.com

Protective Counter Measures and
Consulting, Inc.
70 West Red Oak Lane
White Plains, NY 10604
914-697-4777
www.protectivecountermeausres.com

Sako & Associates
Suite 500 600 West Fulton Street
Chicago, IL 60661
312-879-7230
www.sakosecurity.com

Skybox Security Inc.
Suite 550
2077 Gateway Place
San Jose, CA 95110
408-441-8060
www.skyboxsecurity.com

Templar Titan Inc.
Suite 400
402 West Broadway
San Diego, CA 92101
800-779-0322
www.templartitan.com

BWT Associates
P.O. Box 4515
19 Kenilworth Road
Shrewsbury MA 01545
508-845-6000
www.bwt.com

Homisco Inc.
99 Washington Street
Melrose, MA 02176
781-665-1997
www.homisco.com

ProfitStars, A Jack Henry Company
1025 Central Expressway South
Dallas, TX 75013
913-307-1324
www.profitstars.com

VoiceGard
Suite 206
1951 Old Cuthbert Road
Cherry Hill, NJ 08034-1411
856-669-5100
www.voicegard.com

Table 10.3 Disaster Recovery Web Sites and Other Publications

Organization/Publication	URL
Canadian Security Magazine	www.canadiansecuritymag.com
Computer World	www.computerworld.com
Contingency Planning & Management (CPM)	www.contingencyplanning.com
Continuity Central	www.continuitycentral.com
Continuity Insights Magazine	www.continuityinsights.com
Continuity Insurance & Risk (CIR)	www.cirmagazine.com
Data Center Manager	www.afcom.com
Disaster Recover World	www.disasterrecoveryworld.com
Disaster Recovery Journal	www.drj.com/
Disaster Recovery Planning Forum	www.disasterrecoveryforum.com
Disaster Resource Guide	www.disaster-resource.com
Edwards Information	www.edwardsinformation.com
Enterprise Security Journal	www.esj.com
FacilityZone	www.facilityzone.com
Government Security News Magazine	www.gsnmagazine.com
Homeland Defense Journal	www.homelanddefensejournal.com
Journal of Emergency Management	www.emergencyjournal.com
Journal of Homeland Security	www.homelandsecurity.org/journal
Natural Hazards Observer	www.colorado.edu/hazards/o
NFPA Journal	www.nfpa.org
Occupational Health & Safety	www.ohsonline.com
Risk & Insurance	www.riskandinsurance.com
Risk Management	www.rmmag.org/
Rothstein and Associates (Bookstore)	www.rothstein.com
Safety & Health	www.nsc.org/pubs/sh.htm
SearchSecurity	www.searchsecurity.techtarget.com/#
Security Focus	www.securityfocus.com
Security Magazine	www.securitymagazine.com
Storage Magazine	www.storagemagazine.com
Telecommunications Magazine	www.telecommunications.com
The Institute for Continuity Management	www.drii.org
Today's Facility Manager	www.facilitycity.com

About the Authors

Leo A. Wrobel has more than 30 years of experience with a host of firms engaged in banking, manufacturing, telecom services, and government. An active author and technical futurist, he has published 11 books and more than 500 trade articles on a wide variety of technical subjects. Mr. Wrobel pioneered carrier collocation by being the first in the United States to place a computer disaster recovery center inside a telephone central office 1986. He has lectured throughout the United States and overseas and has appeared on several television programs. Mr. Wrobel is presently the CEO of b4Ci Inc (http://www.b4Ci.com) and TelLAWCom Labs Inc. (http://www.naspa.com). He is also the president of the Network and Systems Professional Association (NaSPA), a nonprofit company specializing in, among other things, information technology and disaster recovery (http://www.napsa.com). He can be reached at leo@b4ci.com.

Sharon M. Wrobel is the vice-president of business development at b4Ci Inc., where she conducts extensive publishing and regulatory research. She has published several trade articles and was a contributor on the book *Business Resumption Planning, Second Edition* (Auerbach Publications, 2008). She can be reached at sharon@b4Ci.com.

Index

Recent Titles in the Artech House Telecommunications Series

Vinton G. Cerf, Senior Series Editor

Wide-Area Data Network Performance Engineering, Robert G. Cole and
   Ravi Ramaswamy

Winning Telco Customers Using Marketing Databases, Rob Mattison

WLANs and WPANs towards 4G Wireless, Ramjee Prasad and
   Luis Muñoz

World-Class Telecommunications Service Development, Ellen P. Ward

For further information on these and other Artech House titles,
including previously considered out-of-print books now available through our
In-Print-Forever® (IPF®) program, contact:

Artech House	Artech House
685 Canton Street	46 Gillingham Street
Norwood, MA 02062	London SW1V 1AH UK
Phone: 781-769-9750	Phone: +44 (0)20 7596-8750
Fax: 781-769-6334	Fax: +44 (0)20 7630-0166
e-mail: artech@artechhouse.com	e-mail: artech-uk@artechhouse.com

Find us on the World Wide Web at: www.artechhouse.com